Elements of Financial Mathematics: From Interest Theory to Options

Elements of Financial Mathematics: From Interest Theory to Options

Edited by:

Stefano Spezia

ARCLER
PRESS

www.arclerpress.com

Elements of Financial Mathematics: From Interest Theory to Options

Stefano Spezia

Arcler Press
224 Shoreacres Road
Burlington, ON L7L 2H2
Canada
www.arclerpress.com
Email: orders@arclereducation.com

ISBN: 978-1-77407-769-6 (Hardcover)

DECLARATION

Some content or chapters in this book are open access copyright free published research work, which is published under Creative Commons License and are indicated with the citation. We are thankful to the publishers and authors of the content and chapters as without them this book wouldn't have been possible.

ABOUT THE EDITOR

Stefano Spezia is Ph.D. holder in Applied Physics at the University of Palermo since April 2012. His major research experience is in noise-induced effects in nonlinear systems, especially in the fields of modeling of complex biological systems and simulation of semiconductor spintronic devices. Associate member of the Italian Physical Society and European Physical Society.

TABLE OF CONTENTS

LIST OF CONTRIBUTORS

Filipe Costa de Souza
Universidade Federal de Pernambuco, Centro de Ciências Sociais Aplicadas, Departamento de Ciências Contábeis e Atuariais, Recife, PE, Brazil

Anne G. Balter
Tilburg University

Bas J.M. Werker
Tilburg University

Ben J. Heijdra
Faculty of Economics and Business, University of Groningen, P. O. Box 800, 9700 AV Groningen, The Netherlands

Laurie S. M. Reijnders
Faculty of Economics and Business, University of Groningen, P. O. Box 800, 9700 AV Groningen, The Netherlands

Mussa Juma
Department of Mathematical and Actuarial Sciences, Universiti Tunku Abdul Rahman, Selangor, Malaysia.

Min Cherng Lee
Department of Mathematical and Actuarial Sciences, Universiti Tunku Abdul Rahman, Selangor, Malaysia.

Seong Tah Chin
Department of Mathematical and Actuarial Sciences, Universiti Tunku Abdul Rahman, Selangor, Malaysia.

Kian Wah Liew
Department of Applied Mathematics, Nottingham University Malaysia Campus, Semenyih, Malaysia.

Kein Joe Lau
Centre for Matheamtical Sciences, Universiti Tunku Abdul Rahman, Bandar Sungai Long, Kajang, Selangor, Malaysia,

Yong Kheng Goh
Centre for Matheamtical Sciences, Universiti Tunku Abdul Rahman, Bandar Sungai Long, Kajang, Selangor, Malaysia,

An Chow Lai
Department of Electrical and Electronic Engineering, Universiti Tunku Abdul Rahman, Bandar Sungai Long, Kajang, Selangor, Malaysia

Guojun Gan
Department of Mathematics, University of Connecticut, Storrs, CT 06269-1009, USA

Sheng Fan
School of Economic Mathematics, Southwestern University of Finance and Economics, Chengdu 611130, China

Jorge Brusa
Division of International Banking & Finance Studies, Sanchez School of Business, Texas A&M International University, Laredo, USA

Jenny Gu
College of Business, University of Dallas, Irving, USA

Grace Yaru Liu
Department of Supply Chain and Information Systems, Pennsylvania State University, University Park, USA

Hong-Ming Yin
Department of Mathematics and Statistics, Washington State University, Pullman, WA 99164, USA; hyin@wsu.edu

Jin Liang
Department of Mathematics, Tongji University, Shanghai 200092, China

Yuan Wu
School of Statistics and Mathematics, Shanghai Lixin University of Accounting and Finance,

Hokuto Ishii
Graduate School of Economics, Nagoya University, Furo-cho, Chikusa-ku, Nagoya 464-8601, Japan;

Florin Turcas
Bucharest Academy of Economic Studies, Bucharest, Romania;

Florin Dumiter
Accountancy and Finance, Vasile Goldis Western University of Arad, Arad, Romania;

Petre Brezeanu
Finance, The Bucharest Academy of Economic Studies, Bucharest, Romania;

Pavel Farcas
Accountancy and Finance, Vasile Goldis Western University of Arad, Arad, Romania;

Sorina Coroiu
University of Oradea, Oradea, Romania

Stanislaus Maier-Paape
Institut für Mathematik, RWTH Aachen University, D-52062 Aachen, Germany

Andreas Platen
Institut für Mathematik, RWTH Aachen University, D-52062 Aachen, Germany

Qiji Jim Zhu
Department of Mathematics, Western Michigan University, Kalamazoo, MI 49008, USA

Ebenezer Fiifi Emire Atta Mills
School of Mathematical Sciences, Dalian University of Technology, Dalian, China.

Dawen Yan
School of Mathematical Sciences, Dalian University of Technology, Dalian, China.
School of Business Management, Dalian University of Technology, Dalian, China.
Bank of Dalian, Dalian, Liaoning, China.

Bo Yu
School of Mathematical Sciences, Dalian University of Technology, Dalian, China.

Xinyuan Wei
School of Mathematical Sciences, Dalian University of Technology, Dalian, China.

Danilo Delpini
Dept. of Economics and Business, Università degli Studi di Sassari, Sassari, Italy,
IMT Institute for Advanced Studies, Lucca, Italy,

Stefano Battiston
Dept. of Banking and Finance, University of Zurich, Zurich, Switzerland,

Guido Caldarelli
IMT Institute for Advanced Studies, Lucca, Italy,
ISC-CNR Uos "Sapienza", Rome, Italy,

Massimo Riccaboni
IMT Institute for Advanced Studies, Lucca, Italy,
Dept. of Managerial Economics, Strategy and Innovation, KU Leuven, Leuven, Belgium

Liurui Deng
College of Economics and Management, Hunan Normal University, Changsha 410081, China;

Traian A. Pirvu
Department of Mathematics and Statistics, McMaster University, 1280 Main Street West, Hamilton, ON L8S 4K1, Canada

Marinella Cadoni
Department of Political Science, Communication, Engineering and Information Technologies, University of Sassari, Sassari, Italy,

Roberta Melis
Department of Economics, Universitas Mercatorum, Rome, Italy,

Alessandro Trudda
Department of Economics and Business, University of Sassari, Sassari, Italy,

Amenawo I. Offiong
Department of Banking and Finance, University of Calabar, Calabar, Nigeria.

Hodo B. Riman
Department of Banking and Finance, University of Calabar, Calabar, Nigeria.

Eyoanwan E. Eyo
Department of Mathematics and Statistics, University of Calabar, Calabar, Nigeria.

Yunyu Zhang
School of Law, Xiamen University, Xiamen, China

Shuhua Chang
Research Center for Mathematics and Economics, Tianjin University of Finance and Economics, Tianjin 300222, China,
Institute of Policy and Management, Chinese Academy of Sciences, Beijing 100190, China,

Xinyu Wang
Research Center for Mathematics and Economics, Tianjin University of Finance and Economics, Tianjin 300222, China,

Yijuan Liang
Agricultural Education and Development Research Center, Southwest University, Chongqing 400715, China
School of Economics and Management, Southwest University, Chonging 400715, China

Xiuchuan Xu
School of Economics and Management, Southwest University, Chonging 400715, China

Nneka Umeorah
Unit for Business Mathematics and Informatics, North-West University, Potchefstroom, South Africa

Phillip Mashele
Unit for Business Mathematics and Informatics, North-West University, Potchefstroom, South Africa

Ho-Chun Herbert Chang
Department of Mathematics, Dartmouth College, Hanover, NH, United States

Kevin Li
Department of Mathematics, Dartmouth College, Hanover, NH, United States

Mingjia Li
Jinan University, Guangzhou, China

Javier Villarroel
Facultad de Ciencias, Universidad de Salamanca, Plaza Merced, 37008 Salamanca, Spain

Kazem Nouri
Department of Mathematics, Faculty of Mathematics, Statistics and Computer Sciences, Semnan University, P.O. Box 35195-363, Semnan, Iran

Behzad Abbasi
Department of Mathematics, Faculty of Mathematics, Statistics and Computer Sciences, Semnan University, P.O. Box 35195-363, Semnan, Iran

Farahnaz Omidi
Department of Mathematics, Faculty of Mathematics, Statistics and Computer Sciences, Semnan University, P.O. Box 35195-363, Semnan, Iran

Leila Torkzadeh
Department of Mathematics, Faculty of Mathematics, Statistics and Computer Sciences, Semnan University, P.O. Box 35195-363, Semnan, Iran

LIST OF ABBREVIATIONS

AIR	Assumed interest rate
CDC	Collective defined contribution
AI	Asymmetric information
FPVA	Flexible premium variable annuity
GBM	Geometric Brownian motion
GLB	Guaranteed living benefits
GMIB	Guaranteed minimum income benefits
SPVA	Single premium variable annuity
VA	Variable annuity
MCMC	Markov chain monte carlo
SDE	Stochastic differential equation
GMDB	Guaranteed minimum death benefit
IDW	Inverse distance weighting method
TFMC	Truncated fuzzy *c*-means algorithm
AFNS	Arbitrage-free Nelson–Siegel models
DNS	Dynamic Nelson–Siegel
CAPM	Capital asset pricing modeltheory
MPT	Modern portfolio theory
CAPM	Capital asset pricing model
CED	Conditional expected Drawdown
KKT	Karush-Kuhn-Tucker
NYSE	New York Stock Exchange
DPR	Degree-preserving randomization
GFC	Global financial crisis
CDF	Cumulative distribution function
CPT	Cumulative prospect theory
EUT	Expected utility theory
CML	Capital market line

CER	Constant expected return
CIR	Cox-Ingersoll-Ross
GHG	Anthropogenic greenhouse gas
CCX	Chicago climate exchange
CDM	Clean development mechanism
PDE	Partial differential equation
EVM	Ethereum virtual machine
CEV	Constant elasticity of variance
FEM	Finite element method
FLOP	Floating strike look-back put
ASE	American stock exchange
FROs	Fixed return options

PREFACE

In the last half-century, financial derivatives have become progressively central in finance being actively traded on many exchanges throughout the world and entered into by financial institutions, fund managers, and corporate treasurers. However, their use can be efficient and positively encourage financial and social sustainability, only if there is a better knowledge of the mathematical relationships among bond prices and interest rates.

Section 1 of this book treats the annuities, i.e., those series of payments made at equal time intervals. In particular, it focuses on bounds for annuities and life insurance from incomplete mortality data, on the effect of the assumed interest rate and smoothing on variable annuities, and the role of mandatory social annuitization. Moreover, it deals with the evaluation of variable annuity and the effect of jumps in the asset price process.

Section 2 discusses the bonds, i.e., loans made by investors to borrowers. In detail, it considers the analysis of corporate bond valuation under an infinite dimensional compound Poisson framework and of the time passage effect on bond prices. Moreover, it takes into account a new corporate bond pricing model with potential credit rating change and stochastic interest rate and forecasting term structure of interest rates in Japan.

Section 3 treats the portfolio theory focusing on its optimization. In particular, it deals with multi-period markets and modular approach, regularized mean–variance portfolio selection strategy with modified Roy safety-first principle, and a new measure for risk of pension fund portfolios.

Finally, the last Section 4 reviews the financial derivatives and their valuation with special attention to the options. In detail, it examines the benefit of stochastic modeling and variance reduction technology based on the Monte Carlo model in the valuation of derivatives.

SECTION 1
ANNUITIES

CHAPTER

1

Upper and Lower Bounds for Annuities and Life Insurance from Incomplete Mortality Data

Filipe Costa de Souza[1]

[1]Universidade Federal de Pernambuco, Centro de Ciências Sociais Aplicadas, Departamento de Ciências Contábeis e Atuariais, Recife, PE, Brazil

ABSTRACT

This study aimed to set upper and lower bounds for the expected present value of whole life annuities and whole life insurance policies from incomplete mortality data, generalizing previous results on life expectancy. Since its inception, in the 17th century, actuarial science has been devoted to the study of annuities and insurance plans. Thus, setting intervals that provide an initial idea about the cost of these products using incomplete mortality data represents a theoretical contribution to the area and this may have major

Citation: Souza, Filipe Costa de. (2019) "Upper and lower bounds for annuities and life insurance from incomplete mortality data". Revista Contabilidade & Finanças, 30(80), 282-291. Epub March 25, 2019. https://doi.org/10.1590/1808-057x201807320

applications in markets lacking historical records or those having little reliability of mortality data, as well as in new markets still poorly explored. For both the continuous and discrete cases, upper and lower bounds were constructed for the expected present value of whole life annuities and whole life insurance policies, contracted by a person currently aged x, based on information about the expected present value of these respective financial products subscribed to by a person of age $x + n$ and the probability that an individual of age x survives to at least age $x + n$. Through the bounds of a continuous annuity, in an environment where the instantaneous interest rate is equal to zero, the results shown also set bounds for the complete life expectancy, which implies that the contribution of this research generalizes previous results in the literature. It was also found that, for both annuities and insurance plans, the length of constructed intervals increases as the data gap size increases and it decreases as the survival curve becomes more rectangular. Illustratively, bounds for life expectancy at 40 and 60 years of age, for the 10 municipalities showing the highest life expectancy at birth in Brazil in 2010, were constructed by using data available in the Atlas of Human Development in Brazil.

Keywords:-Actuarial mathematics; actuarial science; annuities; life insurance; mortality table

INTRODUCTION

According to Pitacco, Denuit, Haberman and Olivieri (2009), actuarial science flourished in the mid-17th century, based on compound interest rate theory and probability theory, as well as observations on mortality. Also according to these authors, one of the first themes to be addressed by this new science was calculating the expected present value of annuities. Such interest arose because governments used to sell whole life annuities as a way to finance public enterprises.

Pitacco et al. (2009) also indicate that Jan de Witt, in 1671, was the precursor in the calculation of annuities using a hypothetical mortality table and a constant interest rate. However, as well emphasized by Haberman and Sibbet (1995), de Witt's contribution had little repercussion at the time. The same, in turn, cannot be said of Edmund Halley's work, in 1693, which, in addition to developing a mortality table by means of actual observations, also introduced a method for calculating the cost of annuities that reverberates up to the present day. The reader interested in historical aspects of actuarial science will benefit from the reading of Hald (1990) and Haberman and Sibbet (1995).

Thus, it can be noticed that actuarial science, since its inception, has taken as one of its central foci the study of annuities. To calculate the expected present value in an annuity for any given individual of age x, it is necessary to have access to a complete mortality table (having information from at least age x) or a survival function representative of the population to which the person belongs. In some instances, it is also possible to generate a complete table by means of abridged life tables and use the resulting complete table to calculate the cost of desired annuity or insurance (Baili, Micheli, Montanari, & Capocaccia, 2005; Ibrahim, 2008).

However, when there is no data on the mortality probabilities (age after age) since age x, calculating the expected present value of desired annuity is compromised. The impossibility to obtain detailed mortality data for a sequence of ages may occur due to lack of historical records or little reliability of existing data or also because this is a new market that is still poorly explored, for instance.

The absence of complete mortality tables also hinders the calculation of longevity measures, as in the case of the complete life expectancy. In this way, Cohen (2011) established upper and lower bounds on life expectancy at a given age x, knowing only detailed mortality data from age $x + n$ (and, indeed, life expectancy at age $x + n$), as well as the probability that a person aged x survives to at least age $x + n$.

This study aims to extend the results of Cohen (2011), by setting bounds (upper and lower) for the expected present value of annuities and life insurance policies subscribed to by a person of age x, both in the continuous and discrete cases. Based on the definition of these bounds, it might be possible to estimate the actuarial premium of these financial products lacking detailed knowledge of the mortality profile between age x and a given age $x + n$.

It is worth emphasizing that, although the absence of data for more advanced ages is more common, it is understood that rather atypical cases also deserve careful discussion. In practice, the data gap for younger ages could occur, for instance, in the so-called Special Regimes of Social Security (Regimes Próprios de Previdência Social - RPPS) - or also in pension funds - which have few employees in the early working ages (or even a sequence of early ages without employees) and those in need to estimate mortality tables and/or calculate the cost of annuities for retirement purposes. So, this study brings a theoretical contribution to the literature, by generalizing the results of Cohen (2011), with a potential practical applicability.

In order to achieve the goal outlined, the remainder of the text is organized as follows: section 2 presents the theoretical framework, focusing on the calculation of net single premiums for whole life annuities and life insurance, as well as an exposition of the results of Cohen (2011) for life expectancy. Section 3 shows the bounds, upper and lower, for the expected present value of annuities and insurance policies; and section 4 shows numerical results in an illustrative way, also promoting discussions on the theme. In turn, section 5 brings the final remarks of this study. It is also worth mentioning that, throughout the text, the standard actuarial notation is adopted. For more details, see Bowers, Gerber, Hickman, Jones and Nesbitt (1997).

THEORETICAL FRAMEWORK

Suppose a complete mortality table that has ω as its maximum age, i.e. no individual is supposed to be alive at age ω. Being $q\ x$ the probability that a person of age x dies along that age, this implies that $q\ \omega\text{-}1 = 1$. In addition, $t\ p\ x$, for $0 \leq x \leq \omega$ an $t \geq 0$, indicates the probability that a person of age x survives to at least age $x + t$. In the continuous case, $t\ p\ x$ is named as survival function. Of course, as a survival function, $t\ p\ x$ is a non-increasing function of t, i.e. as t increases, $t\ p\ x$ decreases or remains at the same value. In addition, $t\ p\ x = 1$ and $t\ p\ x = 0$ whenever $t \geq \omega - x$. In addition, being $\mu_x(t) = \lim_{dt \to 0^-} \frac{{}_{dt}q_{x+t}}{dt}$ the force of mortality (or instantaneous mortality rate) at age $x + t$, for $t > 0$, then, the probability that a person of age x survives to at least age $x + t$ and dies instantaneously thereafter is defined by $t\ p\ x\ \mu\ x\ (t)dt$ (Dickson, Hardy, & Waters, 2013). Finally, $i \geq 0$ is the annual effective interest rate in a compound capitalization regime and $\delta \geq 0$ is the instantaneous interest rate in the continuous capitalization regime, so that $ln(1 + i) = \delta$. In this way, the financial decapitalization factor is defined as $v = 1/(1 + i) = e\ \text{-}\delta$.

So, as well taught by Dickson, Hardy and Waters (2013), the expected present value of whole life continuous annuity subscribed to by a person aged x is given by:

$$\bar{a}_x = \int_0^{\omega - x} {}_tp_x \cdot e^{-\delta t}\,dt. \tag{1}$$

The discrete case, i.e. the net single premium for a whole life annuity-due subscribed to by a person aged x is defined as:

$$\ddot{a}_x = \sum_{t=0}^{\omega - x - 1} {}_tp_x \cdot v^t. \tag{2}$$

Being $\mu_x(t) = \lim_{dt \to 0^+} \frac{{}_{dt}q_{x+t}}{dt}$ the expected present value of a whole life insurance policy in the continuous case subscribed to by a person aged x, and being A_x the discrete counterpart, i.e. paying a monetary unit in the end of the contracting party's death year, then such values are formally defined as:

$$\overline{A}_x = \int_0^{\omega - x} {}_tp_x \mu_x(t) . e^{-\delta t} dt \qquad (3)$$

$$A_x = \sum_{t=0}^{\omega - x - 1} {}_tp_x q_{x+t} . v^t . \qquad (4)$$

It is possible to associate the actuarial present value of a whole life insurance benefit (continuous and discrete) to the present value of the respective annuities as follows:

$$\overline{A}_x = 1 - \delta . \bar{a}_x \qquad (5)$$

$$A_x = 1 - (1 - v) . \ddot{a}_x . \qquad (6)$$

Further details about the relation between annuities and insurance are presented by Dickson, Hardy and Waters (2013).

To calculate both the value of $\ddot{a}_x$ and A_x (when $i > 0$), it is necessary to know the entire survival history since age x. The same requirement would also be needed if we wished to calculate the complete life expectancy of a person aged x, which is defined as:

$$\mathring{e}_x = \int_0^{\omega - x} {}_tp_x dt . \qquad (7)$$

In this spirit, Cohen (2011) set bounds, upper and lower, on life expectancy of a person aged x, , even lacking detailed knowledge of the mortality history between ages x and x + n, as long as the probability that an individual aged x survives to at least age x + n is known, ${}_np_x$, as well as $\mathring{e}_{x+n}$. Formally, according to the author, such bounds are defined as:

$${}_np_x \left(n + \mathring{e}_{x+n} \right) \leq \mathring{e}_x < n + {}_np_x . \mathring{e}_{x+n} . \qquad (8)$$

So, using these bounds, a researcher could have better knowledge about, or even to approximate, the value of $\mathring{e}_x$. An example of the application of this result is the study by Rabbi (2013), who calculated the upper and lower bounds for life expectancy in various age groups, as a way of analyzing trends in life expectancy in Bangladesh.

Besides the clear practical relevance, this result was also used by the author in the development of other theoretical results. Cohen (2015) showed sharper bounds for Markov and Chebyshev's inequalities when information on conditional expectation is available. According to the author, Markov's inequality provides an upper bound for the probability that a non-negative random variable takes a value greater than n (for x > 0). Thus, for instance, being such a random variable the time lived by a newborn up to her/his death, Markov's inequality would indicate that:

$$_{n}p_{0} \leq \frac{\mathring{e}_{0}}{n}. \tag{9}$$

However, from (8), once $\mathring{e}_n$ (conditional expectation) is known, it is possible to refine inequality as follows:

$$_{n}p_{0} \leq \frac{\mathring{e}_{0}}{n + \mathring{e}_{n}}. \tag{10}$$

So, these results only reinforce the importance of the study by Cohen (2011), stimulating the extension of these ideas to other areas.

UPPER AND LOWER BOUNDS FOR ANNUITIES AND LIFE INSURANCE

In this section, the idea proposed by Cohen (2011) is extended to the case of continuous annuities, as set out in Proposition 1. In addition, the discrete case is also presented in Proposition 2. However, it is first necessary to define the present value of a given annuity-certain, immediate, temporary per n years and paying one monetary unit per year, for both the continuous and the discrete (in advance) cases, as expressed in equations (11) and (12), respectively:

$$\bar{a}_{\overline{n|}} = \int_{0}^{n} e^{-\delta t} dt = \begin{cases} n, \delta = 0 \\ \frac{1-e^{-\delta t}}{\delta}, \delta > 0. \end{cases} \tag{11}$$

$$\ddot{a}_{\overline{n|}} = \sum_{t=0}^{n-1} v^{t} = \begin{cases} n, i = 0 \\ \frac{1-v^{n}}{1-v}, i > 0. \end{cases} \tag{12}$$

Proposition 1 (continuous case): being x and $x + n$ two ages such that $0 \leq x \leq x + n < \omega$, it is possible to state that

$$\left(\bar{a}_{\overline{n|}} + \bar{a}_{x+n} \cdot e^{-\delta n}\right) \cdot {}_{n}p_{x} \leq \bar{a}_{x} \leq \bar{a}_{\overline{n|}} + {}_{n}p_{x} \cdot e^{-\delta n} \cdot \bar{a}_{x+n} \tag{13}$$

This proposition sets upper and lower bounds on ex using known values of $\bar{e}_{x+n}$ and $n\,p\,x$. The proof follows the same line as Cohen (2011).

Proof: it is known that

$$\bar{a}_x = \int_0^{\omega-x} {}_tp_x \cdot e^{-\delta t}\,dt. = \int_0^{n} {}_tp_x \cdot e^{-\delta t}\,dt + \int_n^{\omega-x} {}_tp_x \cdot e^{-\delta t}\,dt$$

$$= \int_0^{n} {}_tp_x \cdot e^{-\delta t}\,dt + {}_np_x \cdot e^{-\delta n} \cdot \int_0^{\omega-x-n} {}_tp_{x+n} \cdot e^{-\delta t}\,dt.$$

Since it is a probability, one must have that $t\,p\,x \leq 1$, for every $t \geq 0$, so

$$\int_0^{n} {}_tp_x \cdot e^{-\delta t}\,dt \leq \int_0^{n} e^{-\delta t}\,dt = \bar{a}_{\overline{n}|}.$$

Therefore,

$$\bar{a}_x \leq \int_0^{n} e^{-\delta t}\,dt + {}_np_x \cdot e^{-\delta n} \cdot \int_0^{\omega-x-n} {}_tp_{x+n} \cdot e^{-\delta t}\,dt = \bar{a}_{\overline{n}|} + {}_np_x \cdot v^n \cdot \bar{a}_{x+n}.$$

Now, it remains to prove the lower bound. Thus, return to the point where

$$\bar{a}_x = \int_0^{n} {}_tp_x \cdot e^{-\delta t}\,dt + {}_np_x \cdot e^{-\delta n} \cdot \int_0^{\omega-x-n} {}_tp_{x+n} \cdot e^{-\delta t}\,dt.$$

Since $t\,p\,x$ is a non-increasing function of t, so $n\,p\,x \leq t\,p\,x$, for $0 \leq t \leq n$. Indeed,

$$\int_0^{n} {}_tp_x \cdot e^{-\delta t}\,dt \geq \int_0^{n} {}_np_x \cdot e^{-\delta t}\,dt = {}_np_x \cdot \bar{a}_{\overline{n}|}$$

Therefore,

$$\bar{a}_x \geq \left(\bar{a}_{\overline{n}|} + \bar{a}_{x+n} \cdot e^{-\delta n}\right) \cdot {}_np_x,$$

which ends the proof.

After this demonstration, two points deserve to be highlighted. First, Proposition 1 is, in fact, an extension of the findings by Cohen (2011). To verify this, it is enough to assume that $\delta = 0$ in (13), from there it naturally follows that

$$\left(n + \overset{\circ}{e}_{x+n}\right) \cdot {}_{n}p_{x} \leq \overset{\circ}{e}_{x} \leq n + {}_{n}p_{x} \cdot \overset{\circ}{e}_{x+n}$$

that is, the result of Cohen (2011) is a particular case of Proposition 1 when $\delta = 0$.

Second, for $\delta > 0$, and based on the relation between $\overline{A}_x$ and $\bar{a}_x$ presented in Equation (5), it is also possible to establish bounds for $\overline{A}_x$ using an a priori knowledge about $\overline{A}_{x+n}$ and $_np_x$. In this way, we have:

$$e^{-\delta n}\left(1 - {}_{n}p_{x}\right) + {}_{n}p_{x} \cdot e^{-\delta n} . \overline{A}_{x+n} \leq \overline{A}_{x} \leq 1 - {}_{n}p_{x} \quad {}_{n}p_{x} \cdot e^{-\delta n} . \overline{A}_{x+n}$$

(14)

To develop this result, realize initially that replacing $\bar{a}_x$ by the upper bound, $\bar{a}_{\overline{n}|} + {}_np_x . e^{-\delta n} . \bar{a}_{x+n}$, in Equation (5), would imply that:

$$\overline{A}_{x} \geq 1 - \delta\left(\frac{1 - e^{-\delta n}}{\delta} + {}_{n}p_{x} . e^{-\delta n} . \bar{a}_{x+n}\right)$$

and how, also using Equation (5), we know that

$$\bar{a}_{x+n} = \frac{1 - \overline{A}_{x+n}}{\delta},$$

so,

$$\overline{A}_{x} \geq 1 - \delta\left[\frac{1 - e^{-\delta n}}{\delta} + {}_{n}p_{x} . e^{-\delta n} . \left(\frac{1 - \overline{A}_{x+n}}{\delta}\right)\right] = e^{-\delta n}\left(1 - {}_{n}p_{x}\right) + {}_{n}p_{x} . e^{-\delta n} . \overline{A}_{x+n};$$

thus establishing the lower bound for $\overline{A}_x$.

Similarly, replacing $\bar{a}_x$ by the lower bound, ($\left(\bar{a}_{\overline{n}|} + \bar{a}_{x+n} . e^{-\delta n}\right) . {}_np_x$, in Equation (5), would imply that:

$$\overline{A}_{x} \leq 1 - \delta\left(\frac{1 - e^{-\delta n}}{\delta} + e^{\delta n} . \bar{a}_{x+n}\right){}_{n}p_{x} = 1 - \delta\left[\frac{1 - e^{-\delta n}}{\delta} + e^{-\delta n} . \left(\frac{1 - \overline{A}_{x+n}}{\delta}\right)\right] . {}_{n}p_{x} = 1 - {}_{n}p_{x} + {}_{n}p_{x} . e^{-\delta n} . \overline{A}_{x+n},$$

which, in turn, sets the upper bound for $\overline{A}_x$.

Once the result is known for the continuous case, the discrete case is developed in a similar way, as set out in Proposition 2.

Proposition 2 (discrete case): being x and x + n two integer ages, such that $0 \leq x \leq x + n < \omega$, it is possible to state that

$$\left(\ddot{a}_{\overline{n}|} + \ddot{a}_{x+n} . v^{n}\right) . {}_{n}p_{x} \leq \ddot{a}_{x} \leq \ddot{a}_{\overline{n}|} + {}_{n}p_{x} . v^{n} . \ddot{a}_{x+n} \qquad (15)$$

Proof: it is known that

$$\ddot{a}_x = \sum_{t=0}^{\omega-x-1} {}_tp_x \cdot v^t = \sum_{t=0}^{n-1} {}_tp_x \cdot v^t + \sum_{t=n}^{\omega-x-1} {}_tp_x \cdot v^t \quad \sum_{t=0}^{n-1} {}_tp_x \cdot v^t + {}_np_x \cdot v^n \cdot \sum_{t=0}^{\omega-x-n-1} {}_tp_{x+n} \cdot v^t .$$

Since it is a probability, one must have that tpx ≤ 1, for t = 0,1,..., n - 1, so

$$\sum_{t=0}^{n-1} {}_tp_x \cdot v^t \leq \sum_{t=0}^{n-1} v^t = \ddot{a}_{\overline{n|}} .$$

Therefore,

$$\ddot{a}_x \leq \sum_{t=0}^{n-1} v^t + {}_np_x \cdot v^n \cdot \sum_{t=0}^{\omega-x-n-1} {}_tp_{x+n} \cdot v^t = \ddot{a}_{\overline{n|}} + {}_np_x \cdot v^n \cdot \ddot{a}_{x+n}$$

Now, it remains to prove the lower bound. Thus, return to the point where

$$\ddot{a}_x = \sum_{t=0}^{n-1} {}_tp_x \cdot v^t + {}_np_x \cdot v^n \cdot \sum_{t=0}^{\omega-x-n-1} {}_tp_{x+n} \cdot v^t .$$

Since tpx is a non-increasing function of t, so npx ≤ tpx, for t = 0,1,...,n - 1. Indeed,

$$\sum_{t=0}^{n-1} {}_tp_x \cdot v^t \geq \sum_{t=0}^{n-1} v^t \cdot {}_np_x = {}_np_x \cdot \ddot{a}_{\overline{n|}} .$$

Therefore,

$$\ddot{a}_x \geq \left(\ddot{a}_{\overline{n|}} + \ddot{a}_{x+n} \cdot v^n\right) \cdot {}_np_x ,$$

which ends the proof.

Again, for i > 0, and using Equation (6), the bounds are set for Ax by means of prior knowledge about Ax+n and npx. In this way, we have:

$$v^n\left(1 - {}_np_x\right) + {}_np_x \cdot v^n \cdot A_{x+n} \leq A_x \leq 1 - {}_np_x + {}_np_x \cdot v^n \cdot A_{x+n} \quad (16)$$

The development of this result is omitted, since it follows the same argument introduced for the continuous case and, therefore, it can be easily replicated by the reader.

APPLICATION AND DISCUSSIONS

This section suggests applications and provides numerical examples about the results demonstrated in section 3. In addition, the effects of variation in the interest rate and the data gap (i.e. the value of n) are discussed, as well as the retangularization of the survival curve.

Initially, for illustrative purposes, intervals for the values of $\bar{a}_{20}$ and A 20 are calculated by considering the complete availability of data from the ages of 20, 30, 40, 50, 60, and 70 years, i.e. the intervals were calculated by means of different values for the data gap. In addition, still for illustrative purposes, the central point of the interval is considered an estimate of the actuarial fair value of the respective financial products and, based on this estimate, the error (estimated value less actual value, which uses the complete data since 20 years) is also computed. For the calculations, the 'IBGE 2015 mortality table' was considered for both sexes - extrapolated for ages over 80 years, and an effective interest rate of 3% per year. The results for are summarized in Table 1, while the results for A 20 are set out in Table 2.

Table 1. Bounds, estimates, and errors for $\bar{a}_{20}$ based on complete data on mortality from the ages of 20, 30, 40, 50, 60, and 70 years

Age	**Data gaps (n) - in years**	**Lower bound**	**Upper bound**	**Central value**	**Length of the interval**	**Error**	**Error percentage**
20	0	27.21419	27.21419	27.21419	0.00000	0.00000	0.00000%
30	10	27.13380	27.27153	27.20267	0.13773	-0.01152	0.04234%
40	20	26.88249	27.42508	27.15379	0.54259	-0.06040	0.22194%
50	30	26.24663	27.65945	26.95304	1.41283	-0.26115	0.95960%
60	40	24.72109	28.00480	26.36294	3.28371	-0.85124	3.12794%
70	50	21.47332	28.50447	24.98890	7.03115	-2.22529	8.17695%

Source: Prepared by the author.

Table 2. Bounds, estimates, and errors for A20 based on complete data on mortality from the ages of 20, 30, 40, 50, 60, and 70 years

Age	**Data gaps (n) - in years**	**Lower bound**	**Upper bound**	**Central value**	**Length of the interval**	**Error**	**Error percentage**
20	0	0.20735	0.20735	0.20735	0.00000	0.00000	0.00000%
30	10	0.20568	0.20969	0.20768	0.00401	0.00033	0.16183%
40	20	0.20121	0.21701	0.20911	0.01580	0.00175	0.84840%

50	30	0.19438	0.23553	0.21496	0.04115	0.00760	3.66823%
60	40	0.18551	0.27382	0.22966	0.08831	0.02231	10.7621%
70	50	0.16977	0.37456	0.27216	0.20479	0.06481	31.2578%

Source: Prepared by the author.

It is noticed in Table 1 that the estimate of $\bar{a}_{20}$ using the central value of the interval for the age of 30 years underestimates the actual value of the annuity concerned by less than 0.05%. In turn, if we used data referring to 70 years, where there is a 50-year data gap, the value of would be underestimated by just over 8%. In turn, in Table 2, the estimate of A 20 using the central value of the interval for the age of 30 years might overestimate the actual value of the annuity concerned by only 0.16%. However, if data referring to 70 years were used, the value would be overestimated by over 31%.

As a way of visualizing the effects that a variation in the interest rate would produce on the results, using the same mortality table, intervals were constructed for the value of and A 20 considering data from 30 years and varying the value of interest rate. The results are detailed in tables 3 and 4, respectively. In these latter, given the choices made, it is observed that, in both cases, as the interest rate increases, the error also tends to increase.

Table 3. Bounds, estimates, and erros for $\bar{a}_{20}$ based on data from the age of 30 years in the face of various interest rates

Interest rate	**Lower bound**	**Upper bound**	**Central value**	**Length of the interval**	**Error**	**Error percentage**
1%	43.27120	43.42116	43.34618	0.14996	-0.01012	0.02335%
2%	33.71009	33.85372	33.78191	0.14363	-0.01086	0.03214%
3%	27.13380	27.27153	27.20267	0.13773	-0.01152	0.04234%
4%	22.46578	22.59801	22.53189	0.13223	-0.01211	0.05371%
5%	19.05272	19.17982	19.11627	0.12710	-0.01263	0.06604%

Source: Prepared by the author.

Table 4. Bounds, estimates, and erros for A 20 based on data from the age of 30 years in the face of various interest rates

Interest rate	Lower bound	Upper bound	Central value	Length of the interval	Error	Error percentage
1%	0.57008	0.57157	0.57083	0.00148	0.00010	0.01756%
2%	0.33620	0.33901	0.33761	0.00281	0.00021	0.06312%
3%	0.20568	0.20969	0.20768	0.00401	0.00033	0.16183%
4%	0.13084	0.13593	0.13338	0.00508	0.00046	0.35037%
5%	0.08667	0.09272	0.08970	0.00605	0.00060	0.67511%

Source: Prepared by the author.

At this point, two aspects deserve to be highlighted: first, it would only be possible to calculate the error of having the actual values of financial products; but, surely, such values would not be available in situations involving incomplete mortality data. Second, the error would depend on the value used as an estimate. If, for instance, a researcher decided to conservatively use values greater than the central value of the interval as a way of analyzing the expected present value of annuities and life insurance policies, then the results on error might be different. Therefore, it is useful to evaluate how the length of the interval produced behaves in face of variations in the parameters, as the next subsection does.

The Length of Intervals

Once some illustrative results have been shown, it is worth formally discussing the impact of certain parameters on the length of the interval (upper bound minus lower bound) produced. Being *T a* the length of the interval formed by the bounds in (15), so *T a* is defined by the expression:

$$T_a = \ddot{a}_{\overline{n|}}(1 - {}_n p_x). \qquad (17)$$

Using Equation (17), it is noticed that, everything else being constant, *T a* increases with increases in *n*, since a¯20a¯20 is an increasing function of *n* and *n p x* is a non-increasing function of *n* (see example in Table 1); just as we observe that *T a* decreases with increases in *i*, since a¯20a¯20 is a decreasing function of *i* (Faro, 2006). This result is shown in Table 3. Finally, *T a* decreases with increases in *n p x* , since, as stated, *n p x* is a non-increasing function of *n*.

This last result refers to the phenomenon of retangularization of the survival curve. The rectangularization process, as well shown by Wilmoth and Horiuchi (1999), is characterized by high survival rates in childhood and adulthood, and rapid mortality in advanced ages. Thus, in a hypothetical case where $n\,p\,x = 1$, then the lower and upper bounds in (15) are equal and, indeed, $\ddot{a}_x = \ddot{a}_{\overline{n|}} + \ddot{a}_{x+n}.\, v^n$. This fact reinforces the result that the more rectangular the survival curve, the smaller distance between the lower and upper bounds of ax, i.e. the smaller interval is produced.

Just as in the discussion of the bounds of an annuity, it is also relevant to analyze the bounds for a life insurance. For this product, being TA the length of the interval formed by the bounds in (16), then TA is defined by:

$$T_A = (1 - v^n).(1 - {}_n p_x). \tag{18}$$

For (18), it is observed that TA increases with increases in n (see Table 2) and decreases with increases in ${}_n p_x$, similarly to what happens with Ta. However, TA increases with increases in i, since (1 - vn) is an increasing function of i (see Table 4).

Defining Bounds for e40 and e60 Using the Atlas of Human Development in Brazil

This subsection presents an actual application of the results of Cohen (2011) and, consequently, of this research. This is the calculation of bounds for e40 and e60 using information from the Atlas of Human Development in Brazil (Programa das Nações Unidas para o Desenvolvimento Humano, 2013a).Life expectancy at birth is the main longevity measure and it is commonly used as an indicator of human development (Mayhew & Smith, 2015). Due to their importance, data on life expectancy at birth are typically set out in research on demography, public health, etc. However, when there is no detailed data on mortality, such availability does not occur as easily for life expectancy at other ages. Mathers, Stevens, Boerma, White and Tobias (2015) point out that life expectancy at the age of 60 years, for instance, is a relevant indicator of longevity for older people and knowing it for a given population is key to public planning in social security and health, among other areas.Originally using the result of Cohen (2011), it has been found that we can estimate upper and lower bounds for complete life expectancy at age x, only knowing the probability that a person aged x is alive at age $x + n$ and the complete life expectancy at age $x + n$. Of course, using the same argument, bounds for $\mathring{e}_{x+n}$ are constructed by means of prior knowledge about $n\,p\,x$ and ex. That is, it follows directly from (8) that:

$$\frac{\mathring{e}_x - n}{{}_n p_x} \leq \mathring{e}_{x+n} \leq \frac{\mathring{e}_x}{{}_n p_x} - n. \tag{19}$$

The Atlas of Human Development in Brazil (Programa das Nações Unidas para o Desenvolvimento Humano, 2013b) is a tool for accessing the Municipal Human Development Index, for the years 1991, 2000, and 2010 and it is available at Programa das Nações Unidas para o Desenvolvimento Humano (2013a). Using this tool, it is possible, for instance, for the mentioned years, to obtain information about life expectancy at birth, the probability that a newborn survives until the age of 40 years, and the probability that a newborn survives until the age of 60 years, for all Brazilian municipalities. It is worth noticing that, in 2010, Brazil had 5,565 municipalities. So, it is possible to construct intervals for life expectancy values at ages 40 and 60 years for Brazilian municipalities in the respective years.Illustratively, Table 5 displays the bounds for $\mathring{e}_{40}$ and $\mathring{e}_{60}$ for the 10 municipalities with the highest life expectancy at birth in Brazil, in the year 2010. Curiously, all these municipalities belong to the state of Santa Catarina. Data were collected in the Atlas of Human Development in Brazil Programa das Nações Unidas para o Desenvolvimento Humano, 2013a) on December 22, 2017.

Table 5. Bounds for e40 and e40 for the 10 municipalities with the highest life expectancy at birth in Brazil, in the year 2010

Municipality	Lower bound for e40	Upper bound for e40	Length of the interval for e60	Lower bound for e60	Upper bound for e60	Length of the interval for e60
Blumenau	40.29197	42.00209	1.710115	21.15297	29.24194	8.08897
Brusque	40.29197	42.00209	1.710115	21.15297	29.24194	8.08897
Balneário Camboriú	40.26272	41.96414	1.701418	21.11829	29.16865	8.050357
Rio do Sul	40.22294	41.89395	1.671007	21.05918	28.95553	7.896345
Rancho Queimado	40.20629	41.88164	1.675349	21.04131	28.95303	7.911715
Rio do Oeste	40.14591	41.83429	1.688379	20.98346	28.96442	7.980965
Joaçaba	40.07924	41.78501	1.705766	20.91652	28.97459	8.058076
Iomerê	40.07924	41.78501	1.705766	20.91652	28.97459	8.058076
Porto União	40.06881	41.77458	1.705766	20.90992	28.98344	8.073519
Nova Trento	40.06881	41.77458	1.705766	20.90992	28.98344	8.073519

Source: Prepared by the author.

Taking, for instance, the case of Blumenau and if the central value of the interval (nearly 41.15 years) was used as an estimate for e40; even without knowing the actual value of this life expectancy, it is certain that, in this context, the error caused by using such estimate would be, at most, 312 days, i.e. a little more than 10 months.Thus, using these intervals could provide relevant information for research on the heterogeneity of the Brazilian population's mortality and promote discussions on the adoption of a minimum retirement age, particularly in the current political-economic context that the country has experienced (Souza, 2018).

FINAL REMARKS

In this study, and inspired by the work of Cohen (2011), the upper and lower bounds were demonstrated for the expected present value of annuities and life insurance policies. In addition, through the analysis of bounds for continuous annuities, it may be seen that the results shown are an extension of those provided by Cohen (2011), and this extended the discussion to an actuarial context.

For both annuities and life insurance, the length of constructed intervals increases with increases in the data gap and it decreases as the survival curve becomes more rectangular. The interest rate, in turn, has a different effect on the intervals, increasing them when the interest rate increases, in the case of insurance plans and decreasing them in the case of annuities.

Besides the theoretical contribution, it is worth pointing out that the definition of these intervals may have a significant potential for applicability in markets where there is no historical record of mortality for certain ages or where data are unreliable or even in new markets that are still poorly explored. And, in addition to be used for pricing these financial products, they can provide information for research on demography, public health, social security, and public planning, for instance.

REFERENCES

1. Baili, P., Micheli, A., Montanari, A., & Capocaccia, R. (2005). Comparison of four methods for estimating complete life tables from abridged life tables using mortality data supplied to EUROCARE-3. *Mathematical Population Studies, 12*, 183-198. [
2. Bowers, N. L., Gerber, H. U., Hickman, J. C., Jones, D. A., & Nesbitt, C. J. (1997). *Actuarial mathematics* (2nd ed.). Schaumburg, IL: Society of Actuaries.
3. Cohen, J. E. (2011). Life expectancy: lower and upper bounds from surviving fractions and remaining life expectancy. *Demographic Research, 24*, 251-256.
4. Cohen, J. E. (2015). Markov's inequality and Chebyshev's inequality for tail probabilities: a sharper image. *The American Statistician, 69*(1), 5-7.
5. Dickson, D. C., Hardy, M. R., & Waters, H. R. (2013). *Actuarial mathematics for life contingences risks* (2nd ed.). Cambridge: Cambridge University Press.
6. Faro, C. (2006). *Fundamentos da matemática financeira: uma introdução ao cálculo financeiro e à análise de investimento de risco*. São Paulo: Saraiva. [
7. Haberman, S., & Sibbett, T. A. (1995). The history of actuarial science. In Haberman, S. & Sibbett, T. A. (Ed.), *History of actuarial science* (Vol. 1, pp. xix-lxxxiii). London: William Pickering.
8. Hald, A. (1990). *A history of probability and statistics and their applications before 1750*. New York: John Wiley & Sons. [
9. Ibrahim, R. I. (2008). Expanding an abridged life table using the Heligman-Pollard model. *Matematika, 24*(1), 1-10.
10. Mathers, C. D., Stevens, G. A., Boerma, T., White, R. A., & Tobias, M. I. (2015). Causes of international increases in older age life expectancy. *Lancet, 385*, 540-548.
11. Mayhew, L., & Smith, D. (2015). On the decomposition of life expectancy and limits to life. *Population Studies, 69*(1), 73-89.
12. Pitacco, E., Denuit, M., Haberman, S., & Olivieri, A. (2009). *Modelling longevity dynamics for pension and annuity business*. Oxford: Oxford University Press. [
13. Programa das Nações Unidas para o Desenvolvimento Humano.

(2013a). Atlas do Desenvolvimento Humano no Brasil (homepage). Retrieved from http://www.atlasbrasil.org.br/2013/

14. Programa das Nações Unidas para o Desenvolvimento Humano. (2013b). *O Índice de Desenvolvimento Humano Municipal Brasileiro*. Brasília, DF: PNDU/IPEA/FJD. (Série Atlas do Desenvolvimento Humano no Brasil 2013).
15. Rabbi, A. M. F. (2013). Lower and upper bounds of life expectancies and total expected longevity of a developing country. *South East Asia Journal of Public Health*, 3(2), 52-57.
16. Souza, F. C. (2018). A heterogeneidade da mortalidade da população brasileira e aspectos distributivos na previdência social: uma análise atuarial da proposta de idade mínima de aposentadoria. *Administração Pública e Gestão Social*, *10*(1), 2-11. [
17. Wilmoth, J. T., & Horiuchi, S. (1999). Rectangularization revisited: variability of age at death within human population. *Demography*, *36*(4), 475-495. [

CHAPTER 2

The Effect of the Assumed Interest Rate and Smoothing on Variable Annuities

Anne G. Balter and Bas J. M. Werker

Department of Econometrics and Operations Research; Department of Finance Tilburg University and Netspar

INTRODUCTION

Over the past few decades, defined benefits pension schemes have largely been converted into defined contributions pension schemes without or with lower guarantees. Especially the recent financial crisis and increasing life expectancies affect the sustainability of pension systems that include guarantees. Therefore, there is a rising1 number of products available in the market that explicitly let these risks be borne by the individual rather than

Citation: Balter, A., & Werker, B. (2020). "The Effect Of The Assumed Interest Rate And Smoothing On Variable Annuities". Astin Bulletin, 50(1), 131-154. https://doi.org/10.1017/asb.2019.27

the employer or insurer. If the pension payments in the decumulation phase include risk, we refer to these designs as variable annuities. Fixed annuities are those for which the payments are not uncertain.

There is a wide literature on variable annuities including investigating different embedded guarantees (Mahayni and Schneider, 2012; Chen et al., 2015), pricing variable annuities (Bauer et al., 2008; Bacinello et al., 2011; Nirmalendran et al., 2014), hedging variable annuities (Coleman et al., 2006; Trottier et al., 2018) or combinations of these (Kling et al., 2011; Bernard et al., 2014). Moreover, optimal demand for different annuity products is also investigated (Horneff et al., 2009; Blake et al., 2014; Peijnenburg et al., 2016). Designs in which equity exposure is incorporated in the annuity product is shown to increase welfare by, for example, Koijen et al. (2011).

We investigate variable annuities, where the variability arises due to risky investment returns. We study the relationship between the so-called assumed interest rate (AIR) and the (expected) annuity payments. The AIR effectively determines the decumulation speed of financial wealth over the payout phase: a larger AIR leads to higher early payments and lower later payments. In case the AIR equals the expected return on the underlying portfolio, the income during the payout phase is, in expectation, constant. See Dellinger (2006) and Horneff et al. (2010) for more details on the usage of the AIR concept in insurance pricing. We provide a novel way to solve the optimal consumption problem and, in doing so, derive the AIR that optimizes lifetime utility of consumption. We also analyze the utility loss of investors with constant relative risk aversion (CRRA) preferences who allocate their wealth suboptimally over their life cycle and/or have a suboptimal risk exposure. Under the assumption of an optimally chosen risk exposure, we find that a restriction to a (suboptimal) constant expected pension income does not lead to large utility losses. We also show that pension payments with a horizon of "only" 10 years are fairly insensitive to the choice of the AIR. As a result, unlike common practice, communication about the effect of choosing an AIR is preferably based on results for horizons closer to 20 years. We also investigate how financial shocks can be smoothed and what the effect is on the variable annuity. In that case, we find that a horizon-dependent AIR ensures a constant expected pension income.

It is well known that CRRA investors prefer to absorb financial shocks immediately in their consumption rather than smoothing the effect across future consumption. This follows as the optimal consumption is a constant fraction of the wealth level and thus a shock in wealth leads one-to-one to a shock in consumption. Nonetheless, we investigate the impact of smoothing in variable annuities and show how a stable expected pension payment can be attained within this setting. Maurer et al. (2016) and Guillén et al. (2006) investigate smoothing in collective pension plans with both intergenerational risk transfers and risk-sharing between pension providers and holders. We abstain from wealth transfers between retirees and insurers, and focus on the method to smooth within the retiree's own available wealth. Bruhn and Steffensen (2013) show that habit formation preferences can explain the demand for such collective smoothing designs.

Our approach to smoothing financial market shocks is novel. More precisely, we consider strategies that lead to smooth consumption paths. The term smoothing refers to a preference for smooth consumption paths in habit formation like specifications. Such preferences lead to spreading out financial shocks and, thus, being less exposed to large upside and downside risk on short horizons. As mentioned above, we explicitly separate the allocation of initial (pension) wealth to pension payments at various horizons. We call these "pension buckets" for the given horizons. Given the allocation of pension wealth to the individual pension buckets, each investment problem is a standard final-wealth problem whose properties are well studied that started with the seminal work of Merton (1969) and Cox and Huang (1989). Smoothing is then implemented by reducing the stock allocation for the pension bucket in the final years before the horizon. Up to our knowledge, this paper is new in its approach toward smoothness. The possibility of smoothing financial market returns has been incorporated in many products in order to allow for preferences that exhibit habit formation. Smoothing financial market returns essentially means that a decrease of, say, 10% in pension wealth needs not be translated immediately into a decrease of 10% in pension payments. This implies that pension payments will decrease, due to this adverse financial market development, at a slower pace, but ultimately by more than 10%. We explicitly model this possibility and show how it influences the risk return trade-off for variable pension payments. In order to design the annuity such that constant nominal expected pension payments are obtained, we show that smoothing leads to a horizon-dependent AIR.

The rest of this paper is organized as follows. In Section 2, we investigate the risk–return trade-off for variable annuities by deriving the AIR that leads to constant expected pension payments, the optimal AIR for a CRRA investor and the utility loss between these. In Section 3, we consider how to smooth financial market shocks. And Section 4 concludes the paper. As numerical illustration, we provide a running example throughout the paper.

Keywords
Optimal consumption, certainty equivalent loss, variable annuities, assumed interest rates, conversion risk, smoothing financial market shocks.

VARIABLE ANNUITIES

The financial market that we consider is described by the seminal work of Merton (1971). This implies that in the standard Black–Scholes/Merton setting there is a risk-free asset with a constant interest rate r, there is a risky asset with price St at time t that evolves by the diffusion process.

$$dS_t = \mu S_t dt + \sigma S_t dZ_t \quad (2.1)$$

$$= (r + \lambda\sigma)\, S_t dt + \sigma S_t dZ_t. \quad (2.2)$$

Thus, we assume that the stock price S_t follows a geometric Brownian motion, where μ stands for the expected return, σ is the stock volatility, λ is the Sharpe ratio

$$\lambda = (\mu - r)\,/\sigma, \quad (2.3)$$

and Z is a standard Brownian motion on the probability space (Ω, F, P)

Moreover, we assume the isoelastic (power) function for utility that exhibits a CRRA and is given by

$$u(x) = \begin{cases} \frac{x^{1-\gamma}}{1-\gamma} & \text{if } \gamma > 0, \gamma \neq 1, \\ \ln(x) & \text{if } \gamma = 1, \end{cases} \quad (2.4)$$

where γ is the relative risk aversion level. The more risk averse the investor is, the higher γ. We exclude negative risk aversion levels which would imply risk loving preferences. Since additive constant terms in objective functions do not affect optimal decisions, a term of minus one is omitted from the numerator which would be needed to show that the limiting case of γ to one converges to logarithmic utility.

In general, the investor is endowed with initial wealth W_0 which can be used for consumption and the remainder is invested in the financial market. The wealth process is given by

$$dW_t = ((r + w_t(\mu - r))\, W_t - c_t)\, dt + \sigma w_t W_t dZ_t, \quad (2.5)$$

where w_t is the fraction invested in the risky asset and c_t is the withdrawal (consumption) rate. For the CRRA utility function, the optimal time-varying risk exposure w_t is known to be

$$w^* = \frac{\lambda}{\gamma\sigma}; \tag{2.6}$$

see, for example Theorem 3.8.8. in Karatzas and Shreve (1998). That is, the optimal exposure is state- and time-independent. Concerning the optimal consumption choice, we represent the withdrawal via the AIR which determines the allocation of initial wealth to the (optimal) consumption at various horizons. We formulate this problem using a discrete number of consumption dates which effectively means that we solve H separate terminal wealth problems. The novelty in this setup allows us to directly cast optimal consumption questions into AIRs in variable annuities.

As an example, consider a retiree who enters retirement with total wealth W_0 at time 0 and who needs to finance H annual pension payments at times. For ease of exposition, we assume H to be given; that is, we consider fixed term instead of lifelong variable annuities. Think of H as the remaining life expectancy at retirement age.[2]

The pension payments at each horizon $h = 0, \dots, H - 1$ have to be financed from the initial total pension wealth W_0. This simple idea is formalized by the notation in the next definition.

Definition 2.1

Let $W_0(h)$ be the amount reserved at time 0 that finances (through a self-financing strategy) the consumption in year h.

The budget constraint now implies

$$W_0 = \sum_{h=0}^{H-1} W_0(h). \tag{2.7}$$

Stated otherwise, at time 0 we consider an amount of wealth $W_0(h)$ that is available to finance the pension payment at time h. The actual pension payment will, of course, depend on the investment strategy that is followed and the financial market returns. We can, conceptually, allow for a different investment strategy for the wealth allocated to each horizon $h = 0, \dots, H - 1$. This is indeed precisely what happens when financial market returns are smoothed; see Section 3.

The way total pension wealth W_0 is allocated over all pension buckets $W_0(h)$ determines implicitly the so-called AIR $a_0(h)$.

Definition 2.2

The AIR at time 0 for horizon h, a0(h), is defined by

$$\frac{W_0(h)}{W_0(0)} = \exp\left(-ha_0(h)\right). \tag{2.8}$$

Jointly with the total wealth W0 available in (2.7), this implies

$$\frac{W_0(h)}{W_0} = \frac{W_0(h)}{\sum_{k=0}^{H-1} W_0(k)} = \frac{\exp\left(-ha_0(h)\right)}{\sum_{k=0}^{H-1} \exp\left(-ka_0(k)\right)}. \tag{2.9}$$

Intuitively, this relation determines the fractions of total wealth allocated to the various horizons h. Note that $a_0(h)$ generally does not need to be constant in h; that is, there can be an AIR term structure or, equivalently, a horizondependent AIR. This will become particularly relevant when discussing the possibility of smoothing in Section 3.

We introduce a running example that is used as illustration throughout this paper with the following parameters: the expected return is $\mu = 6.00\%$, the stock volatility is $\sigma = 20.00\%$ and the risk-free interest rate is $r = 2.00\%$, which leads to a Sharpe ratio of $\lambda = 0.20$. We fix the horizon at $H = 20$ years. The initial wealth is assumed to be equal to W0 = e100, 000. Figure 1 shows, for two different AIRs, the percentage of total wealth W0 that is allocated to each horizon; that is, we depict $W_0(h)/W0$ as given by (2.9)

Let us, for the sake of illustration and before turning to variable annuities, briefly review fixed annuities. We assume that the wealth $W_0(h)$ for each horizon h is completely invested in a risk-free bank account. Using an AIR of 2.00%, that is equal to the prevailing risk-free rate r, implies that each subsequent bucket that is allocated to a specific year contains 2.00% less of the initial wealth than the previous bucket; see the solid bars in Figure 1. Since each bucket will grow each year at the risk-free rate, we ultimately get a constant stream of pension payments. The initial allocation to the first bucket will be largest since it will not earn any interest income, while the second bucket can contain 2.00% less since the additional income from interest for 1 year is also 2.00%.

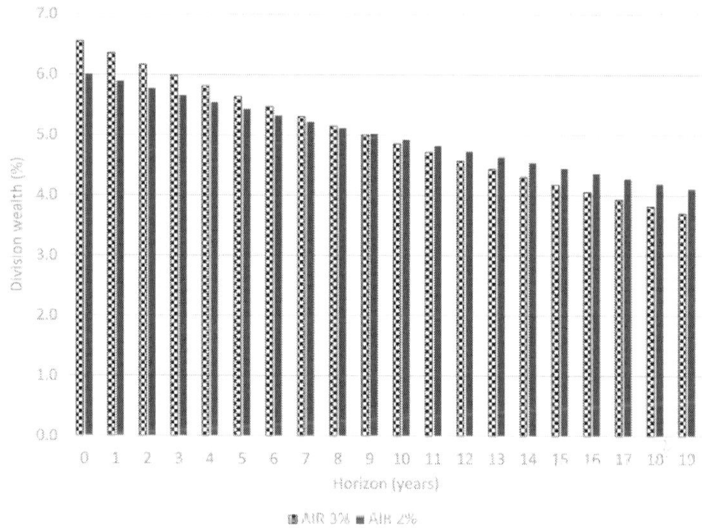

Figure 1: Wealth division for a0(h) = 2.00% and a0(h) = 3.00% for all h ∈ [0, ..., 19].

Now consider the blocked bars in Figure 1, which is based on an AIR of $a_0(h)$ = 3.00%. In this case, each subsequent bucket will contain 3.00% fewer initial wealth than the previous one. As a result, more wealth is allocated to earlier payments and less to later payments. Stated differently, since the initial total wealth W_0 is given, the higher earlier payments will lead to lower later payments. This is visible in Figure 1, where the blocked bar compared to solid bars imply that the first payment is 9.1% higher while the last payment is 9.8% lower. In terms of pension payments, an entirely risk-free investment would lead to a r = 2.00% return every year. As a result, pension payments will decrease by 3.00% − 2.00% = 1.00% every year.

Now, let us consider what happens in case each of the buckets is not invested risk free but partly in the risky stock St. That is, we invest each $W_0(h)$ in a continuously re-balanced strategy with a stock exposure w. Standard calculations then show that wealth $W_t(h)$, for the pension payment at time h, evolves as

$$dW_t(h) = (r + w\lambda\sigma)\, W_t(h)dt + w\sigma\, W_t(h)dZ_t, \qquad (2.10)$$

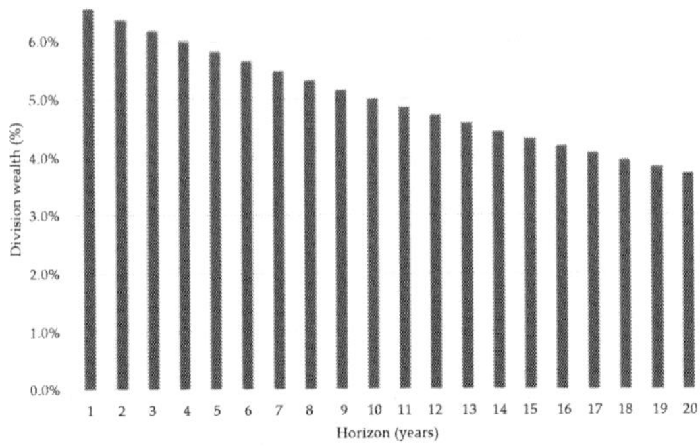

Figure 2: Sensitivity AIR.

for $0 \leq t \leq h$. For an investment mix of w, the expected pension payment of a variable annuity at horizon h is given by

$$\mathbb{E}_0[W_t(h)] = W_0(h) \exp\left(t\left(r + w\lambda\sigma\right)\right). \tag{2.11}$$

The quantile at level α is given by

$$Q_0^{(\alpha)}\left(W_t(h)\right) = W_0(h) \exp\left(t\left(r + w\lambda\sigma - \frac{1}{2}w^2\sigma^2\right) + z_\alpha\sqrt{t}w\sigma\right), \tag{2.12}$$

where z_α denotes the corresponding quantile of the standard normal distribution. Note that the risk exposure that is constant per bucket can easily be extended to time-varying risk exposure, though optimality is obtained for a constant w.

Remark 2.1. Figure 2 shows the sensitivity of the AIR on the expected pension payment at h = 9 and h = 19. We observe that an increase in the AIR has hardly any effect on the distribution of the 10th pension payment, but all the more on the 20th payment. The higher the AIR, the larger the expected payments in the near future and the lower the expected payments in the distant future, with the expected payments in the middle of the horizon being hardly influenced by the choice of AIR.

Remark 2.2. Contrary to popular belief, a variable annuity does not provide protection against interest rate risk upon conversion of initial pension wealth W0 into an annuity. Conversion risk is also known as annuity risk, that is, the risk that if financial market conditions are unfavorable at retirement, annuities are relatively expensive. Both for fixed and variable annuities, the level of the interest rates at the moment of conversion plays an important role. Although we do not explicitly consider a market with interest rate risk, we can assess the conversion risk effect by studying the implications of changes in the risk-free rate r.

In general $W_0(h) = \frac{W_0 \exp(-ha_0(h|w))}{\sum_{k=1}^{H-1} \exp(-ka_0(k|w))}$ holds, which we obtain by slightly rewriting (2.9). For the fixed annuity, the pension payments are known to be equal to $W_h(h) = W_0(h) \exp(rh)$, while for the variable annuity, $E_0[W_h(h)] = W_0(h) \exp((r + w\lambda\sigma)h)$. And thus a change in the risk-free rate from r to r' and in the AIR from a to a' leads to an equal shock in all payments of

$$\frac{W_h(h)|_r}{W_h(h)|_{r'}} = \frac{E_0[W_h(h)]|_r}{E_0[W_h(h)]|_{r'}} = \exp((r - r' + a' - a)h)\frac{\sum_{k=1}^{H-1} \exp(-ka')}{\sum_{k=1}^{H-1} \exp(-ka)}. \quad (2.13)$$

The only[3] viable way of dealing with conversion risk is to use appropriate lifecycle investment strategies (in terms of bond duration) during the accumulation phase (Koijen et al., 2011). Products such as guaranteed annuity options (Kling et al., 2014) guarantee the policyholder at the start of the deferred annuity a minimum conversion factor or a minimum pension income both circumventing the risk of expensive lifelong annuities at the retirement age. However, the rate is fixed at an earlier date at which market conditions can also be unfortunate.

Constant expectation AIR

One may be interested to choose the AIR $a_0(h)$ in such a way that the expected pension payments are constant with respect to h, that is such that $E_0[W_h(h)] = W0(0)$ (recall that the first pension payment W0(0) is without investment risk).

Proposition 2.1. The AIR $AIR\ \bar{a}_0(h|w)$ that leads to constant expected payments for variable annuities equals

$$\bar{a}_0(h|w) = r + w\lambda\sigma. \quad (2.14)$$

Proof. From (2.11), we find that $E_0[W_h(h)] = W_0(0)$ implies

$$\frac{W_0(h)}{W_0(0)} = \exp(-h(r + w\lambda\sigma)), \quad (2.15)$$

or (2.14), using (2.8).

This constant AIR leads to nominally constant expected pension payments. In case our financial market would exhibit interest rate risk (that is, a horizondependent risk-free term structure) and/or stock market predictability, we would need horizon-dependent AIRs to obtain expected constant pension payments. We will see that, even in the present financial market, also smoothing financial market returns leads to a horizon-dependent AIR if annuity payments are required to be constant in expectation.

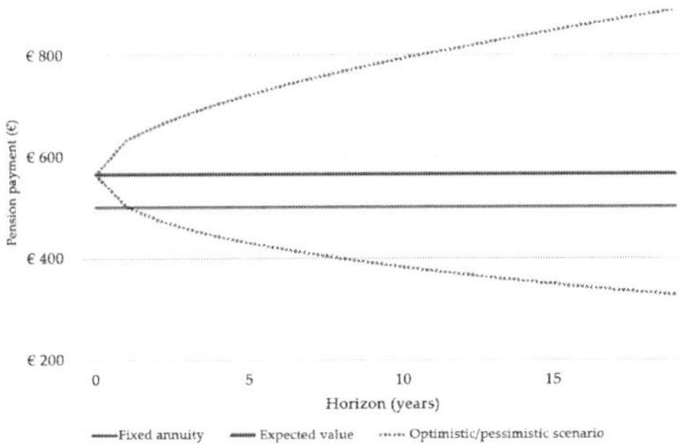

Figure 3: Variable annuity with w = 35% and $\bar{a}_0(h|w)$ = 3.4%.

Figure 3 shows, in solid black, the expected yearly pension payment with a stock exposure w = 35% and the 5% and 95% quantiles. If the AIR equals the expected return as in (2.14), then a constant stream of expected pension payments is obtained. The median is (slightly) below the expected pension payments since the log-normal distribution is skewed. The difference between the variable and the fixed annuity is caused by the risk included in the variable annuity. As a result, the variable annuity leads to higher expected payments, but there is the risk that the realized payments are actually lower than the fixed annuity. The dash-dotted line is the fixed annuity in which $W_0(h)$ is fully invested in the risk-free asset.

Optimal AIR

For given preferences, we derive the optimal AIR that maximizes the expected utility from all the pension payments subject to the budget constraint of the total available pension wealth. The optimal withdrawal is determined by the optimal allocation strategy $a_0^*(h|w)$. The retiree has to determine how much of his wealth he allocates to each horizon for a given investment strategy w. Thus, a retiree who maximizes the expected utility subject to the budget constraint solves the following optimization problem:

Problem 2.1.

$$\max_{\{W_0(h)\}} \mathbb{E}_0 \left[\sum_{h=0}^{H-1} \exp\left(-\beta h\right) u\left(W_h(h)\right) \right] \tag{2.16}$$

$$s.t. \quad W_0 = \sum_{h=0}^{H-1} W_0(h), \tag{2.17}$$

where β is the subjective discount rate that reflects time preferences, that is, impatience.

Proposition 2.2. The optimal AIR that solves Problem 2.1 with utility function (2.4) is

$$a_0^*(h|w) = \begin{cases} r + \frac{1}{\gamma}(\beta - r) - \left(\frac{1}{\gamma} - 1\right) w\sigma\left(\lambda - \frac{1}{2}\gamma w\sigma\right) & if\ \gamma > 0, \gamma \neq 1, \\ \beta & if\ \gamma = 1. \end{cases} \tag{2.18}$$

Proof. Using Itô's lemma and (2.10), we find

$$dW_t(h)^{1-\gamma} - \left(r + w\lambda\sigma - \frac{1}{2}\gamma w^2\sigma^2\right)(1-\gamma)W_t(h)^{1-\gamma}dt + w\sigma(1-\gamma)W_t(h)^{1-\gamma}dZ_t. \tag{2.19}$$

This leads to the optimization problem

$$\max_{\{W_0(h)\}} \sum_{h=0}^{H-1} \exp(-\beta h)\frac{W_0(h)^{1-\gamma}}{1-\gamma}\exp\left(\left(r + w\lambda\sigma - \frac{1}{2}\gamma w^2\sigma^2\right)(1-\gamma)h\right) \tag{2.20}$$

$$\text{s.t. } W_0 = \sum_{h=0}^{H-1} W_0(h). \tag{2.21}$$

The Lagrangian is

$$\mathcal{L}(W_0(h)) = \sum_{h=0}^{H-1} \exp(-\beta h)\frac{W_0(h)^{1-\gamma}}{1-\gamma}\exp\left(\left(r + w\lambda\sigma - \frac{1}{2}\gamma w^2\sigma^2\right)(1-\gamma)h\right) - \kappa\left(\sum_{h=0}^{H-1} W_0(h) - W_0\right).$$

For ease of notation, let
$f(h) - \exp\left(-\beta h + \left(r + w\lambda\sigma - \frac{1}{2}\gamma w^2\sigma^2\right)(1-\gamma)h\right)$. Then the first-order condition with respect to W0(h) becomes

$$\frac{\partial \mathcal{L}(W_0(h))}{\partial W_0(h)} = W_0(h)^{-\gamma} f(h) - \kappa = 0, \tag{2.22}$$

$$W_0^*(h) = \kappa^{-\frac{1}{\gamma}} f(h)^{\frac{1}{\gamma}}. \tag{2.23}$$

The Lagrangian multiplier can be solved by plugging the optimal wealth allocation in the budget constraint, implying

$$(\kappa^*)^{-\frac{1}{\gamma}} = \frac{W_0}{\sum_{h=0}^{H-1} f(h)^{\frac{1}{\gamma}}}. \tag{2.24}$$

Combining (2.23) and (2.24) gives

$$W_0^*(h) = f(h)^{\frac{1}{\gamma}} \frac{W_0}{\sum_{h=0}^{H-1} f(h)^{\frac{1}{\gamma}}}. \tag{2.25}$$

By the definition of the AIR in Equation (2.9), the optimal AIR equals

$$a_0^*(h|w) = -\frac{1}{h} \ln\left(f(h)\right)^{\frac{1}{\gamma}}. \tag{2.26}$$

Note that both $\bar{a}_0(h|w)$ and $a_0^*(h|w)$are independent of h. Thus because the problem is time consistent, it is never optimal to transfer money from year to year; that is, re-optimization leaves the strategies unchanged.

The optimal AIR given the optimal risk exposure leads to the following corollary.

Corollary 2.1. The optimal AIR for the optimal Merton stock exposure is

$$a_0^*(h|w^*) = \begin{cases} r + \frac{1}{\gamma}(\beta - r) - \frac{1}{2\gamma}\left(\frac{1}{\gamma} - 1\right)\lambda^2 & if\, \gamma > 0, \gamma \neq 1, \\ \beta & if\, \gamma = 1, \end{cases} \tag{2.27}$$

and the AIR that leads to constant expectations is

$$\bar{a}_0(h|w^*) = \begin{cases} r + \frac{\lambda^2}{\gamma} & if\, \gamma > 0, \gamma \neq 1, \\ r + \lambda^2 & if\, \gamma = 1. \end{cases} \tag{2.28}$$

Proof. Plugging the optimal exposure w∗ from (2.6) into (2.27) and (2.14) gives $a_0^*(h|w^*)$ and $\bar{a}_0(h|w^*)$, respectively.

Remark 2.3. Both the optimal risk exposure and the optimal AIR can be solved for other utility functions as well. Karatzas and Shreve (1998)'s Theorem 3.7.3 shows the general representation of the optimal consumption. For the required technical assumptions, we refer to their Chapter 3. Both the optimal risk exposure and the AIR are implied by the optimal consumption that is obtained via the martingale method. We have, essentially, used the

Hamilton–Jacobi–Bellman method in the proof of Proposition 2.2 for the special case of CRRA utility and a geometric Brownian motion for the stock price process.

A main difference between a^* and $\bar{a}$ in Corollary 2.1 is that the optimal one includes the individualistic risk aversion level and subjective discount rate which are part of one's individual preferences. If the subjective discount rate is higher than the risk-free rate, then the cost of waiting is higher than the reward of waiting. Therefore, a higher utility is obtained by immediate consumption which coincides with allocating less to future payments, that is, a higher AIR and vice versa. For logarithmic utility, the income effect and substitution effect cancel out the risk-free rate term of the optimal AIR. Moreover, it can be seen that for large risk aversion the optimal exposure converges to zero and thus the variable annuity with either AIR boils down to a fixed annuity.

By comparing the two AIRs, we can determine for which values the individual prefers increasing, decreasing or constant expected pension payments. By definition if $a_0^*(h|w^*) = \bar{a}_0(h|w^*)$, then the expected pension payments are constant, while for $a_0^*(h|w^*) < \bar{a}_0(h|w^*)$, less is allocated to early buckets implying an increase in expectation throughout the lifetime and vice versa. Thus, the optimal expected pension payments for the CRRA investor are either decreasing, constant or increasing for the following parameter values:

$$\frac{1+\gamma}{2\gamma}\lambda^2 \begin{cases} > \beta - r & \text{increasing,} \\ = \beta - r & \text{constant,} \\ < \beta - r & \text{decreasing.} \end{cases} \tag{2.29}$$

Proof. $a_0^*(h|w^*) < \bar{a}_0(h|w^*) \Leftrightarrow r + \frac{1}{\gamma}(\beta - r) - \frac{1}{2\gamma}\left(\frac{1}{\gamma} - 1\right)\lambda^2 < r + \frac{\lambda^2}{\gamma}.$

If $\beta = r$, then $a_0^*(h|w^*) < \bar{a}_0(h|w^*)$ for any $\gamma > 0$ and thus the expected pension payments are increasing through time. The more impatient, that is the higher β, the more likely the expectation is constant or decreasing

Munk (2017) derives the optimal consumption in the Merton model as given by (2.5) and finds that instantaneous consumption is expected to increase with age, decrease with age or to be age-independent depending on which case of (2.29) holds. Realistic parameters appear to imply that the consumption increases, on average, over life. Figure 4 depicts[4] several AIRs; the dashed line shows $\bar{a}_0(h|w^*)$ which generates constant expected payments for different risk aversion levels of γ , and the solid line shows $a_0^*(h|w^*)$ for $\beta = 2\%$ and the dotted line for $\beta = 7\%$ which both determine the preferred wealth allocation for a CRRA investor.

If we fixed the risk aversion level to $\gamma = 2.9$, the optimal risk exposure coincides with the exogenously chosen strategy of the running example which was w = 35%. Since the optimal AIR based on $\beta = 2\%$ is lower than the AIR that determines constant expectations, the implied expected pension stream will be increasing through time, while the expected payments of the CRRA investor with a subjective discount factor of 7% will be decreasing every year. Figure 3 shows the expected pension payments for the $\bar{a}_0(h|w^*)$ and Figures 5 and 6 show this for the two optimal AIRs $a_0^*(h|w^*)$ for $\gamma = 2.9$.

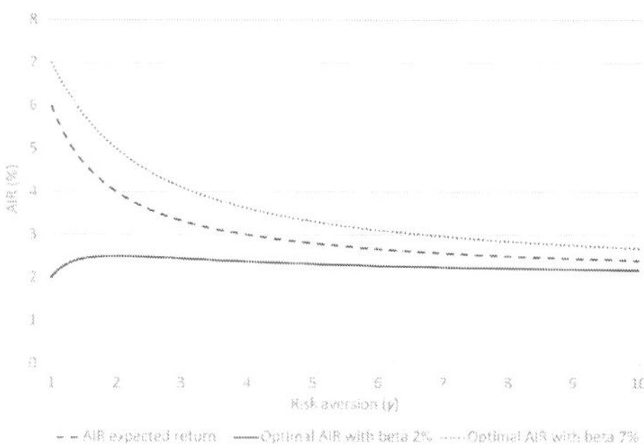

Figure 4: AIRs $a_0^*(h|w^*)$ with $\beta = 2\%$ or $\beta = 7\%$ and $\bar{a}_0(h|w^*)$.

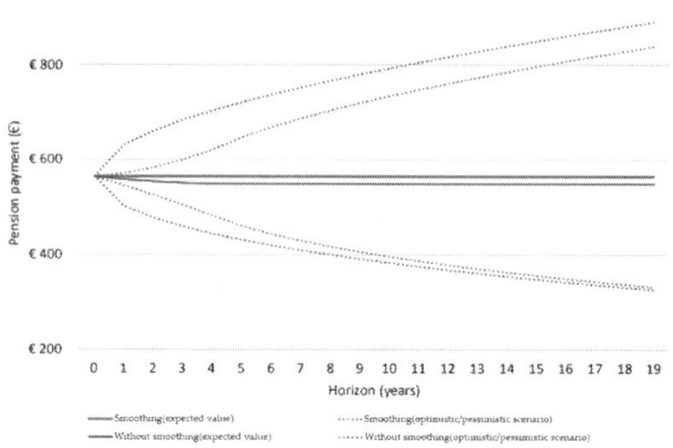

Figure 5: Expected annuity payments for $\gamma = 2.9$, where $a_0^*(h|w^*)$ with $\beta = 2\%$.

Utility loss

We define the total utility that the CRRA investor obtains by U which is a function of the AIR and total initial wealth and given by

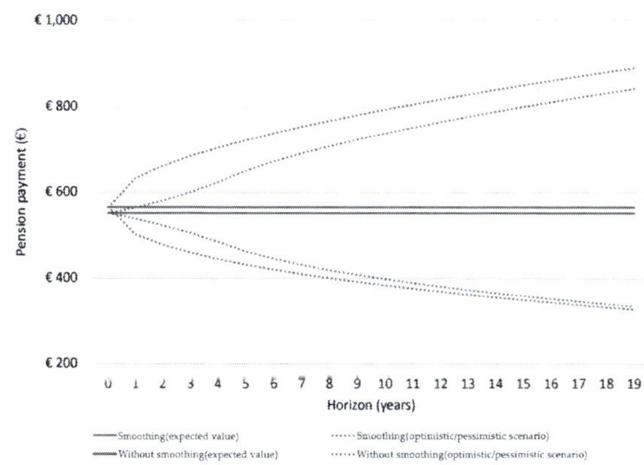

Figure 6: Expected annuity payments for γ =2.9, where $a_0^*(h|w^*)$ with β =7%.

$$U(a_0(h|w), W_0) = \mathrm{E}_0\left[\sum_{h=0}^{H-1} \exp\left(-\beta h\right) u\left(W_h(h)\right) \middle| W_0(h) = W_0(0) \exp\left(-h a_0(h|w)\right), \right.$$

$$\left. W_0 = \sum_{h=0}^{H-1} W_0(h) \right]. \tag{2.30}$$

A total wealth of W0 and an AIR of $a_0(h|w)$ give the investor the same utility as a total – lower – wealth of αW0 and the optimal AIR of $a_0^*(h|w)$. We call α the wealth loading, which is the reduction on the total wealth the investor accepts when his wealth is distributed by the optimal AIR rather than a suboptimal one. Thus the wealth loading is implied by

$$U(a_0(h|w), W_0) - U(a^*(h|w), \alpha W_0). \tag{2.31}$$

If α = 1, the investor is indifferent between the two AIRs. So the utility loss is zero. Therefore, we define the certainty equivalent utility loss as =l=1 − α, which represents the percentage of wealth reduction due to suboptimality. By definition 0 ≤ l ≤ 1, and the higher l, the more the investor looses due to the suboptimal AIR.

Comparing the CRRA certainty equivalents gives a closed-form expression for the utility loss. The definition of a certainty equivalent is $CE(a_0(h|w), W_0) = u^{-1}(U(a_0(h|w), W_0))$. Indifference between utilities is one-to-one related to indifference between certainty equivalents for inverse utility functions that are monotone transformations only. Therefore, (2.31) can also be stated by

$$CE(a_0(h|w), W_0) = CE(a^*(h|w), \alpha W_0). \tag{2.32}$$

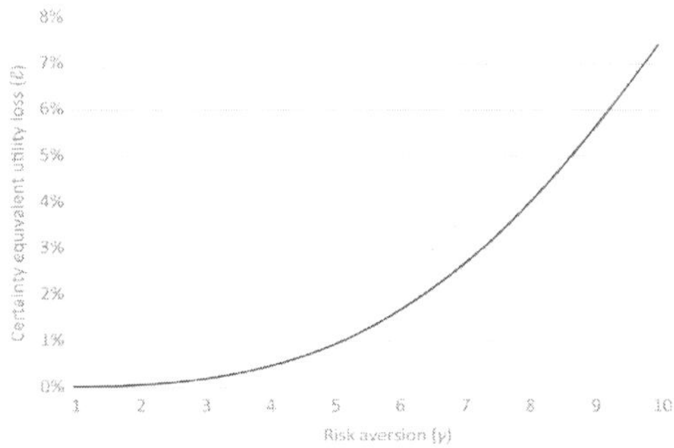

Figure 7: Certainty equivalent utility loss $\ell\left(\bar{a}_0(h|w), a_0^*(h|w)\right)$ for w = 35%.

For CRRA, the certainty equivalent is given by

$$CE(a_0(h|w), W_0) = \frac{W_0}{\sum_{k=0}^{H-1} \exp\left(-k a_0(k|w)\right)} \left(\sum_{h=0}^{H-1} \exp\left(h\left(-\beta - \left(a_0(h|w) - \left(r + w\lambda\sigma - \frac{1}{2}\gamma w^2 \sigma^2 \right)\right)(1-\gamma)\right)\right)\right)^{\frac{1}{1-\gamma}}. \tag{2.33}$$

When comparing two general AIRs $a_{0,1}(h|w_1)$ with wealth W_0 and $a_{0,2}(h|w_2)$ with wealth αW0 with each other, the certainty equivalent utility loss for a CRRA investor is given by

$$\begin{aligned} \ell\left(a_{0,1}(h|w_1), a_{0,2}(h|w_2)\right) &= 1 - \alpha\left(a_{0,1}(h|w_1), a_{0,2}(h|w_2)\right) \\ &= 1 - \frac{CE(a_{0,1}(h|w_1), W_0)}{CE(a_{0,2}(h|w_2), W_0)}, \end{aligned} \tag{2.34}$$

and the utility loss for logarithmic utility is obtained by the limit of $\lim_{\gamma \to 1} \frac{CE(a_{0,1}(h|w_1), W_0)}{CE(a_{0,2}(h|w_2), W_0)}$. Thus $\ell\left(a_{0,1}(h|w_1), a_{0,2}(h|w_2)\right)$ represents the percentage of wealth that the investor with $a_{0,2}(h|w_2)$ requires less – to be indifferent in terms of utility – when the AIR is $a_{0,1}(h|w_1)$ with 100% total wealth.

Several losses of suboptimality can be analyzed since optimality is obtained for two parameters, the AIR and the exposure. In general, the following losses occur, $U(a_0^*(h|w^*)) \geq U(a_0^*(h|w))$, $U(a_0^*(h|w^*)) \geq U(\bar{a}_0(h|w^*))$, $U(a_0^*(h|w^*)) \geq U(\bar{a}_0(h|w))$, $U(a_0^*(h|w)) \geq U(\bar{a}_0(h|w))$ and $U(\bar{a}_0(h|w^*)) \geq U(\bar{a}_0(h|w))$.

Figures 7 and 8 show the utility losses by comparing the optimal AIR $a_0^*(h|w)$ with the AIR that implies constant expectations $\bar{a}_0(h|w)$ for r = 2%, λ = 0.20, σ = 20%, β = 3% and an exposure of w = 35% and w = 100%, respectively. For example, an investor with a risk aversion of about γ = 6.2 and a risk exposure of w = 100% receives equal utility from W_0 = €100, 000 and a constant pension in expectation or from W_0 = €50, 000 and the optimal AIR while if w = 35%, the utility loss implies W_0 = €98, 200

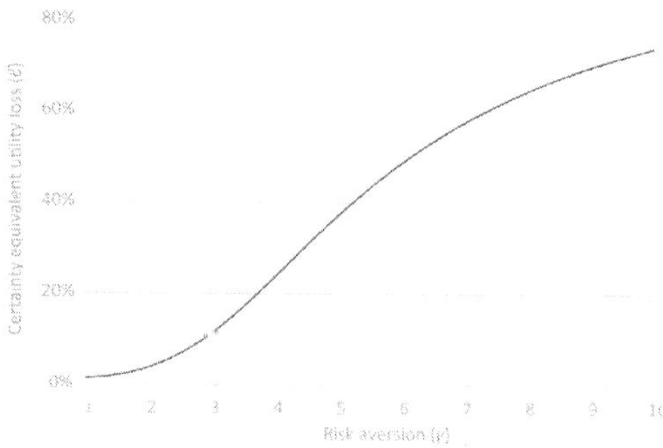

Figure 8: Certainty equivalent utility loss $\ell\left(\bar{a}_0(h|w), a_0^*(h|w)\right)$ for w = 100%.

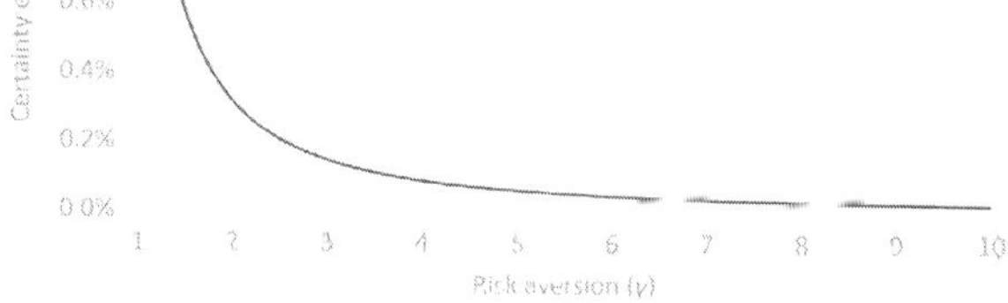

Figure 9: Certainty equivalent utility loss $\ell\left(\bar{a}_0(h|w^*), a_0^*(h|w^*)\right)$.

Figure 9 shows the utility loss for the optimal risk exposure w∗ for different levels of risk aversion γ , that is, by comparing $a_0^*(h|w^*)$ with $\bar{a}_0(h|w^*)$. The optimal risk exposure, the constant Merton fraction, for each different level of risk aversion is shown in Figure 10. For very high levels of risk aversion, the optimal Merton strategy converges to no risk exposure and thus both AIRs converge to the risk-free rate implying no loss because of suboptimality. While for fixed risk exposure, a higher risk aversion implies a higher utility loss. Moreover, when $\gamma \to 1$, the utility loss does not converge zero but to the loading of logarithmic preferences as depicted in Figure 11, which is shown for different risk exposures.

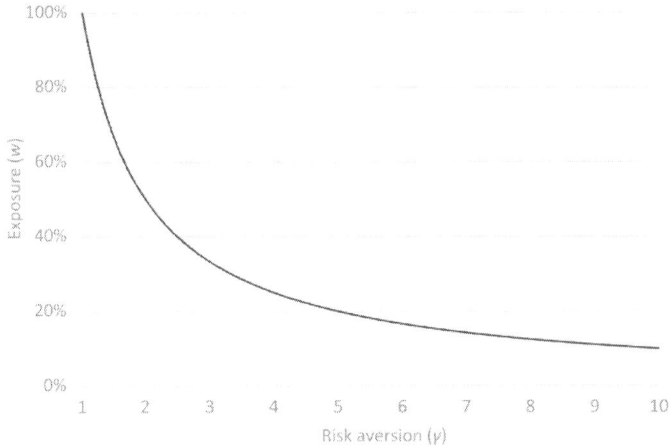

Figure 10: Wealth exposure w*.

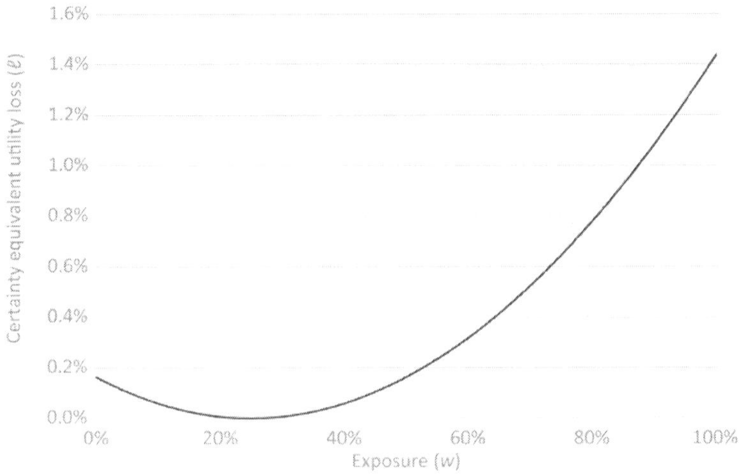

Figure 11: Certainty equivalent utility loss $\ell\left(\bar{a}_0(h|w), a_0^*(h|w)\right)$ for logarithmic utility.

Thus for optimal risk exposure, the utility loss due to the allocation that leads to constant expected payments is small compared to the optimal AIR. Therefore, abstracting from individual measurements of the risk aversion level and subjective discount rate can be justified based on its minor impact on utility.

SMOOTHING FINANCIAL RETURNS

If agents have habit-formation preferences, they may want to reduce yearto-year volatility in the pension payments. Utility functions of this type capture individuals who receive utility from consumption relative to a habit. It rationalizes the demand for smooth consumption as shown by Abel (1990), Constantinides (1990), Fuhrer (2000), Carroll (2000), Crawford (2010) and Davidoff et al. (2005).

The traditional view to achieve smooth consumption, that is lower yearto-year volatility, is to "smooth" financial market returns. That is, in case portfolio returns are −20%, instead of reducing the pension payment immediately by 20%, it is only reduced by a fraction, say, 20%/5 = 4%. This implies that pension payments later in the retirement phase need to be cut by more than 20% to fulfill the budget constraint. Smoothing thus leads to a smaller year-to-year volatility, but the long-term volatility is larger. We derive the conceptual implications of smoothing in the framework of the discrete pension buckets and show the change in the design via the AIR that generates constant expectations.

The reduced year-to-year volatility can be achieved as follows. Recall that the initial pension payment at time 0 is given by $W_0(0)$. In order to have limited risk in the pension payment W+(1), we do not invest it according to a stock exposure w, as in Section 2, but with a stock exposure w0(1) = w/N, where N denotes the smoothing period, say, N = 5 years. Subsequently, the pension wealth $W_0(2)$ for the pension payment $W_2(2)$ is invested with exposure $w_0(2)$ − 2w/N the first year and $w_1(2)$ − w/N the second year. Different smoothing mechanisms can be chosen as long as the exposure is decreased. All results, that is, formulas, below hold for general wj−1(h), which is the stock exposure from year j − 1 to j for the pension wealth that generates the payment in year h. For illustration, we provide figures based on the exposures

$$w_{j-1}(h) = w \min\left\{1, \frac{1+h-j}{N}\right\}, \quad j = 1, \ldots, h, \tag{3.1}$$

for given smoothing period N and long-term stock exposure w. Figure 12 shows these stock exposure $w_{j-1}(h)$ as a function of j for h = 17.

The horizon-dependent stock exposure $w_{j-1}(h)$ induces a life-cycle-type investment strategy. That is, with smoothing, the investment mix is no longer constant over time. Note that the classical reasoning behind a life-cycle argument does not hold for the retirement phase. The idea is that human capital – future wealth from income that is still to be earned – decreases with age. As such, the fraction of savings that is allocated to investments is higher for younger workers because the absolute value of savings is relatively low. In the retirement phase, the future income does not decrease since the income at old age continues until death. However, smoothing causes a horizon-dependent investment strategy, as illustrated above. This decreasing risk exposure per bucket is the favored design that is in line with the finding of Blake et al. (2014), who show that it is optimal to gradually exchange equity for fixed annuities in the decumulation phase for Epstein–Zin preferences. Note here that the deterministic risk exposure per bucket implies a path-dependent risk exposure in terms of the total pension wealth and thus a stochastic smoothing mechanism.

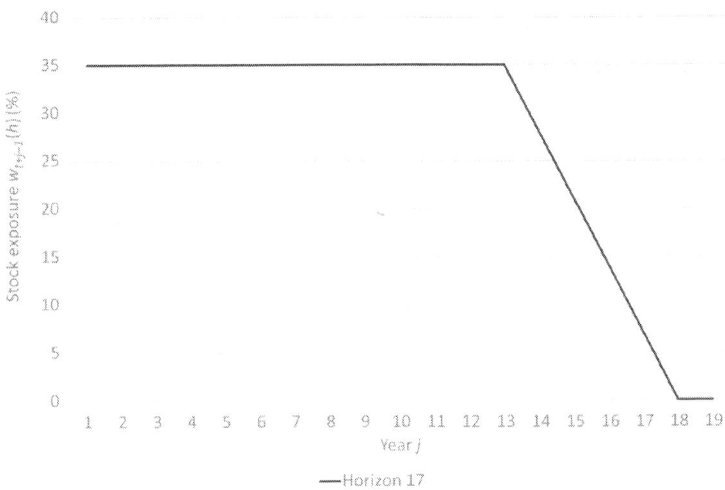

Figure 12: Smoothing stock exposure.

We can now calculate the distribution of the pension payment at time h. Again, this distribution is log-normal, but now with parameters

$$\sum_{j=1}^{h}\left(r + w_{j-1}(h)\lambda\sigma - \frac{1}{2}w_{j-1}^2(h)\sigma^2\right) \tag{3.2}$$

and

$$\sum_{j=1}^{h} w_{j-1}^2(h)\sigma^2. \tag{3.3}$$

The expected nominal pension payments, and their quantiles, can be calculated as before.

A variable annuity as specified by Equation (2.27) in Corollary 2.1 reflects the optimal design for a CRRA investor. Thus both smoothing and embedding guarantees in the design will not be preferred. Utility functions with a reference level could drive demand for smoothing or mixed annuities, where the latter combines the fixed and variable annuity according to constant proportion portfolio insurance (CPPI) to guarantee a minimum pension payment.

In practice, smoothing mechanisms are often implemented for individual products as well as among generations. The collective product with smoothing and intergenerational risk transfers, that we do not consider here, is often based on a solidarity argument. Moreover, to protect retirees from poverty due to impatience it is for instance by Dutch national law5 imposed that the AIR can never be chosen such that it leads to decreasing expectations. From a communicative perspective, constant expectations are usual too. Thus for the remainder of the paper, we illustrate the impact of smoothing within in the "bucket" setting for an AIR that leads to constant expectations.

Constant expectations with smoothing

The present setting allows for an exact derivation of the AIR that leads to pension payments that have a constant expectation. The AIR $\tilde{a}_0(h|w)$ leads to a pension payment that is constant in expected nominal terms and is given by the next proposition.

Proposition 3.1. The AIR $\tilde{a}_0(h|w)$ that leads to constant expected payments in nominal terms when financial shocks are smoothed according to (3.1) equals

$$\tilde{a}_0(h|w) = r + \lambda\sigma\frac{1}{h}\sum_{j=1}^{h} w_{j-1}(h). \tag{3.4}$$

Proof. With smoothing, the expected nominal pension payment at time h is given by

$$\mathbb{E}_0[W_h(h)] = W_0(h) \exp\left(\sum_{j=1}^{h} (r + w_{j-1}(h)\lambda\sigma)\right). \tag{3.5}$$

In order to have a constant expected nominal pension payment, we must choose the AIR $\tilde{a}_0(h|w)$ such that this expectation equals $W_0(0)$ for all h. Recall, see (2.8),

$$\frac{W_0(h)}{W_0(0)} = \exp(-ha_0(h)). \tag{3.6}$$

Thus, together with $\mathbb{E}_0[W_h(h)] = W_0(0)$, it follows that a˜ 0(h|w) is implied by $\tilde{a}_0(h|w)$ is implied by the $a_0(h)$ that solves exp $(-ha_0(h)) \exp\left(\sum_{j=1}^{h} (r + w_{j-1}(h)\lambda\sigma)\right) = 1$ for a given exposure w.

The dash-dotted gray line in Figure 13 shows the expected pension payment and the dashed gray lines show the 5% and 95% quantiles with smoothing period N = 5 years for a stock exposure w = 35% and the AIR equal to $\tilde{a}_0(h|w)$ of Proposition 3.1. The black dotted and black solid lines are obtained without smoothing, similar to Figure 3.

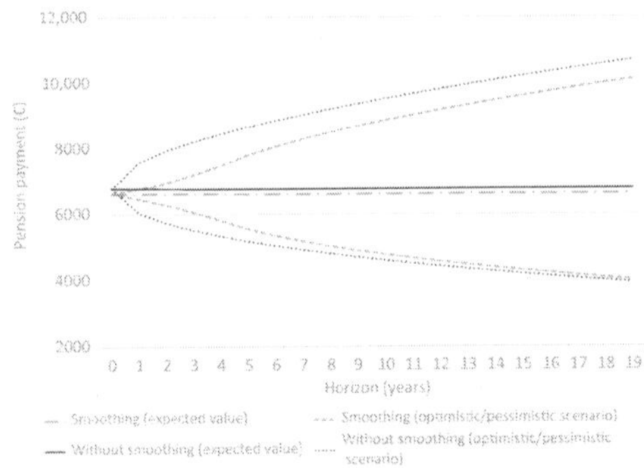

Figure 13: Smoothing with w = 35%, N = 5.

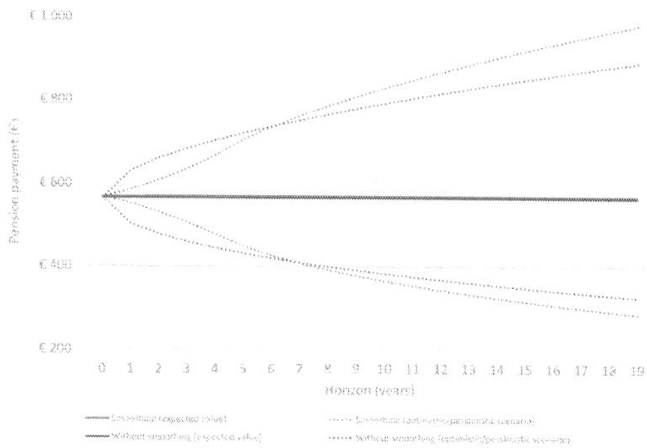

Figure 14: Smoothing with w = 46.5%, N = 5.

Because the total risk exposure is lower due to smoothing, the expected payments are lower than without smoothing. Therefore, a comparison of risk for the same level of expectations is accomplished by a stock exposure of w = 46.5% with smoothing and w = 35% without smoothing, as shown in Figure 14.

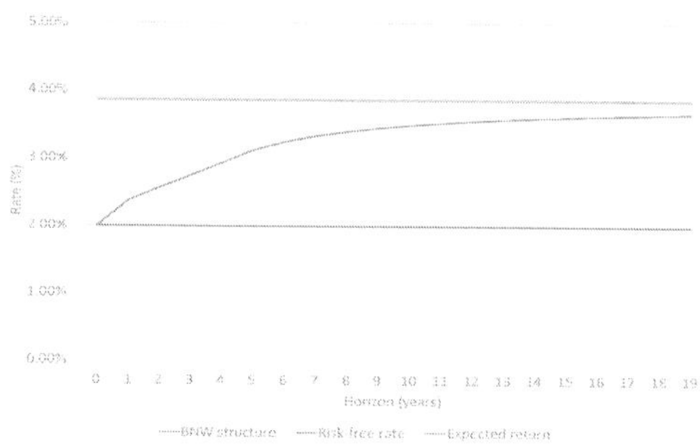

Figure 15: Smoothing structure.

The dotted line in Figure 15 shows the AIR $\tilde{a}_0(h|w)$ as a function of the horizon such that the expected pension payments are constant when smoothing is applied, which is equal to the curve from Bovenberg et al. (2012). If the AIRs are set equal to the risk free rate, as given by the dash-dotted line, the fixed annuity is obtained. The expected return with a stock exposure of w = 46.5% is the solid line.

CONCLUSION

This paper provides analytical expressions for the risk–return trade-off of variable annuities, with a special focus on the explicit allocation of initial wealth across the "pension buckets" reserved for future payments. The latter is completely determined by the AIR and relates to the consumption portfolio problem. This conceptual division provides useful insights by analyzing different AIRs. We derive the AIR that leads to constant expected pension payments as well as the optimal AIR for an investor with CRRA preferences. The utility loss between these two is small when the risk exposure is optimal, that is for the Merton fraction. We also consider the situation where financial market returns may be smoothed over the remaining retirement period. In order to obtain, in a contract with smoothing, a constant expected nominal pension, the AIR has to be horizon dependent. Other insights that we obtain from investigating the effect of the AIR on the variable annuity is that the allocation's sensitivity on the payments toward the end of life is large while early or intermediate expected payments are hardly affected by the AIR. Moreover, we show that variable annuities do not solve conversion risk since we find that shocks in the risk-free rate have the same effect for variable and fixed annuities.

Notes

1. For instance, variable annuities entered the Dutch market since the Improved Defined Contribution Scheme Act of 2016 and the Australian market has introduced Comprehensive Income Products for Retirement (CIRPs).
2. Mortality could be incorporated by multiplying each "bucket" $W0(h)$ reserved for the pension payment in year h by the survival probability from 0 to h. For the impact of longevity risk, see Balter *et al.* (2019).
3. There is only a small effect that *may* lead to some protection. Under the empirically hard to defend assumption that lower interest rates r imply increased prices-of-risk , some protection is provided. But even if this assumption is true, such protection only pertains to the risky investment in the variable annuity.
4. For the parameter assumptions of the running example, that is, μ=6%, r=2%, =20% and =0.2.
5. Source: http://www.toezicht.dnb.nl/2/50-235784.jsp

REFERENCES

1. Abel, A.B. (1990) Asset prices under habit formation and catching up with the Joneses. *American Economic Review*, 80(2), 38–42.
2. Bacinello, A.R., Millossovich, P., Olivieri, A. and Pitacco, E. (2011) Variable annuities: A unifying valuation approach. *Insurance: Mathematics and Economics*, 49(3), 285–297.
3. Balter, A.G., Rangvid, J. and Kallestrup-Lamb, M. (2019) *The move towards riskier pensions: The importance of mortality*. Working Paper.
4. Bauer, D., Kling, A. and Russ, J. (2008) A universal pricing framework for guaranteed minimum benefits in variable annuities. *ASTIN Bulletin: The Journal of the IAA*, 38(2), 621–651.
5. Bernard, C., Hardy, M. and MacKay, A. (2014) State-dependent fees for variable annuity guarantees. *ASTIN Bulletin: The Journal of the IAA*, 44(3), 559–585.
6. Blake, D., Wright, D. and Zhang, Y. (2014) Age-dependent investing: Optimal funding and investment strategies in defined contribution pension plans when members are rational life cycle financial planners. *Journal of Economic Dynamics and Control*, 38, 105–124.
7. Bovenberg, A., Nijman, T.E. and Werker, B. (2012) *Voorwaardelijke pensioenaanspraken: Over waarderen, beschermen, communiceren en beleggen*. Netspar Occasional Paper.
8. Bruhn, K. and Steffensen, M. (2013) Optimal smooth consumption and annuity design. *Journal of Banking & Finance*, 37(8), 2693–2701.
9. Carroll, C.D. (2000). Solving consumption models with multiplicative habits. *Economics Letters*, 68(1), 67–77.
10. Chen, A., Hentschel, F. and Klein, J.K. (2015) A utility-and CPT-based comparison of life insurance contracts with guarantees. *Journal of Banking & Finance*, 61, 327–339.
11. Coleman, T.F., Li, Y. and Patron, M.-C. (2006) Hedging guarantees in variable annuities under both equity and interest rate risks. *Insurance: Mathematics and Economics*, 38(2), 215–228.
12. Constantinides, G.M. (1990) Habit formation: A resolution of the equity premium puzzle. *Journal of Political Economy*, 98(3), 519–543.
13. Cox, J.C. and Huang, C.-f. (1989) Optimal consumption and portfolio policies when asset prices follow a diffusion process. *Journal of Economic Theory*, 49(1), 33–83.

14. Crawford, I. (2010) Habits revealed. *The Review of Economic Studies*, 77(4), 1382–1402.
15. Davidoff, T., Brown, J.R. and Diamond, P.A. (2005). Annuities and individual welfare. *American Economic Review*, 95(5), 1573–1590.
16. Dellinger, J.K. (2006). *The Handbook of Variable Income Annuities*. Hoboken: John Wiley & Sons.
17. Fuhrer, J.C. (2000). Habit formation in consumption and its implications for monetary-policy models. *American Economic Review*, 90(3), 367–390.
18. Guillén, M., Jørgensen, P.L. and Nielsen, J.P. (2006). Return smoothing mechanisms in life and pension insurance: Path-dependent contingent claims. *Insurance: Mathematics and Economics*, 38(2), 229–252.
19. Horneff, W.J., Maurer, R.H., Mitchell, O.S. and Stamos, M.Z. (2009). Asset allocation and location over the life cycle with investment-linked survival-contingent payouts. *Journal of Banking & Finance*, 33(9), 1688–1699.
20. Horneff, W.J., Maurer, R.H., Mitchell, O.S. and Stamos, M.Z. (2010). Variable payout annuities and dynamic portfolio choice in retirement. *Journal of Pension Economics & Finance*, 9(2), 163–183.
21. Karatzas, I. and Shreve, S.E. (1998). *Methods of Mathematical Finance*. New York: Springer-Verlag.
22. Kling, A., Ruez, F. and Russ, J. (2011). The impact of stochastic volatility on pricing, hedging, and hedge efficiency of withdrawal benefit guarantees in variable annuities. *ASTIN Bulletin: The Journal of the IAA*, 41(2), 511–545.
23. Kling, A., Russ, J. and Schilling, K. (2014). Risk analysis of annuity conversion options in a stochastic mortality environment. *ASTIN Bulletin: The Journal of the IAA*, 44(2), 197–236.
24. Koijen, R.S., Nijman, T.E. and Werker, B.J. (2011) Optimal annuity risk management. *Review of Finance*, 15(4), 799–833.
25. Mahayni, A. and Schneider, J.C. (2012) Variable annuities and the option to seek risk: Why should you diversify? *Journal of Banking & Finance*, 36(9), 2417–2428.
26. Maurer, R., Mitchell, O.S., Rogalla, R. and Siegelin, I. (2016) Accounting and actuarial smoothing of retirement payouts in participating life annuities. *Insurance: Mathematics and Economics*, 71, 268–283.

27. Merton, R.C. (1969) Lifetime portfolio selection under uncertainty: The continuous-time case. *The Review of Economics and Statistics*, 51(3), 247–257.
28. Merton, R.C. (1971) Optimum consumption and portfolio rules in a continuous-time model. *Journal of Economic Theory*, 3(4), 373–413.
29. Munk, C. (2017) *Dynamic Asset Allocation. Lecture Notes*, Copenhagen Business School, February 2017.
30. Nirmalendran, M., Sherris, M. and Hanewald, K. (2014). Pricing and solvency of value-maximizing life annuity providers. *ASTIN Bulletin: The Journal of the IAA*, 44(1), 39–61.
31. Peijnenburg, K., Nijman, T. and Werker, B.J. (2016). The annuity puzzle remains a puzzle. *Journal of Economic Dynamics and Control*, 70, 18–35
32. Trottier, D. A., Godin, F. and Hamel, E. (2018). Local hedging of variable annuities in the presence of basis risk. *ASTIN Bulletin: The Journal of the IAA*, 48(2), 611–646.

CHAPTER

3

Adverse Selection in Private Annuity Markets and the Role of Mandatory Social Annuitization

Ben J. Heijdra and Laurie S. M. Reijnders

Faculty of Economics and Business, University of Groningen, P. O. Box 800, 9700 AV Groningen, The Netherlands

ABSTRACT

We study the effects on the macroeconomic equilibrium, the wealth distribution, and welfare of adverse selection in private annuity markets in a closed economy inhabited by overlapping generations of heterogeneous agents who are distinguished by their health status. If an agent's health type is private information there will be a pooling equilibrium in the private

Citation: Heijdra, B.J., Reijnders, L.S.M. "Adverse Selection in Private Annuity Markets and the Role of Mandatory Social Annuitization". De Economist 160, 311–337 (2012). https://doi.org/10.1007/s10645-012-9191-z

annuity market. We also study the implications for the macro-economy and welfare of a social security system with mandatory contributions that are constant across health types. These social annuities are immune to adverse selection and therefore offer a higher rate of return than private annuities do. However, they have a negative effect on the steady-state capital intensity and welfare. The positive effect of a fair pooled rate of return on a fixed part of savings and a higher return on capital in equilibrium is outweighed by the negative consequences of increased adverse selection in the private annuity market and a lower wage rate.

Keywords:-Annuity markets, adverse selection, overlapping generations, Demography

INTRODUCTION

In a seminal paper, Yaari (1965) argues that in the face of life span uncertainty nonaltruistic individuals will fully annuitize their financial wealth. That is, they will invest all their savings in the annuity market, thereby insuring themselves against the risk of outliving their assets. However, empirical evidence has revealed that, despite the theoretical attractiveness of annuities, in practice people tend not to invest much in private annuity markets.

Several explanations for the low participation in annuity markets have been given in the literature. First of all, individuals may have a bequest motive, in that they wish to leave an inheritance to those they leave behind. If so, they will want to keep part of their financial assets outside the annuity market. Secondly, psychological factors may play a role. According to Cannon and Tonks (2008), people might feel uncomfortable to "bet on a long life". Investing in annuities only seems attractive if you expect to live long enough, as most of us would hate to die before having received at least our initial outlay back in periodic payments. Third, private annuity demand may be crowded out by a system of social security benefits.

A fourth explanation is that in reality annuities may not be actuarially fair, in the sense that individuals are insufficiently compensated for their risk of dying. This may be due to administrative costs and taxes or monopoly profits as a result of imperfect competition among annuity firms. The implications for macroeconomic growth and welfare of a loading factor on annuities proportional to the mortality rate are investigated in Heijdra and Mierau (2012). Another reason for annuity market imperfection is adverse selection. The healthier someone believes herself to be, the more likely she is to buy an annuity. As a consequence, low-mortality (and thus high-risk) individuals are overrepresented in the annuity market. Annuity firms will have to take this selection effect into account when pricing their products, as they will incur a loss if they offer a rate based on average survival probabilities in the population. The resulting higher prices (or lower return) will induce high-mortality (low-risk) individuals to invest less in the annuity market.

In this paper we abstract from bequest motives, administrative costs, and imperfect competition and focus on the adverse selection channel and the role of social annuities. Our work mainly builds on the foundations laid out by Heijdra and Reijnders (2009). They consider a continuous-time endogenous growth model with two types of agents differing in their health status acquired at birth, which is assumed to be private information. The equilibrium in the annuity market is then characterized by risk pooling among health types which induces the unhealthy agents to drop out of the market in the last stages of their lives. This pooling equilibrium is slightly dominated in welfare terms by a hypothetical full-information equilibrium (in which each health type receives its actuarially fair rate of return).

We augment the work by Heijdra and Reijnders (2009) in several directions. First, rather than distinguishing only two types of agents, we model a continuum of health types. Second, instead of a employing a continuous-time model with endogenous growth we work with a discrete-time framework in which long-run growth is exogenously determined. This allows us to easily study what happens during the transition from one steady state to another.

Our main findings are as follows. First, we note that in the absence of private and social annuities there exist accidental bequests that must be redistributed somehow. When such bequests flow to the young, private saving is boosted as this constitutes an intergenerational transfer from dissavers to savers. In this so-called TY equilibrium the unhealthiest agents typically experience a binding borrowing constraint. Second, we demonstrate that when a private annuity market is opened up and information is perfect (the FI case), then all agents will purchase positive amounts of annuities. Third, in the more realistic case with asymmetric information (the AI scenario), the equilibrium in the market for annuities will be a pooling equilibrium in which the unhealthiest individuals face a self-imposed borrowing constraint and the other agents receive a common yield on their annuities. Fourth, both in the FI and AI cases the opening up of a private annuity market causes a 'tragedy of annuitization', as described in Heijdra et al. (2010). Intuitively, whilst it is individually optimal for agents to invest in annuities, it is not socially optimal. Agents of all health types are in the long run worse off compared to the benchmark case in which annuities are absent and accidental bequests are redistributed to the newly arrived young (TY). Fifth, the introduction of a mandatory social annuity system, while providing a 'fairer' rate of return than the private annuity market, reduces steady-state welfare even more. It aggravates the degree of adverse selection in the private annuity market and reduces the overall level of savings in the economy.

Other papers closely related to ours are Abel (1986), Walliser (2000), and Palmon and Spivak (2007). All three find that the introduction of social annuities accentuates the problem of adverse selection in the private annuity market. In Abel (1986) a two-period exogenous growth model is developed in which agents have privately known heterogenous mortality profiles and a bequest motive. Due to adverse selection, the rate of return on private annuities is less than actuarially fair. In this context, the introduction of a mandatory (demographically fair1) social security system further decreases the return on annuities in the steady state. Walliser (2000) builds on the work of Abel (1986), but calibrates his model with 75 instead of only 2 periods. The paper investigates the effects of pay-as-you-go social security benefits on private annuity demand and shows that privatization (i.e. elimination) of social security lowers the loading factor on annuities resulting from adverse selection. Finally, Palmon and Spivak (2007) argue that a modest social security system may reduce welfare in an adverse selection economy. The positive effect of providing agents with social annuities at an demographically fair pooling rate is outweighed by the negative impact of increased adverse selection in the private annuity market.In contrast to our work, however, both Walliser (2000) and Palmon and Spivak (2007) focus on the features of private annuity markets in isolation and do not take general equilibrium effects into account. Moreover, the latter fail to specify what happens to accidental bequests in the absence of annuities and therefore incorrectly conclude that private annuities are always welfare improving. Abel (1986) on the other hand does model a production sector with potentially endogenous factor prices and provides signs for the responses of key economic variables following a change in the rate of contribution to the social security system. Yet he does not give any insight in the magnitude of the effect on consumer welfare, nor how it may differ among risk types. Our contribution lies in providing a consistent general equilibrium framework for studying the main aspects related to life annuitization and social security. We are able to both analytically characterize the underlying mechanisms and to quantify their relative importance through a simulation with realistic parameter values.

The remainder of this paper is structured as follows. Section 2 describes the key features of the model in terms of the decisions made by households, firms, and the government sector. In Sect. 3 we introduce private annuity markets, while Sect. 4 shows the effects on general equilibrium and welfare when mandatory social annuities are added to the model. Section 5 concludes.

MODEL

Consumers

The population consists of overlapping generations of finitely-lived agents who are identical in every respect except for their health type. Agents live for a maximum of two periods, termed 'youth' (superscript y) and 'old age' (o). At birth each agent learns her health status as proxied by the survival probability, μ. This is where the difference between health types comes in: unhealthy agents have a higher risk of dying, and therefore a shorter expected life span (which equals 1 + μ periods). We assume that cohorts are sufficiently large such that there is no aggregate uncertainty and probabilities and frequencies coincide. For example, the fraction of young agents of type μ who die after the first period equals exactly 1−μ. Note that from the perspective of an individual agent, lifetime uncertainty is resolved at the start of the second period. When still alive, the agent will live for exactly one additional period.

Labour supply is exogenous. During youth the agent is fully employed while during old age labour supply is only a fraction λ of the unit time endowment as a result of mandatory retirement (0 <λ< 1). The expected lifetime utility of a representative agent of health type μ who is born in period t is given by:

$$\mathbb{E}\Lambda_t(\mu) = U(C_t^y(\mu)) + \mu\beta U(C_{t+1}^o(\mu)), \tag{1}$$

where $C_t^y(\mu)$ and $C_{t+1}^o(\mu)$ are consumption during youth and old age, respectively, β is a parameter capturing pure time preference (0 <β< 1), and U (·) is the felicity function:

$$U(x) \equiv \frac{x^{1-1/\sigma} - 1}{1 - 1/\sigma}, \qquad \sigma > 0. \tag{2}$$

This functional form is chosen for analytical convenience and it implies a constant intertemporal substitution elasticity, σ. We assume that the agent does not have a bequest motive such that she does not derive any utility from wealth that remains after her death.

The agent's periodic budget identities are given by:

$$C_t^y(\mu) + S_t(\mu) = w_t + Z_t, \tag{3}$$

$$C_{t+1}^o(\mu) = \lambda w_{t+1} + (1 + r_{t+1})S_t(\mu), \tag{4}$$

where w_t is the wage rate, r_{t+1} is the rental rate of capital, $S_t(\mu)$ is the amount saved, and Z_t denotes a lump-sum income transfer received from the government during youth. Since the government cannot observe an individual's health type, the transfer is the same for everybody. All workers are equally productive so the wage rate is common to all agents.

If annuity markets do not exist, agents cannot insure themselves against life span uncertainty. Their only option is to invest their savings in the capital market at a net rate of return r_{t+1}. Since there is a risk of dying after youth the agent may pass away before being able to consume her savings, thereby leaving an unintended bequest. It is not possible to borrow in the capital market, as the agent is not allowed to die indebted. Hence we impose the borrowing constraint $S_t(\mu) \geq 0$. Individuals who face a binding borrowing constraint have no better option than to consume their current and future endowments.

For unconstrained agents we can combine the two budget identities to obtain the consolidated budget constraint:

$$C_t^y(\mu) + \frac{C_{t+1}^o(\mu)}{1 + r_{t+1}} = w_t + Z_t + \frac{\lambda w_{t+1}}{1 + r_{t+1}}. \tag{5}$$

That is, the present value of total consumption (left-hand side) should equal lifetime income or human wealth at birth (right-hand side). The representative agent maximizes life-time utility (1) subject to the budget constraint (5). The agent's optimal plans are fully characterized by:

$$C_t^y(\mu) = \Phi(\mu, 1 + r_{t+1})\left[w_t + Z_t + \frac{\lambda w_{t+1}}{1 + r_{t+1}}\right], \tag{6}$$

$$S_t(\mu) = \left[1 - \Phi(\mu, 1 + r_{t+1})\right][w_t + Z_t] - \Phi(\mu, 1 + r_{t+1})\frac{\lambda w_{t+1}}{1 + r_{t+1}}, \tag{7}$$

$$\frac{C_{t+1}^o(\mu)}{1 + r_{t+1}} = \left[1 - \Phi(\mu, 1 + r_{t+1})\right]\left[w_t + Z_t + \frac{\lambda w_{t+1}}{1 + r_{t+1}}\right], \tag{8}$$

Where $\phi(\mu, x)$ is the marginal propensity to consume out of lifetime wealth during youth:

$$\Phi(\mu, x) \equiv \frac{1}{1 + (\mu\beta)^{\sigma} x^{\sigma-1}}, \qquad (0 < \Phi(\mu, x) < 1). \tag{9}$$

We find that consumption during youth and the present value of consumption during old age are proportional to human wealth at birth. The level of savings depends positively on the total income received during youth and negatively on the discounted value of future wage income.

The optimal consumption choices can be illustrated with the aid of Fig. 1. The line labeled HBC is the household budget constraint as given in (5). It passes through the income endowment point A which is the same for all individuals. Agents wish to consume such that the marginal rate of substitution between current and future consumption is equated to the survival-adjusted gross interest rate:

$$\frac{U'(C_t^y(\mu))}{\beta U'(C_{t+1}^o(\mu))} = \mu(1 + r_{t+1}). \tag{10}$$

For the felicity function employed in this paper, the Euler equation is a straight line from the origin with a slope that depends positively on the agents survival probability, μ. In the diagram, $EE(\mu_h)$ and $EE(\mu_l)$ denote the Euler equations for, respectively, the healthiest and unhealthiest types of agents. The former types consume at point B which is attained by saving during youth. The latter types, however, would like to consume at point C but this point is unattainable because it violates the borrowing constraint. The best such an agent can do is to consume at the income endowment point A.[2]

With a continuum of health types, it thus follows that in general not all agents are net savers in the absence of annuities. In the case drawn in Fig. 1, there exists a marginal health type $\mu_{bc,t}$ such that $EE(\mu_{bc,t})$ passes through the income endowment point A. Agents of this type optimally choose not to save anything during youth. Hence, all relatively healthy agents ($\mu_{bc,t} \leq \mu \leq \mu_h$) are unconstrained whereas all relatively unhealthy agents ($\mu_l \leq \mu < \mu_{bc,t}$).

Demography

Let h (μ) denote the probability density function of health types in a given cohort upon its arrival. Then the distribution of agents in the cohort born at time t can be written as:

$$L_t(\mu) \equiv h(\mu) L_t, \tag{11}$$

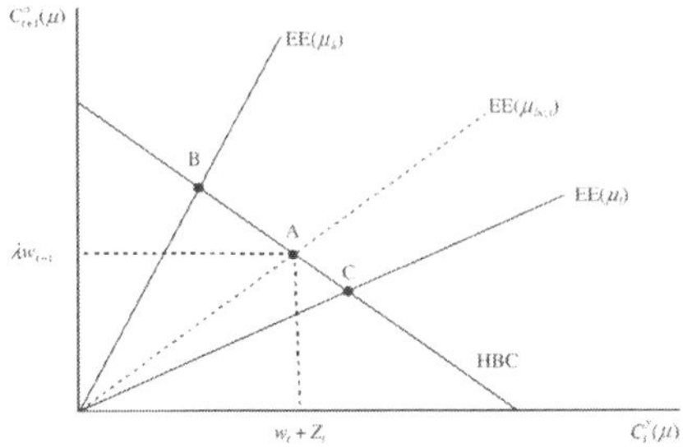

Figure 1. Choices of a μ-type individual in the absence of annuities.

where L_t is the total size of the cohort. By definition, $\int_{\mu_l}^{\mu_h} h(\mu)\,d\mu = \int_{\mu_l}^{\mu_h} dH(\mu) = 1$ where H (μ) is the cumulative density function. The density of μ-type agents alive at time t is given by $P_t(\mu) \equiv \mu L_{t-1}(\mu) + L_t(\mu)$. Assuming that newborn cohorts evolve according to $L_t = (1 + n) L_{t-1}$ (with $n > -1$) we thus find that:

$$P_t(\mu) = \frac{1 + \mu + n}{1 + n} L_t(\mu). \qquad (12)$$

The total population alive in period t is obtained by aggregating over health types:

$$P_t \equiv \int_{\mu_l}^{\mu_h} P_t(\mu)\,d\mu = \frac{1 + \bar{\mu} + n}{1 + n} L_t, \qquad (13)$$

where $\bar{\mu} \equiv \int_{\mu_l}^{\mu_h} \mu h(\mu)\,d\mu$ is the average survival rate in the population as a whole.

Government

In the absence of annuity markets we have to make an assumption about how the accidental bequests left by the dead are redistributed among the agents who are still alive. We therefore introduce a government sector which collects the bequests and uses them to finance lump-sum income transfers Z_t to the young.[3] The government budget constraint is given by:

$$(1 + r_t) \int_{\mu_{bc,t}}^{\mu_h} (1 - \mu) L_{t-1}(\mu) S_{t-1}(\mu)\,d\mu = L_t Z_t. \qquad (14)$$

That is, the total amount of accidental bequests (left-hand side) equals the sum of income transfers (right-hand side).

Production

0The production side of this closed economy features a large number of perfectly competitive firms who produce a homogeneous commodity. The technology is represented by the following Cobb–Douglas production function:

$$Y_t = \Omega_0 K_t^{\varepsilon} N_t^{1-\varepsilon}, \quad 0 < \varepsilon < 1, \tag{15}$$

where Y_t is total output, $\Omega_0 > 0$ is an exogenously given index of general factor productivity, K_t is the aggregate capital stock, and $N_t \equiv L_t + \bar{\mu}\lambda L_{t-1}$ is the labour force. By defining $y_t \equiv Y_t / N_t$ and $k_t \equiv K_t / N_t$ we can write the intensive-form production function as:

$$y_t = \Omega_0 k_t^{\varepsilon}. \tag{16}$$

Profit-maximizing behaviour of firms yields the following factor demand equations:

$$r_t + \delta = \varepsilon \Omega_0 k_t^{\varepsilon - 1}, \tag{17}$$

$$w_t = (1 - \varepsilon)\Omega_0 k_t^{\varepsilon}, \tag{18}$$

where δ is the constant rate of depreciation of the capital stock ($0 < \delta < 1$). The general model without annuities is fully characterized by the following fundamental difference equation:

$$k_{t+1} = \phi_1^{TY}(r_{t+1})[w_t + Z_t] - \phi_2^{TY}(r_{t+1}) \frac{\lambda w_{t+1}}{1 + r_{t+1}}. \tag{19}$$

where $\phi_1^{TY}(r_{t+1})$ and $\phi_2^{TY}(r_{t+1})$ are given by:

$$\phi_1^{TY}(r_{t+1}) \equiv \frac{1}{1 + n + \lambda\bar{\mu}} \int_{\mu_{bc,t}}^{\mu_h} \left[1 - \Phi(\mu, 1 + r_{t+1})\right] h(\mu)\, d\mu,$$

$$\phi_2^{TY}(r_{t+1}) \equiv \frac{1}{1 + n + \lambda\bar{\mu}} \int_{\mu_{bc,t}}^{\mu_h} \Phi(\mu, 1 + r_{t+1}) h(\mu)\, d\mu.$$

Equation (19) is obtained by imposing equilibrium in the savings market and using the cohort size evolutions described in Sect. 2.2. At time t the predetermined capital intensity, k_t, pins down r_t, w_t, and Z_t, so that (19) in combination with (17) and (18) constitutes an implicit function determining k_{t+1}, r_{t+1}, and w_{t+1}.

Table 1. Structural parameters and functions.

β	Time preference factor	0.4128
σ	Intertemporal substitution elasticity	0.7000
n	Population growth factor	0.4889
ε	Capital efficiency parameter	0.2750
Ω_0	Scale factor production function	2.4312
δ	Capital depreciation factor	0.9158
λ	Mandatory retirement parameter	0.6250
$h(\mu)$	PDF of the survival probability	Uniform on $[\mu_l, \mu_h]$
μ_l	Survival rate of the unhealthiest	0.0500
μ_h	Survival rate of the unhealthiest	0.9500

The cut-off value $\mu_{bc,t}$ is such that $St(\mu_{bc,t}) = 0$. From these variables Z_{t+1} can be computed. In the steady state we have $k_{t+1} = k_t = \hat{k}$ such that the level of output per worker $\hat{y}$ is constant. As a consequence, the long-run rate of economic growth over and above population growth equals zero, which is a standard result in an exogenous growth model without technological process.

Throughout this paper we visualize and quantify the properties of the model by means of simulations with realistic parameter values. The main structural parameters are summarized in Table 1. The calibration was done as follows. In step 1, we postulate plausible values for the intertemporal substitution elasticity ($\sigma = 0.7$), the efficiency parameter of capital ($\varepsilon = 0.275$), the annual capital depreciation rate ($\delta_a = 0.06$), the annual growth rate of the population ($n_a = 0.01$) and choose a realistic target annual steady-state interest rate ($\hat{r}_a = 0.05$). Taken together these parameters ensure that the annual steady-state capital-output ratio is $\hat{K}/\hat{Y} = \varepsilon/(\hat{r}_a + \delta_a) = 2.5$. In step 2, we fix the time dimension by assuming that each period lasts for 40 years. The values for n, δ, and r^ follow readily from this.[4] We assume that mandatory retirement takes place at age 65, which yields the value for λ in Table 1. Furthermore, we postulate a distribution for the probability of surviving into the second period and note that life expectancy at birth of a μ-type agent is given by $40(1 + \mu)$ in years. In step 3 we choose Ω_0 such that $\hat{y} = 1$ in the initial steady state. This also pins down the steady-state values for $\hat{w}$ and $\hat{k}$. Finally, in step 4 we solve for steady-state transfers $\hat{z}$, the discount factor β, and the cut-off survival level for the borrowing constraint $\hat{\mu}_{bc}$ [satisfying $S(\hat{\mu}_{bc}) = 0$] such that a consistent general equilibrium solution is obtained. To interpret the resulting value for β we note that $\rho_a = \beta^{-1/40} - 1 = 0.0237$ represents the annual rate of time preference. In view of the fact that the annual rate of interest is $\hat{r}_a = 0.05$agents in this economy are relatively patient.

The key features of the steady-state equilibrium that is attained in the absence of private and social annuities and with accidental transfers flowing to the young (abbreviated as TY) are reported in column (a) of Table 2. As was enforced by the calibration procedure, output per worker is equal to unity and the steady-state interest rate is five percent on an annual basis.

Table 2. Allocation and welfare

	(a) TY	(b) FI	(c) AI	(d) SA $\theta = 0.01$	(e) SA $\theta = 0.02$	(f) SA $\theta = 0.03$
$\hat{y}$	1.0000	0.9638	0.9473	0.9395	0.9318	0.9247
$\hat{k}$	0.0395	0.0346	0.0325	0.0315	0.0306	0.0297
%Q1	3.62	9.10	5.86	3.25	1.23	0.11
%Q2	19.66	20.51	20.66	17.25	13.43	9.14
%Q3	32.96	30.66	32.05	28.94	25.46	21.52
%Q4	43.76	39.73	41.43	38.56	35.35	31.69
%SAS				12.00	24.53	37.54
$\hat{r}$	6.0400	6.7500	7.1079	7.2840	7.4632	7.6348
$\hat{r}^a$	0.0500	0.0525	0.0537	0.0543	0.0548	0.0554
$\hat{w}$	0.7250	0.6988	0.6868	0.6811	0.6756	0.6704
$\hat{Z}$	0.1146					
$H(\hat{\mu}_{bc})$	0.1029	0.0000	0.0533	0.0999	0.1556	0.2217
$\hat{\mu}^p$			0.6418	0.6597	0.6802	0.7035
$\hat{c}^y$	0.6351	0.5261	0.5193	0.5161	0.5128	0.5098
%Q1	26.98	26.56	26.78	26.70	26.57	26.35
%Q2	25.50	25.44	25.40	25.43	25.48	25.56
%Q3	24.26	24.45	24.34	24.37	24.41	24.49
%Q4	23.26	23.56	23.47	23.49	23.54	23.60
$\hat{c}^o$	0.3094	0.3892	0.3824	0.3792	0.3760	0.3731
%Q1	4.05	8.80	3.88	3.98	4.21	4.66
%Q2	14.98	20.16	14.88	14.87	14.83	14.77
%Q3	30.81	30.65	30.81	30.77	30.70	30.56
%Q4	50.16	40.40	50.44	50.38	50.26	50.02
$E\hat{\Lambda}(\mu_l)$	−0.2010	−0.3913	−0.4287	−0.4439	−0.4608	−0.4784
$E\hat{\Lambda}(\mu_1)$	−0.2717	−0.4107	−0.4858	−0.4931	−0.4994	−0.5042
$E\hat{\Lambda}(\mu_2)$	−0.2884	−0.4335	−0.4830	−0.4911	−0.4984	−0.5043
$E\hat{\Lambda}(\mu_3)$	−0.2789	−0.4595	−0.4555	−0.4643	−0.4726	−0.4795
$E\hat{\Lambda}(\mu_h)$	−0.2540	−0.4884	−0.4135	−0.4231	−0.4322	−0.4401

Here %Q j denotes the share accounted for by population quartile j of the variable directly above it. %SAS is the share owned by the social annuity system. μ_j is the upper bound for μ in quartile j, i.e. $H(\mu_j) = j/4$ for j ∈ {1, 2, 3, 4} and $\mu_4 = \mu_h$. $E\hat{\Lambda}(\mu_j)$ gives expected utility for an agent with health type μ_j

Steady-state transfers to the young amount to $\hat{Z} = 0.1146$ which is almost sixteen percent of wage income during youth. Steady-state consumption per worker by the young and surviving old are given by, respectively $\hat{c}^y \equiv [\int_{\mu_l}^{\mu_h} L_t(\mu)\hat{C}^y(\mu)d\mu]/N_t$ and $\hat{c}^o = [\int_{\mu_l}^{\mu_h} \mu L_{t-1}(\mu)\hat{C}^o(\mu)d\mu]/N_t$. Whilst the distribution over quartiles is rather even for youth consumption, it is very skewed for old-age consumption with the healthiest quartile accounting for half of consumption by the old. More than ten percent of the agents encounter a binding borrowing constraint $[H(\hat{\mu}_{bc}) = 0.1029]$. The wealth distribution by quartiles (as evidenced by capital ownership) is very uneven, with the unhealthiest and healthiest quartiles owning, respectively, 3.62 and 43.76 % of the capital stock.

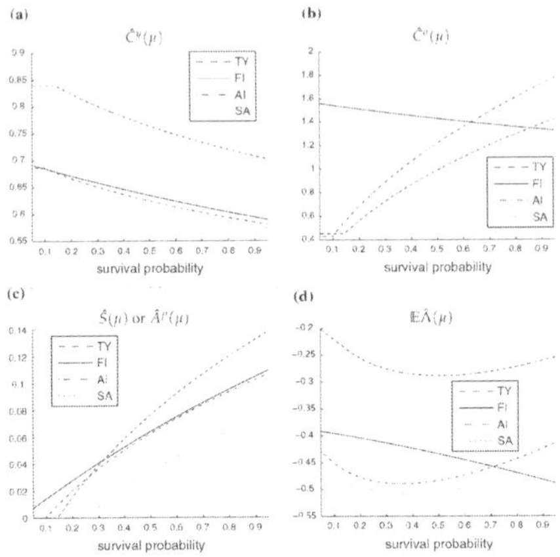

Figure 2. Steady-state profiles. (a) Youth consumption $\hat{C}^y(\mu)$, (b) old-age consumption $\hat{C}^o(\mu)$, (c) saving/annuity demand $\hat{S}(\mu)$ or $\hat{A}^p(\mu)$, (d) expected utility $\mathbb{E}\hat{\Lambda}(\mu)$.

In Fig. 2 the steady-state health profiles for youth- and old-age consumption, saving, and expected utility at birth are illustrated—see the dashed lines. The horizontal segments in panels (a–c) are the realizations for the (borrowing-constrained) endowment consumers. As expected from our theoretical discussion above, for unconstrained agents youth consumption is falling in health whilst the opposite holds for old-age consumption which, like saving, is increasing in health. Interestingly, expected utility is a non-monotonic function of health.[5]

PRIVATE ANNUITY MARKETS

Now consider the introduction of a private annuity market in this economy. A (life) annuity is an asset which pays a stipulated return contingent upon survival of the annuitant. In order to compete with other investment products, annuities have to provide a rate of return which exceeds the market rate of interest to compensate for the risk of death. The annuity firm is willing to pay this additional return on savings as it is held free of any obligation after the annuitant has died. Conversely, an agent who sells an actuarial note gets a life-insured loan and will have her debts acquitted when she dies prematurely. Hence, an agent who exclusively uses the annuity market for financial transactions will never die indebted or leave an (unintended) bequest. It is important to note that an annuity market constitutes just another type of redistribution scheme. Instead of channeling funds from the dead to the newly arrived young by means of lump sum transfers as above, an annuity market redistributes money from the dead to those that survive by offering a premium over and above the rental rate of capital. Therefore, the macroeconomic and welfare effects of introducing a private annuity market are not a priori clear.

Full Information Equilibrium

We make the following assumptions about the private market for annuities:

(A1) The annuity market is perfectly competitive. There is a large number of risk neutral firms offering annuities to individuals, and firms can freely enter or exit the market.

(A2) Annuity firms do not use up any real resources.

As a consequence of these assumptions the expected profit of each annuity firm is zero. We start by considering the benchmark case in which health status is public information. In the Full Information equilibrium in the annuity market (abbreviated as FI) each health type receives its actuarially fair rate of return and achieves perfect insurance against life span uncertainty. There is market segmentation in the sense that there is a separate market for each type of agent.

Let $A_t^p(\mu)$ denote the private annuity holdings of an agent of health type μ. As the net rate of return on annuities exceeds the rental rate of capital, all agents will completely annuitize their wealth. This is the famous result found by Yaari (1965).

An annuity firm sells annuities to agents of health type μ that pay a net rate of return $r_{t+1}^p(\mu)$ contingent upon survival of the annuitant. The firm invests the proceeds in the capital market, thus earning a rate of return r_{t+1}. Since some of its clients will die young and will subsequently lose their claim at an early stage, the annuity firm can redistribute their assets among the surviving clientele. The zero-profit conditions for the annuity market are therefore given by

$$(1+r_{t+1})L_t(\mu)A_t^p(\mu) = \mu(1+r_{t+1}^p(\mu))L_t(\mu)A_t^p(\mu), \quad \mu \in [\mu_L, \mu_H]. \tag{20}$$

That is, the gross return earned on the amount of savings collected from clients of health type μ in period t (left-hand side) should equal the (expected) claim originating from surviving clients of health type μ in period t + 1. It follows that:

$$1 + r_{t+1}^{p}(\mu) = \frac{1 + r_{t+1}}{\mu}, \quad \mu \in [\mu_l, \mu_h]. \tag{21}$$

Note that $r_{t+1}^{p}(\mu)$ is decreasing in the survival probability, i.e. unhealthy agents earn a higher gross rate of return on their investments in the annuity market than do healthy agents.

Since $0 < \mu_l < \mu_h < 1$ it follows that the interest rate on annuities exceeds the rate of return on capital for all agents, i.e. $r_{t+1}^{p}(\mu) > r_{t+1}$ for all $\mu \in [\mu_l, \mu_h]$. As a result rational agents will fully annuitize their savings (as was asserted above) so that the lifetime budget constraint is given by:

$$C_t^y(\mu) + \frac{C_{t+1}^o(\mu)}{1 + r_{t+1}^{p}(\mu)} = w_t + \frac{\lambda w_{t+1}}{1 + r_{t+1}^{p}(\mu)}. \tag{22}$$

Comparing the TY and FI scenarios [as described by, respectively, (5) and (22)], we find that for the latter case accidental bequests (and thus transfers from the government) are absent and the relevant return is type dependent. The consumption Euler equation for the FI case is given by:

$$\frac{U'(C_t^y(\mu))}{\beta U'(C_{t+1}^o(\mu))} = \mu\left(1 + r_{t+1}^{p}(\mu)\right) = 1 + r_{t+1}. \tag{23}$$

With actuarially fair annuities and full annuitization the marginal rate of substitution between current and future consumption is equated to the gross interest factor on capital. The mortality rate does not feature in this expression because agents are fully insured against the unpleasant aspects of lifetime uncertainty. This celebrated result was first proposed by Yaari (1965) in a continuous-time model.

By combining the lifetime budget constraint (22) and the Euler equation (23) we obtain the following expressions for consumption during youth, the demand for annuities, and old-age consumption:

$$C_t^y(\mu) = \Phi\left(\mu, (1 + r_{t+1})/\mu\right)\left[w_t + \frac{\lambda\mu w_{t+1}}{1 + r_{t+1}}\right]. \tag{24}$$

$$A_t^p(\mu) = \left[1 - \Phi\left(\mu, (1 + r_{t+1})/\mu\right)\right] w_t - \Phi\left(\mu, (1 + r_{t+1})/\mu\right)\frac{\lambda\mu w_{t+1}}{1 + r_{t+1}}, \tag{25}$$

$$\frac{\mu C_{t+1}^o(\mu)}{1 + r_{t+1}} = \left[1 - \Phi\left(\mu, (1 + r_{t+1})/\mu\right)\right]\left[w_t + \frac{\lambda\mu w_{t+1}}{1 + r_{t+1}}\right], \tag{26}$$

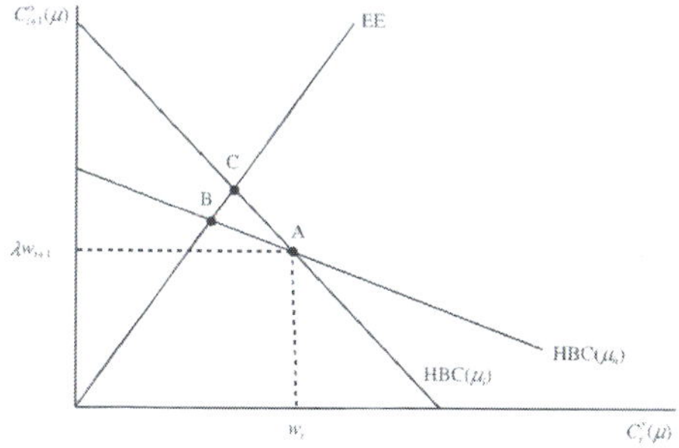

Figure 3. Choices of a μ-type individual with perfect annuities.

where ϕ (μ, x) is defined in (9) above. The optimal consumption choices can be illustrated with the aid of Fig. 3. The line labeled EE is the Euler equation (23) which is the same for all health types. In contrast, there exists a continuum of HBC(μ) lines passing though the income endowment point A. The lines labeled HBC(μ_l) and HBC(μ_h) depict the lifetime budget constraints faced by, respectively, the unhealthiest and healthiest individuals. It follows that the unhealthiest types consume at point C whilst the healthiest choose point B. Throughout the population consumption patterns are located along the line segment BC.

Under the maintained hypothesis that a non-trivial macroeconomic equilibrium exists, two qualitative conclusions can already be drawn on the basis of Fig. 3. First, since the endowment point is type-independent, all individuals must be net savers during youth. (The opposite case with all agents being net borrowers is inconsistent with a positive capital stock in a closed-economy equilibrium.) Whereas the unhealthiest agents in the population tend to face a binding borrowing constraint in a world without annuities (see above), these same agents express a positive demand for annuities in the FI scenario because the rate they receive on such instruments in case they survive into old age is quite high [i.e. HBC(μ_l) is rather steep]. Second, just as the savings function is in the TY case, the demand for annuities in the FI scenario is increasing in health type.

The fundamental difference equation for the FI model is given by:

$$k_{t+1} = \phi_1^{FI}(r_{t+1})\, w_t - \phi_2^{FI}(r_{t+1})\, \frac{\lambda w_{t+1}}{1 + r_{t+1}}, \tag{27}$$

where $\phi_1^{FI}(r_{t+1})$ and $\phi_2^{FI}(r_{t+1})$ are defined as follows:

$$\phi_1^{FI}(r_{t+1}) \equiv \frac{1}{1 + n + \lambda \bar{\mu}} \int_{\mu_l}^{\mu_h} \left[1 - \Phi\left(\mu, (1 + r_{t+1})/\mu\right)\right] h(\mu)\, d\mu.$$

$$\phi_2^{FI}(r_{t+1}) \equiv \frac{1}{1+n+\lambda\bar{\mu}} \int_{\mu_l}^{\mu_h} \mu\Phi(\mu, (1+r_{t+1})/\mu)\, h(\mu)\, d\mu.$$

The existence of a well-defined macroeconomic equilibrium is confirmed by the results reported in column (b) of Table 2. Compared to the TY case, output, the wage rate, and the capital stock are lower and the interest rate is higher under the FI scenario. The intuition behind these results is explained in Sect. 3.3 below.

In Fig. 2 the steady-state health profiles for youth and old-age consumption, annuity demand, and expected utility at birth are depicted—see the solid lines. Youth consumption, old-age consumption, and expected utility are decreasing in health whilst the opposite holds for the demand for annuities.

Asymmetric Information Equilibrium

The following assumptions are added to the ones above:

(A3) Health status is private information of the annuitant. The distribution of health types in the population and the corresponding survival probabilities are common knowledge.

(A4) Annuitants can buy multiple annuities for different amounts and from different annuity firms. Individual annuity firms cannot monitor an annuitant's holdings with other firms.

Under these assumptions, the Asymmetric Information (AI) equilibrium in the annuity market is a pooling equilibrium. The existence of this pooling equilibrium depends critically on (A3): as annuity firms cannot observe their customers' health type, they will offer a single rate of return which applies to all their clients. They can exploit their knowledge about the distribution of health types in the population in setting this common annuity rate. Note that if (A4) would not hold then annuity firms could indirectly deduce an agent's health type by the amount of wealth she has invested. Healthy individuals tend to save more, which exacerbates the degree of adverse selection in the annuity market.

Our conclusion that a pooling equilibrium will prevail in the annuity market appears opposite to that of Rothschild and Stiglitz (1976), who show that a pooling equilibrium does not exist in an insurance market. However, their result relies heavily on the assumption that customers can buy only one insurance contract such that the insurer sets both price and quantity. They admit themselves that this may be an objectionable assumption in some cases, and Walliser (2000) argues that it indeed does not apply to the annuity market. First of all, monitoring the receipts of annuity payments from other annuity firms would be very difficult. Secondly, in contrast to other types of insurance, the death of the annuitant ends the liability of the annuity firm instead of creating it. Hence, withholding payments and investigating compliance is not feasible.

We therefore believe it to be justified to assume that the annuity firm can only control the rate of return on annuities and cannot set prices and quantities simultaneously in order to obtain full information revelation.

An implicit assumption in the model is that annuitants cannot credibly signal their health status to the market. In the absence of cheap and credible medical tests, this is not a strong assumption at all. When asked about her health type, each individual has a clear incentive to claim to be unhealthy in order to get the highest possible return in a separated market. However, firms know this and will therefore not believe the annuitant's claim. Hence, even though part of their clientele tells the truth, the fact that some have an incentive to lie is enough for the annuity firm to assume that everyone will be a fraud.

We now turn to the determination of the pooled annuity rate, denoted by $\bar{r}^p_{t+1}$. As this rate will apply to agents of all health types it does not depend on μ. Assume that all agents with $\mu \in [\mu_{bc,t}, \mu_h]$ purchase annuities, i.e. are net savers. Then the zero-profit condition for the annuity market is given by:

$$(1 + r_{t+1}) \int_{\mu_{bc,t}}^{\mu_h} L_t(\mu) A_t^p(\mu) d\mu = (1 + \bar{r}^p_{t+1}) \int_{\mu_{bc,t}}^{\mu_h} \mu L_t(\mu) A_t^p(\mu) d\mu. \tag{28}$$

As in the case of a full-information equilibrium the gross return earned on the amount of savings collected from clients in period t (left-hand side) should equal the (expected) claim originating from surviving clients in period t + 1. It follows that:

$$1 + \bar{r}^p_{t+1} = \frac{1 + r_{t+1}}{\bar{\mu}^p_t}, \tag{29}$$

where $\bar{\mu}^p_t$ denotes the asset-weighted average survival rate of annuity purchasers:

$$\bar{\mu}^p_t \equiv \int_{\mu_{bc,t}}^{\mu_h} \mu \xi_t(\mu) d\mu, \quad \xi_t(\mu) = \frac{A_t^p(\mu) h(\mu)}{\int_{\mu_{bc,t}}^{\mu_h} A_t^p(\mu) h(\mu) d\mu}, \quad \mu \in [\mu_{bc,t}, \mu_h]. \tag{30}$$

This result has also been found in a partial equilibrium context by Sheshinski (2008) and relates to the linear equilibrium concept of Pauly

(1974). As noted by Walliser (2000), it can alternatively be interpreted as a Nash equilibrium among annuity firms in which each firm that deviates from the zero-profit price incurs a loss. The lifetime budget constraint for annuitants is given by:

$$C_t^y(\mu) + \frac{C_{t+1}^o(\mu)}{1+\bar{r}_{t+1}^P} = w_t + \frac{\lambda w_{t+1}}{1+\bar{r}_{t+1}^P}, \tag{31}$$

where $\bar{r}_{t+1}^P$ is defined in (29) above. Utility maximization results in the following expression for the consumption Euler equation:

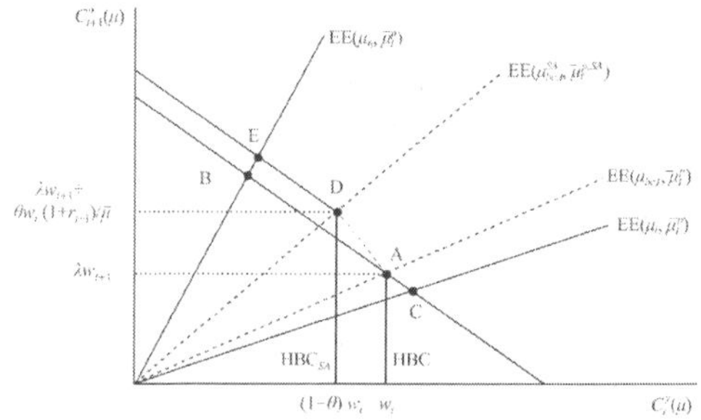

Figure 4. Choices of a μ-type individual with imperfect annuities.

$$\frac{U'(C_t^y(\mu))}{\beta U'(C_{t+1}^o(\mu))} = \mu\left(1+\bar{r}_{t+1}^P\right) = \frac{\mu}{\bar{\mu}_t^P}\left(1+r_{t+1}\right). \tag{32}$$

In the AI case the marginal rate of substitution between current and future consumption is equated to the expected gross return on pooled annuities. The mortality rate still features in this expression because agents are imperfectly insured against the unpleasant aspects of lifetime uncertainty.

By combining the lifetime budget constraint (31) and the Euler equation (32) we obtain the following expressions for consumption during youth, the demand for annuities, and old-age consumption:

$$C_t^y(\mu) = \Phi\left(\mu, (1+r_{t+1})/\bar{\mu}_t^P\right)\left[w_t + \frac{\lambda\bar{\mu}_t^P w_{t+1}}{1+r_{t+1}}\right], \tag{33}$$

$$\begin{aligned} A_t^P(\mu) &= \left[1-\Phi\left(\mu, (1+r_{t+1})/\bar{\mu}_t^P\right)\right] w_t \\ &\quad -\Phi\left(\mu, (1+r_{t+1})/\bar{\mu}_t^P\right)\frac{\lambda\bar{\mu}_t^P w_{t+1}}{1+r_{t+1}}, \end{aligned} \tag{34}$$

$$\frac{\bar{\mu}_t^P C_{t+1}^o(\mu)}{1+r_{t+1}} = \left[1-\Phi\left(\mu, (1+r_{t+1})/\bar{\mu}_t^P\right)\right]\left[w_t + \frac{\lambda\bar{\mu}_t^P w_{t+1}}{1+r_{t+1}}\right], \tag{35}$$

where $\phi(\mu, x)$ is defined in (9) above. In Fig. 4 we illustrate the case (confirmed for the calibrated model) in which not all agents purchase annuities in the AI equilibrium, i.e. $\mu_{bc,t} > \mu_1$. The household budget constraint faced by all agents alike is given by HBC whilst the endowment point is at point A. For given values of $\bar{\mu}_t^p$ and $1 + r_{t+1}$, the Euler equation (32) depends on the agents survival probability, with EE(μ_h) and EE(μ_l) depicting the optimal wealth expansion paths for, respectively, the healthiest and unhealthiest agents. Optimal consumption choices for the two types of agents are located at points B and C, respectively. But point C is unattainable because it would involve going short on annuities, i.e. selling life-insured loans. But the moment an agent attempts to do so she reveals her health type which results in the borrowing rate being set at such a high level that borrowing is unattractive. It follows that there is a binding "self-imposed" borrowing constraint, the HBC is vertical at the endowment wage w_t, and all health types such that $\mu_l \leq \mu < \mu_{bc,t}$ are bunching at the endowment point A.6.

For all agents featuring $\mu_{bc,t} \leq \mu \leq \mu_h$ the demand for annuities is positive and increasing in the survival probability. It thus follows from (30) that $\mu_{bc,t} < \bar{\mu}_t^p < \mu_h$, i.e. the relatively healthy agents obtain a better than actuarially fair rate of return on their private annuity holdings (because $\bar{\mu}_t^p < \mu_h$) whereas the comparatively unhealthy annuitants obtain a less than fair return (as $\bar{\mu}_t^p > \mu_{bc,t}$). The presence of unhealthy agents in the market allows the healthy to earn an informational rent.

Comparing the FI and AI cases we find that the informational asymmetry in the annuity market affects individuals via two related but distinct channels. First, whereas all agents are net purchasers of annuities ($A_t^p(\mu) > 0$ for $\mu \in [\mu_l, \mu_h]$) under symmetric information, the unhealthiest part of the population typically experiences a binding borrowing constraint and is thus absent from this market altogether [$A_t^p(\mu) = 0$ for $\mu \in [\mu_l, \mu_{bc,t})$] when information is asymmetric. Second, the pooling rate under asymmetric information is "demographically unfair" in the sense that it is not based on the average demographic survival rate in the population of annuitants but on the asset-weighted survival rate $\bar{\mu}_t^p$. In order to demonstrate this phenomenon, note that the demographically fair pooling rate is defined as follows:

$$1 + \bar{r}_{t+1}^f = \frac{1 + r_{t+1}}{\bar{\mu}_t^f}, \tag{36}$$

where the superscript f stands for 'demographically fair' and $\bar{\mu}_t^f$ denotes the average survival rate of annuity purchasers among the cohort born in period t:

$$\bar{\mu}_t^f \equiv \frac{\int_{\mu_{bc,t}}^{\mu_h} \mu h(\mu)\, d\mu}{\int_{\mu_{bc,t}}^{\mu_h} h(\mu)\, d\mu}. \tag{37}$$

Clearly, the demographically fair pooling rate is not sustainable in equilibrium because annuity firms would make a loss if they were to offer it to annuitants. To see why this is the case, note that the demand for annuities is increasing in the survival probability.

It follows that $\bar{\mu}_t^p > \bar{\mu}_t^f$ and thus $1+\bar{r}_{t+1}^p < 1+\bar{r}_{t+1}^f$ for all t. 7

The fundamental difference equation for the AI case is given by:

$$k_{t+1} = \phi_1^{AI}\left(r_{t+1}, \bar{\mu}_t^p\right) w_t - \phi_2^{AI}\left(r_{t+1}, \bar{\mu}_t^p\right) \frac{\lambda \bar{\mu}_t^p w_{t+1}}{1+r_{t+1}}, \tag{38}$$

where $\phi_1^{AI}\left(r_{t+1}, \bar{\mu}_t^p\right)$ and $\phi_2^{AI}\left(r_{t+1}, \bar{\mu}_t^p\right)$ are defined as follows:

$$\phi_1^{AI}\left(r_{t+1}, \bar{\mu}_t^p\right) \equiv \frac{1}{1+n+\lambda\bar{\mu}} \int_{\mu_{bc,t}}^{\mu_h} \left[1 - \Phi\left(\mu, (1+r_{t+1})/\bar{\mu}_t^p\right)\right] h(\mu)\, d\mu,$$

$$\phi_2^{AI}\left(r_{t+1}, \bar{\mu}_t^p\right) \equiv \frac{1}{1+n+\lambda\bar{\mu}} \int_{\mu_{bc,t}}^{\mu_h} \Phi\left(\mu, (1+r_{t+1})/\bar{\mu}_t^p\right) h(\mu)\, d\mu.$$

The key features of the steady-state AI equilibrium are summarized in column (c) of Table 2. As far as the macroeconomic variables ($\hat{y}$, $\hat{k}$, $\hat{r}$, $\hat{w}$, $\hat{c}^y$, and $\hat{c}^o$) are concerned the AI and FI equilibria are rather similar. In contrast, the composition of old-age consumption as well as expected lifetime utility differ dramatically for these two cases. The informational asymmetry in annuity markets leads to a redistribution from the unhealthy to the healthy agents in the economy. We elaborate on this point in Sect. 4 below.

Finally, Fig. 2 shows the steady-state health profiles for youth- and old-age consumption, annuity demand, and expected utility at birth—see the dash-dotted lines. Youth consumption is decreasing in health whilst the opposite holds for old-age consumption and the demand for annuities. Expected utility is U-shaped as was the case for the TY scenario.

Transitional Dynamics

We assume that the economy is initially in the steady state of the model without annuities. At a given time t a private annuity market is introduced. The transition paths of the capital intensity and the expected lifetime utility of different health types are computed by means of an iterative procedure.

The results are given in Figs. 5 and 6 for both the full information (FI) equilibrium and the asymmetric information (AI) equilibrium in the private annuity market. The paths for the capital intensity are scaled by the steady-state equilibrium value in the no annuities case (TY).

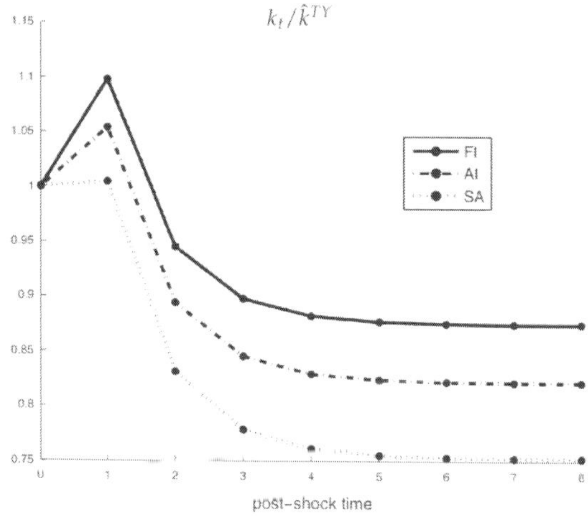

Figure 5. Capital intensity relative to TY.

As such, a value in excess of unity indicates that the capital intensity increases relative to its benchmark scenario, when it falls short of unity there is a decrease. For lifetime utility we report the difference with the steady-state values obtained in the TY case, i.e. $\Delta\mathbb{E}\Lambda_t(\mu_j) \equiv \mathbb{E}\Lambda_t(\mu_j) - \mathbb{E}\hat{\Lambda}^{TY}(\mu_j)$. Table 2 presents a numerical comparison between (unscaled) steady-state values.

We find that the introduction of a private annuity market boosts the capital intensity at first but gives rise to severe capital crowding out in the long run. The effects are more pronounced for the FI equilibrium than for the AI equilibrium. Agents save a larger part of their income in the first period and therefore consume less during youth and more in old age. Comparing the lifetime utility profiles we find that comparatively healthy agents prefer the AI to the FI equilibrium, and vice versa for the relatively unhealthy

At the time the annuity market is opened up there are still accidental bequests in the economy which have been left by the previous generation. Hence, in the first period of the new regime young agents benefit from the higher return on their savings through full annuitization while also receiving the intergenerational transfer. Their expected utility level increases relative to the benchmark. However, from period t+1 (i.e. postshock time 1) onwards the transfers are abolished and all health types have a lower utility level under either equilibrium in the private annuity market than in the absence of annuities. This is an example of the so-called 'tragedy of annuitization', as described in Heijdra et al. (2010). Even though it is individually rational to fully annuitize (as the annuity rate of return exceeds the return on capital), this is not optimal from a social point of view. If all agents invests their financial wealth in the annuity market then the resulting long-run equilibrium leaves future newborns worse off compared to the case where annuities are absent and accidental bequests are redistributed to the young. In other words, the introduction of an annuity market is welfare improving when considered in isolation, but not when the general equilibrium repercussions as a result of endogenous factor prices are taken into account.

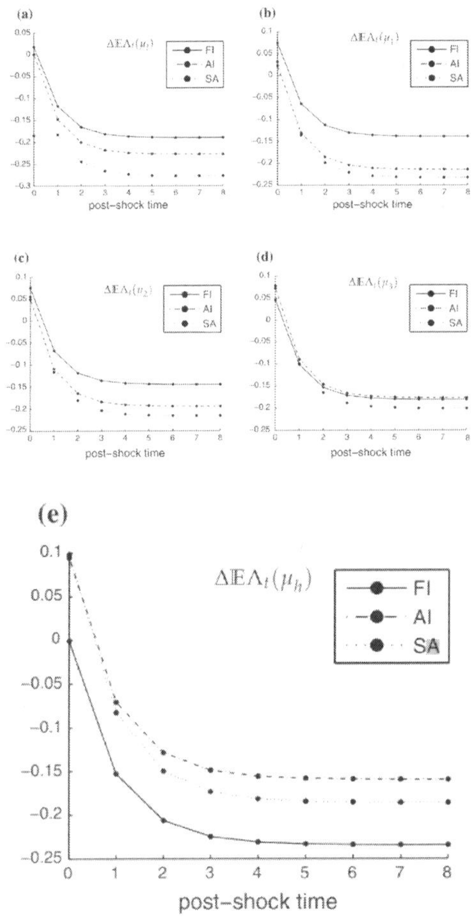

Figure 6. Expected lifetime utility relative to TY. **a** Type μ_l: $\Delta\mathbb{E}\Lambda_t(\mu_l)$, **b** type μ_1: $\Delta\mathbb{E}\Lambda_t(\mu_1)$, **c** type μ_2: $\Delta\mathbb{E}\Lambda_t(\mu_2)$, **d** type μ_3: $\Delta\mathbb{E}\Lambda_t(\mu_3)$, **e** type μ_h: $\Delta\mathbb{E}\Lambda_t(\mu_h)$.

Table 3. Equivalent variations with private and social annuities

	(a) AI	(b) SA $\theta = 0.01$	(c) SA $\theta = 0.02$	(d) SA $\theta = 0.03$
$\Pi(\mu_l)$	0. 1540	0.1615	0.1700	0.1788
$\pi(\mu_l)$	0. 1835	0.1924	0.2025	0.2129
$\Pi(\mu_1)$	0. 1335	0.1376	0.1413	0.1441
$\pi(\mu_1)$	0. 1656	0.1707	0.1753	0.1788
$\Pi(\mu_2)$	0. 1117	0.1160	0.1201	0.1234
$\pi(\mu_2)$	0. 1464	0.1521	0.1573	0.1617
$\Pi(\mu_3)$	0. 0946	0.0992	0.1034	0.1071
$\pi(\mu_3)$	0. 1297	0.1360	0.1418	0.1468
$\Pi(\mu_h)$	0. 0804	0.0851	0.0896	0.0935
$\pi(\mu_h)$	0. 1147	0.1214	0.1277	0.1334

As measures of utility are purely ordinal, the nominal difference in steady-state utility levels as presented in Table 2 cannot be easily interpreted. In order to evaluate the magnitude of utility differences in a more insightful way we use the metric of consumption-equivalent variations. That is, we determine how much additional consumption during youth one would have to give to an agent of a given health type in the private annuity market scenario in order to make her as well off as in the no-annuities case, keeping all prices (i.e. the wage rate and the interest rate) and old-age consumption fixed at their initial level. That is, we find that value of Π (μ) which satisfies the following equation:

$$U(\hat{C}^y(\mu) + \Pi(\mu)) + \mu\beta U(\hat{C}^o(\mu)) = \mathbb{E}\hat{\Lambda}^{TY}(\mu). \tag{39}$$

where $\mathbb{E}\hat{\Lambda}^{TY}(\mu)$ is the steady-state benchmark utility level

The results are given in column (a) of Table 3. We present both unscaled and scaled values for the compensation measure, i.e. $\pi(\mu_j) = \Pi(\mu_j)/\hat{C}^y(\mu_j)^{TY}$ where $\hat{C}^y(\mu_j)^{TY}$ is the steady-state youth consumption level of an agent with health type μj in the TY scenario. For example, to ensure that the unhealthiest type of agent has the same level of utility when there is a pooled private annuity market as when there are no annuities at all one would have to give her 0.1540 in terms of youth consumption. This corresponds to 18.35 % of the consumption level she would have had in the absence of annuities. These numbers are quite significant, showing that the (negative) welfare effects of introducing a private annuity market in this economy are not inconsequential.

SOCIAL ANNUITIES

Following Abel (1986) we now study the consequences of introducing mandatory social annuities. These can be interpreted, for example, as a fully funded pension system. Agents contribute a fixed amount A_t^S during their youth and receive a benefit $(1+\bar{r}_{t+1}^s)A_t^s$ As t if they survive into the second phase of life, where $\bar{r}_{t+1}^S$ is the (implicit) net return on social annuities. All contributions are pooled and invested in the capital market. The gross return is then divided equally among the survivors. This implies that the following resource constraint should be satisfied:

$$(1+r_{t+1})L_t A_t^s = (1+\bar{r}_{t+1}^s)A_t^s \int_{\mu_l}^{\mu_h} \mu L_t(\mu)\,d\mu. \tag{40}$$

Solving for the implied rate of return yields:

$$1+\bar{r}_{t+1}^s = \frac{1+r_{t+1}}{\bar{\mu}}, \tag{41}$$

where $\bar{\mu} \equiv \int_{\mu_l}^{\mu_h} \mu h(\mu)\,d\mu$ is the average survival rate. Intuitively, as contributions are mandatory and independent of health type, the social annuity system is immune to adverse selection and forces everyone to participate. This makes the introduction of social annuities appear like an attractive option for the government: it can offer agents a higher rate of return on part of their savings than the private market can. Note that, in general, $\bar{r}_{t+1}^s > \bar{r}_{t+1}^f$, the demographically fair pooling rate as defined in (37), because all agents are forced to buy social annuities whereas not all agents purchase private annuities (if $\mu_{bc,t} > \mu l$ it follows that $\bar{\mu}_t^f > \bar{\mu}$).

We assume that the agent can invest her remaining financial wealth in a pooled private annuity market. The lifetime budget constraint can be written as

$$C_t^y(\mu) + \frac{C_{t+1}^o(\mu)}{1+\bar{r}_{t+1}^p} = w_t + \frac{\lambda w_{t+1}}{1+\bar{r}_{t+1}^p} + \frac{\bar{\mu}_t^p - \bar{\mu}}{\bar{\mu}} A_t^s. \tag{42}$$

Note that, ceteris paribus, the introduction of social annuities has a positive effect on the level of total human wealth as $\bar{\mu}_t^p > \bar{\mu}$ for all t due to adverse selection in the private annuity market. By using (42) in combination with the Euler equation (32) we find that the agent's optimal plans are fully characterized by:

$$C_t^y(\mu) = \Phi\left(\mu, (1+r_{t+1})/\bar{\mu}_t^p\right)\left[w_t + \frac{\lambda \bar{\mu}_t^p w_{t+1}}{1+r_{t+1}} + \frac{\bar{\mu}_t^p - \bar{\mu}}{\bar{\mu}} A_t^s\right], \tag{43}$$

$$\begin{aligned} A_t^p(\mu) + A_t^s &= \left[1 - \Phi\left(\mu, (1+r_{t+1})/\bar{\mu}_t^p\right)\right] w_t \\ &\quad - \Phi\left(\mu, (1+r_{t+1})/\bar{\mu}_t^p\right)\left[\frac{\lambda \bar{\mu}_t^p w_{t+1}}{1+r_{t+1}} + \frac{\bar{\mu}_t^p - \bar{\mu}}{\bar{\mu}} A_t^s\right]. \end{aligned} \tag{44}$$

$$\frac{\bar{\mu}_t^p C_{t+1}^o(\mu)}{1+r_{t+1}} = \left[1-\Phi\left(\mu, (1+r_{t+1})/\bar{\mu}_t^p\right)\right]\left[w_t + \frac{\lambda\bar{\mu}_t^p w_{t+1}}{1+r_{t+1}} + \frac{\bar{\mu}_t^p - \bar{\mu}}{\bar{\mu}} A_t^s\right]. \quad (45)$$

Compared to the model without social annuities discussed in Sect. 3.2 above there are a few noteworthy differences. Consumption during youth and old age are still proportional to human wealth, but the definition of human wealth has been augmented by a term reflecting the relative return on social annuities. From (44) we see that private annuity demand has been replaced by total annuity demand, the sum of private and social annuity holdings. As such, mandatory social annuities do indeed crowd out their private counterparts. Keeping everything else constant, this crowding out is more than one-for-one as the higher level of total income in the second period reduces the incentive for private saving.

We assume that the compulsory contribution to the social annuity system is proportional to wage income during youth, i.e. $A_t^s = \theta w_t$ with $0 < \theta < 1$. Note that for high values of θ agents might be overannuitized, i.e. they could be forced by the social security system to annuitize a larger portion of their financial wealth than they would voluntarily do. Since there is a borrowing constraint in the pooled annuity market this overannuitization cannot be undone by going short in private annuities. Figure 4 depicts a partial equilibrium representation of the effects of social annuities. In the presence of social annuities the endowment point shifts to point D and the relevant lifetime budget constraint is given by HBCSA. The thin dashed line connecting A and D is steeper than the downward sloping segments of HBC and HBCSA because the rate on mandatory annuities is greater than the pooled private rate ($\bar{r}_{t+1}^s > \bar{r}_{t+1}^p$). The choice set is enlarged for the healthier part of the population, but borrowing constraints become more severe for the unhealthier (as $\mu_{bc,t}^{SA} > \mu_{bc,t}$).

The fundamental difference equation for the SA case is given by:

$$k_{t+1} = \phi_1^{AI}\left(r_{t+1}, \bar{\mu}_t^p\right) w_t - \phi_1^{SA}\left(r_{t+1}, \bar{\mu}_t^p\right) \theta w_t - \phi_2^{AI}\left(r_{t+1}, \bar{\mu}_t^p\right) \frac{\lambda\bar{\mu}_t^p w_{t+1}}{1+r_{t+1}}, \quad (46)$$

where $\phi_1^{AI}\left(r_{t+1}, \bar{\mu}_t^p\right)$ and $\phi_2^{AI}\left(r_{t+1}, \bar{\mu}_t^p\right)$ are defined directly below (38) and $\phi_1^{SA}\left(r_{t+1}, \bar{\mu}_t^p\right)$ is given by:

$$\phi_1^{SA}\left(r_{t+1}, \bar{\mu}_t^p\right) = \frac{1}{1+n+\lambda\bar{\mu}} \times \left[\frac{\bar{\mu}_t^p - \bar{\mu}}{\bar{\mu}} \int_{\mu_{bc,t}}^{\mu_h} \Phi\left(\mu, (1+r_{t+1})/\bar{\mu}_t^p\right) h(\mu)\, d\mu - H(\mu_{bc,t})\right]$$

The key features of the steady-state SA equilibria that emerge for different values of θ are summarized in columns (d–f) of Table 2. The larger is the social annuity system (as measured by the magnitude of θ), the stronger is capital crowding out and the larger is the proportion of borrowing-constrained (overannuitized) agents in the economy. We elaborate on this point in Sect. 4.1 below.

Finally, in Fig. 2 the steady-state health profiles for youth- and old-age consumption, private annuity demands, and expected utility at birth are illustrated—see the dotted lines.

To avoid cluttering these figures, we restrict attention to the case of a "large" social annuity system with $\theta = 0.03$.[8] Apart from the larger range of borrowing-constrained individuals, these profiles are very similar to the ones obtained for the AI case.

Transitional Dynamics

The transition paths of the capital intensity and lifetime utility after introducing both private and social annuities in the economy (denoted by SA) are given in Figs. 5 and 6 for the case that social security contributions amount to three percent of wage income ($\theta = 0.03$). We find that, compared to the benchmark of no annuities and transfers to the young, the capital intensity increases slightly for one period but afterwards decreases sharply. The overall effect on lifetime utility is negative for all health types, except for the impact period in which the transfers have not yet been abolished.

In order to evaluate the magnitude of the relative welfare decrease in the steady state we calculate the consumption-equivalent variations for three different values of θ, see columns (b–d) in Table 3 and Sect. 3.3 above. Recall that column (a) (featuring $\theta = 0$) corresponds to the pooling equilibrium. We find that the equivalent variations increase monotonically in θ for all health types, implying that the equilibrium becomes progressively worse in welfare terms when the contribution to the social annuity system increases. At first view this might seem somewhat counterintuitive, as social annuities offer agents a higher rate of return on a fixed part of their savings than private annuities do, keeping everything else constant. However, it is exactly this ceteris paribus condition which is misleading.

Importantly, the introduction of social annuities has two opposing effects on the rate of return on private annuities. First of all, there is a partial equilibrium effect. The mandatory investment in social annuities results in a decrease in the demand for private annuities. Unhealthy agents reduce their private annuity demand disproportionally more than healthy agents do. As a consequence, the asset share of healthy agents increases and thereby the degree of adverse selection in the private annuity market. This leads to a decrease in the rate of return on private annuities, in line with the findings of by Abel (1986), Walliser (2000), and Palmon and Spivak (2007).

However, there is also a general equilibrium effect which partly offsets the decrease. The total level of savings decreases in the steady state, leading to a rise in the rental rate of capital relative to the scenario without social annuities. This response is not taken into consideration when the focus is restricted to partial equilibrium analysis or when the interest rate is assumed to be exogenously fixed.

The net effect on the private annuity rate is negative in all our simulations. Overall, we find that in the long run the negative effects of introducing social annuities (a lower wage rate and a lower return on private annuities) outweigh the positive effect (a demographically fair rate of return on a part of financial wealth) as is evident from the drop in welfare of future generations.

CONCLUSION

In this paper we have constructed a discrete-time general equilibrium model featuring exogenous growth and overlapping generations of heterogeneous agents who are distinguished by their health status. An agent's health type is assumed to be private information. We show that if a private annuity market is introduced in this economy then it will be characterized by a pooling equilibrium. Due to adverse selection, the resulting annuity rate of return is less than the actuarially fair pooling rate. Even though it is individually optimal to invest in annuities, agents of all health types are in the long run worse off in welfare terms compared to the case where annuities are absent and accidental bequests are redistributed to the young.

We have also studied the welfare implications of a social security system with fixed mandatory contributions for all health types. These social annuities are immune to adverse selection and therefore offer a higher rate of return than private annuities do under asymmetric information. However, they have a negative effect on steady-state welfare. The positive effect of a higher rate of return on a fixed part of savings and a higher return on capital in equilibrium is outweighed by the negative consequences of increased adverse selection in the private annuity market and a lower wage rate.

Our results suggest that privatization of social security may be welfare improving when annuity markets are characterized by asymmetric information. Elimination of social annuities reduces the degree of adverse selection in the private annuity market, increases the level of savings, and has a positive effect on lifetime utility for all health groups in the population. In general, lifetime utility of newborns is highest in the complete absence of annuities provided the redistribution scheme transfers accidental bequests to the newly arrived young agents.

REFERENCES

1. Abel, A. B. (1986). Capital accumulation and uncertain lifetimes with adverse selection. Econometrica, 54, 1079–1098.
2. Cannon, E., & Tonks, I. (2008). Annuity markets. Oxford: Oxford University Press.
3. Heijdra, B. J. & Mierau, J. O. (2012). The individual life-cycle, annuity market imperfections and economic growth. Journal of Economic Dynamics and Control (forthcoming).
4. Heijdra, B. J., Mierau, J. O., & Reijnders, L.S.M. (2010). The tragedy of annuitization. Working paper 3141, München: CESifo.
5. Heijdra, B. J. & Reijnders, L.S.M. (2009). Economic growth and longevity risk with adverse selection. Working paper 2898, München: CESifo.
6. Palmon, O., & Spivak, A. (2007). Adverse selection and the market for annuities. Geneva Risk and Insurance Review, 32, 37–59.
7. Pauly, M. V. (1974). Overinsurance and public provision of insurance: The role of moral hazard and adverse selection. Quarterly Journal of Economics, 88, 44–62.
8. Pecchenino, R. A., & Pollard, P. S. (1997). The effects of annuities, bequests, and aging in an overlapping generations model with endogenous growth. Economic Journal, 107, 26–46.
9. Rothschild, M., & Stiglitz, J. (1976). Equilibrium in competitive insurance markets: An essay on the economics of imperfect information. Quarterly Journal of Economics, 90, 629–649.
10. Shcshinski, E. (2008). The economic theory of annuities. Princeton: Princeton, N.J.: University press.
11. Villeneuve, B. (2003). Mandatory pensions and the intensity of adverse selection in life insurance markets. Journal of Risk and Insurance, 70, 527–548.
12. Walliser, J. (2000). Adverse selection in the annuities market and the impact of privatizing social security. Scandinavian Journal of Economics, 102, 373–393.
13. Yaari, M. E. (1965). Uncertain lifetime, life insurance, and the theory of the consumer. Review of Economic Studies, 32, 137–15

CHAPTER

4

Evaluation of Variable Annuity Guarantees with the Effect of Jumps in the Asset Price Process

Mussa Juma[1], Min Cherng Lee[1], Seong Tah Chin[1] and Kian Wah Liew[2]

[1]Department of Mathematical and Actuarial Sciences, Universiti Tunku Abdul Rahman, Selangor, Malaysia.

[2]Department of Applied Mathematics, Nottingham University Malaysia Campus, Semenyih, Malaysia.

ABSTRACT

Financial crisis in 2007 2008 have caused losses to life insurance companies issuing variable annuities with guarantees. This is partly due to failure of variable annuity (VA) issuers to anticipate the large variations in asset prices

Citation: Juma, M., Lee, M. C., Chin, S. T., & Liew, K. W. (2017). "Evaluation of variable annuity guarantees with the effect of jumps in the asset price process". Cogent Economics & Finance, 5(1), 1326218. https://doi.org/10.1080/23322039.2017.1326218

during the financial crisis times in their pricing framework and also setting a higher guaranteed rate. This study aims to investigate the pricing of the guaranteed minimum death and accumulation benefits embedded in flexible premium VA. We compare the prices from calibrated Black–Scholes model to that of calibrated jump-diffusion model. Although both models assume constant volatility, the fact that Black–Scholes model ignores abnormal asset price changes due to jumps is likely to under-price the VA. We also conduct a case study to analyse the impact on guarantee fees for different stock market performance and regional mortality rates.

Keywords:-Jump-diffusion model; variable annuity

INTRODUCTION

A variable annuity (VA) is an agreement between the life insurance company and the annuitant for the writer to pay periodical income in exchange for premium payment from the holder. They are of two types according to premium payment: single premium variable annuity (SPVA) and flexible premium variable annuity (FPVA). FPVA is a VA with periodical premium payments or instalment during the deferred period or accumulation period (Bernard, Cui, & Vanduffel, 2016; Chi & Lin, 2012). The policyholder makes many payments during the life of the policy (Costabile, 2013). The periodical payment can be constant, increasing or decreasing. In SPVA only one payment is paid during inception. VAs are known by various names like segregated funds (Canada) and Unit- or equity- or fundlinked (UK).

The investments of the VAs are in mutual funds. Equity-linked life insurance products enable the annuitants' contributions to be invested in equity market. This means that the returns on investments are directly linked to a particular equity fund of the annuitants' choice. Upon selection, the premiums are transferred to a third-party fund manager by the insurer. Alternatively the investments can be linked to the performance of an index rather than a stock, in this case they are known as Equity-Indexed Annuities or simply Indexed Annuities (EIA) rather than Equity-Linked or UnitLinked Annuities (ELA).

VAs have embedded options which provide protection against the downside of the stock market performance (during the bear market, poor performance and investment during the bull market, good performance) and outliving the annuitant`s income due to longevity risk. The embedded options include guaranteed minimum death benefits (GMDB) and guaranteed living benefits (GLB). GLB comprise guaranteed minimum income benefits

(GMIB), guaranteed minimum accumulation benefits (GMAB), guaranteed minimum withdrawal benefits (GMWB) and guaranteed lifetime withdrawal benefits (GLWB). For detailed explanation please refer to (Hardy, 2003; Stone, 2003). Many studies on evaluation of guarantees are based on guarantees embedded in SPVA (Bacinello, Millossovich, Olivieri, & Pitacco, 2011; Milevsky & Posner, 2001). The implication of the SPVA is that it is mainly meant for adults in their 50s and above who have earned enough from their work to make only one payment (Bacinello, Millossovich, Olivieri, & Pitacco, 2012; Bauer, Kling, & Russ, 2008).

Insurance companies charge a fee to fund the guarantees. The fee is known as mortality and expense fee (M&E fee) or insurance risk charges or insurance fees. If the fees are too high relative to the fund performance, the annuitant can terminate or surrender the guarantee and if the fee are too low relative to the fund performance then the life insurance company stands to suffer losses. Hence the fee should be fair.

Many writers of VAs suffered losses during the global financial crisis of 2007–2008 and other crisis which affected specific regions in previous times. Equitable life UK is an example of failure to incorporate jumps in the asset prices and time-varying volatility in pricing of the guarantee. The Company guaranteed a very high guaranteed rate when inflation was high in the early 1970s.

After decades of operations, the economic changes slowed down the inflation. This caused the Company to suffer losses and forced to close for new business by December 2000. Fluctuation of asset prices in such times becomes high which made the value of the guarantees more expensive than previously calculated.

Many literature use Black–Scholes model to price the guarantees embedded in VAs. Black–Scholes approach assumes a geometric Brownian motion (GBM) for the asset price which fails to reflect the leptokurtic distribution of asset returns, the volatility smiles and the variation of asset returns due to abnormal changes in asset price or jumps in the price process (Merton, 1976). Because of their longterm nature and importance to the retirees, it is necessary for the writers to incorporate volatility smiles in their evaluation. It has been shown that deterministic volatility including the constant volatility models does not take into account the fat tails and volatility smile behaviours of the asset returns.

Many studies have been done to improve the Black–Scholes model. They include:

- Incorporating jumps in the price process assuming constant volatility as in BS model (Hanson & Westman, 2002; Kou, 2002; Merton, 1976).
- Assuming a constant elasticity of variance (Cox & Ross, 1976).
- Assuming a stochastic volatility without jumps in the asset price process (Amin & Ng, 1993; Heston, 1993; Hull & White, 1987; Stein & Stein, 1991).
- Assuming a stochastic volatility and jumps in the asset price dynamics (Bates, 1996; Scott, 1997).

Black–Scholes model ignores the jumps in the asset prices caused by over or under reaction due to good or bad news coming from the market or an individual company. Hence it doesn't consider the impact of information arrival on the asset price changes. It only considers normal asset price changes but not abnormal.

By assuming the impact of market or company news arrival in specific discrete times which are random (jumps in asset prices), the GBM of the Black–Scholes model is transformed into a leptokurtic distribution with volatility smiles. Jump models have better analytical tractability than stochastic volatility models specifically for path-dependent options. They capture short-term characteristics or behaviour of the financial market better than stochastic volatility models (Yan & Hanson, 2006). In other words, they handle short-term smiles better.

The plots of time series of the market data whether market index prices, refer Figures 1–4 or individual company price movements show such presence of jumps. Also plots of times series of the asset returns whether market index returns, refer Figures 5–8 or individual company returns depict the presence of upward and downward jumps.

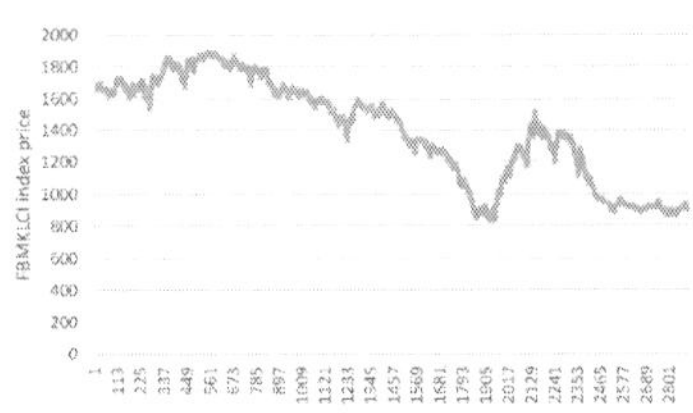

Figure 1. FBMKLCI index price.

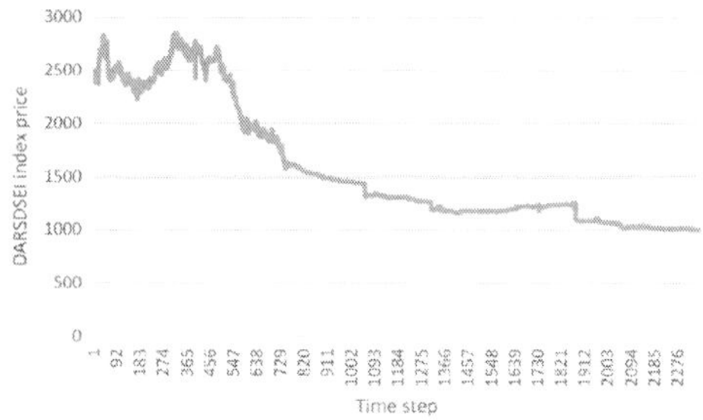

Figure 2. DARSDSEI index price.

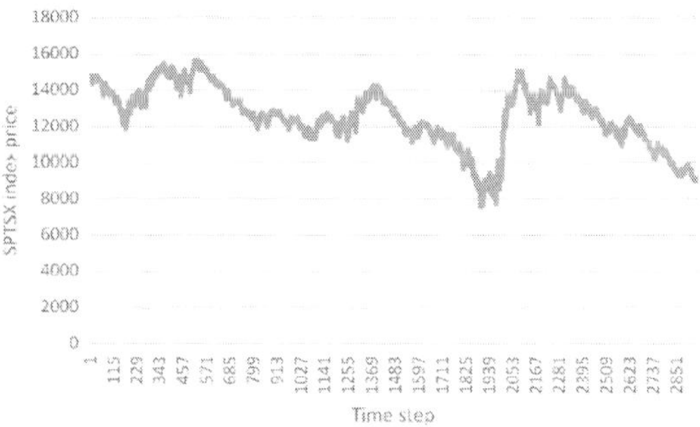

Figure 3. SPTXS index price.

Figure 4. S&P500 index price.

Figures 1–4 shows the stock market index price movements for the Tanzania, Malaysia, Canada and USA markets and their corresponding stock market index returns are shown in Figures 5–8 for the dates 1 January 2005–1 October 2016. Good news from the market or individual company causes upward jumps, while bad news causes downward jumps. Big and very big downward jumps relate to financial crisis and cause the financial market to crash. Big downward jumps in history include 1987, 2000–2002 and 2007–2008. In their papers, Kou (2002, 2007) showed that the addition of jump in the Black–Scholes model generates high peak (under-reaction) in the asset return leptokurtic distribution and heavy tails (over-reaction). In his paper, Merton (1976) assumed that the extra randomness of the underlying asset price due to jumps can be diversified away. By modelling jumps, the smiles can better be fitted.

Because of the 2007–2008 financial crisis and the previous 1987, 2000–2002 we intend to investigate whether a Black–Scholes model is appropriate to calculate the M&E fee for the combined guaranteed minimum death and accumulation benefits (GMDB&GMAB) embedded in a FPVA in such times of financial crashes. We hypothesize that the financial crisis in that period was due to jumps in the asset prices, especially downward jumps, caused by abnormal changes in the asset prices. We use stock market indices from 2005 to 2016 (which includes one of the global financial crisis period) to calibrate the jump-diffusion model and the diffusion model. Then we conduct a case study to determine the impact of investing into various stock markets by the life insurance company writing VAs on the M\&E fees and the impact of regional mortality differences on the combined guarantees fees when the VA writer issues the product to annuitants of different regions (countries).

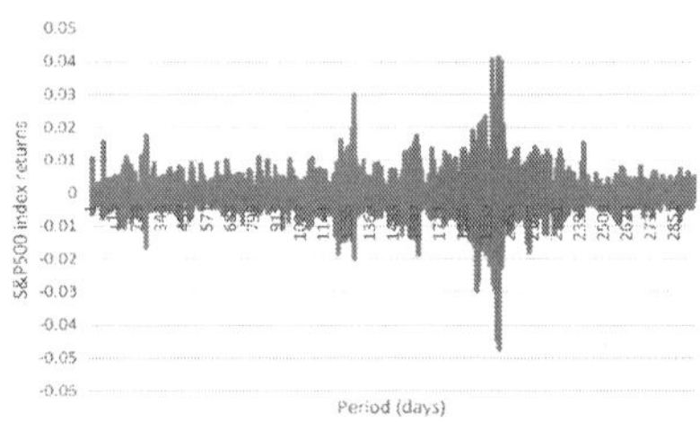

Figure 5. S&P500 index returns.

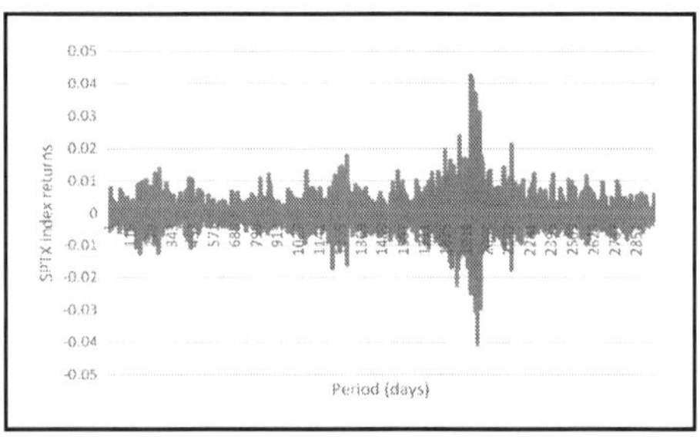

Figure 6. SPTXS index returns.

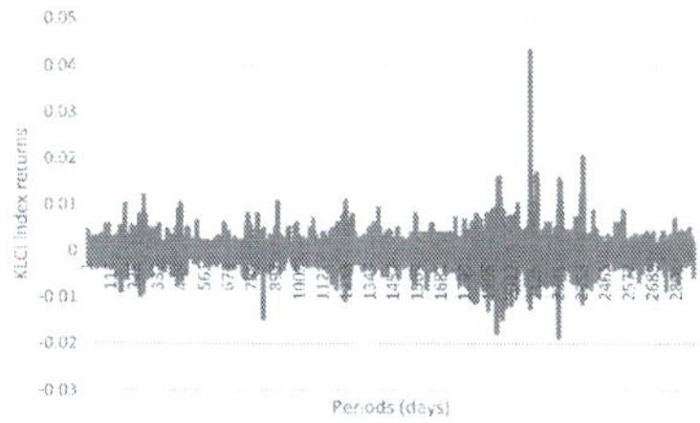

Figure 7. FBMKLCI index returns.

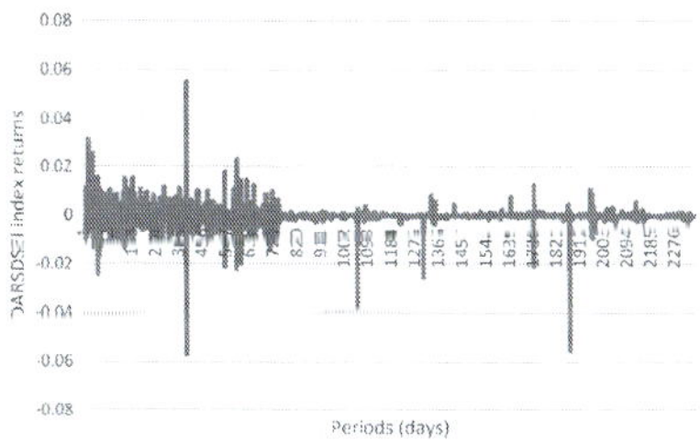

Figure 8. DARSDSEI index returns.

The study seeks to address the following questions:

- What is the impact of jumps in the underlying asset price process on the guarantee fees?
- Will the insurance company charge the same price when the premium is invested in a foreign stock market as in a local stock market?
- Will the insurer charge the local policyholder same price as the foreign policyholder when the insurer issues the VA to both local and foreign annuitants?

The contribution of the study is three fold:

- To evaluate the impacts of jumps in the underlying asset price process on the guarantees fees.
- To analyse the impact of investment in a foreign stock market on the costs of GMDB and GMAB embedded in FPVA.
- To assess and evaluate the impacts of entering a foreign VA market on the price of the GMDB and GMAB embedded in FPVA.

The organization of the remaining part of the paper is as follows: section two explains the methodology while section three addresses the results and discussion of the findings. Section four concludes the study.

METHODOLOGY

In this section we outline the models used and the experiments conducted under different cases, we explain the data used and the calibration procedure of the diffusion and jump-diffusion models.

Underlying asset price dynamics

We propose a new model, jump-diffusion model, which takes into account the large, infrequent and abnormal variations of asset price caused by market news arrival in addition to normal changes of asset price due to disequilibrium in market demand and supply. We extend the work of Milevsky and Posner (2001) by allowing jumps in the underlying asset price process.

Compared to Black–Scholes model which considers only normal asset price changes, jump-diffusion model considers both normal and abnormal asset price changes and hence makes it a better model. Different from stochastic volatility models, jump-diffusion models have better analytical tractability specifically for path-dependent options and capture short term characteristics or behaviour of the financial market better (Yan & Hanson, 2006). In other words, they handle short-term smiles better.

We evaluate the guarantee fees using this proposed model and compare to the fees evaluated using the Black–Scholes model which only considers risk to the life insurer brought by normal changes of asset price caused by the imbalance of market demand and supply.

In the proposed new model, jump-diffusion model, the value of an asset at time t, (S_t) follows a GBM and a jump process outlined by Poisson process, N_t. The asset price follows GBM between jumps.

$$dS_t = \mu S_t dt + \sigma S_t dB_t + JS_t dN_t, \quad S(0) = S_0 \tag{1}$$

where N_t is a Poisson process with an intensity of λ given by:

$$P\{N(t) = n\} = \frac{(\lambda t)^n}{n!} e^{-\lambda t} \tag{2}$$

S_t is the asset price at time t, B_t is a standard Brownian motion, J (is a function which causes a jump of stock price) is the jump size or magnitude which are i.i.d r.vs. The study by Merton (1976) considers the case where the jump sizes are normally distributed ($J \sim N(\mu, \sigma)$).

Under the Black–Scholes model, the dynamics of the stock price (S_t) follows a GBM and is given by the following SDE:

$$dS_t = \mu S_t dt + \sigma S_t dB_t, \quad S(0) = S_0 \tag{3}$$

Account value dynamics

At and c are denoted as the sub account value at time t and the M&E fee payable continuously, respectively, and k is the subsequent contributions. Under the Black–Scholes model assumptions, the sub account is given by:

$$dA_t = (\mu - c)A_t dt + \sigma A_t dB_t + kdt, \quad A(0) = A_0 \tag{4}$$

While in the proposed jump-diffusion model assumptions, the sub account is given by:

$$dA_t = A_t dS_t / S_t - cA_t dt + kdt \tag{5}$$

$$= (\mu - c)A_t dt + \sigma A_t dBt + A_t JdN_t + kdt, \quad A(0) = A_0 \tag{6}$$

Guarantee benefits dynamics

The study considers combined guaranteed minimum death and accumulation benefits (GMDB&GMAB) with roll-up premiums. A pre-agreed guaranteed interest rate $g \geq 0$ is chosen in such a way that $g < r$. The guarantee benefit, G_t, is given by:

$$G(t) = A_0 e^{gt} + \frac{k(e^{gt} - 1)}{g}, \text{ for } g > 0,\ G(0) = A_0,\ G(t) = A_0 + kt, \text{ for } g = 0 \tag{7}$$

The guarantee benefits considered resemble the arithmetic Asian put option where the account value, $A_{t,}$ becomes the underlying asset. Their pay-off function, P(t), is given by:

$$P(t) = [G(t) - A_t]+ = \max\{G(t) - A_t, 0\} \text{ for } t \leq T \tag{8}$$

Guarantee benefits evaluation

Cash outflow less cash inflow constitutes the guarantee benefits loss function. The cash outflow is composed of the GMAB and GMDB pay-out and cash inflow is comprised of the guarantee benefits fee (Stone, 2003). The present value of the loss function (L_0) is given by:

$$L_0 = \int_0^T e^{-rt} P(t)\, {}_t p_x \mu_{x+t} dt + e^{-rT} P(T)\, {}_T p_x - c \int_0^T e^{-rt} A(t)\, {}_t p_x dt \tag{9}$$

The expected present value of the loss function as a function of the M&E fee is given by:

$$L_0(c) = \int_0^T \prod(0,t)_t p_x u_{x+t} dt + \prod(0,T)_T p_x - cA_0 \bar{a}_{x:\bar{T}|c} - \frac{kc}{r-c}\left(\bar{a}_{x:\bar{T}|c} - \bar{a}_{x:\bar{T}|r}\right) \quad (10)$$

Where

$$\prod(0,t) = e^{-rt} EP(t) = e^{-rt} E([G(t) - A(t)]_+), \quad t \le T \quad (11)$$

is the price of the put option and

$$\bar{a}_{x:\bar{T}|r} = \int_0^T e^{rt}\ {}_t p_x dt,\ \bar{a}_{x:\bar{T}|c} = \int_0^T e^{ct}\ {}_t p_x dt \quad (12)$$

is the T-year temporary life annuity payment payable continuously for a life aged x with rate r and c. The fair M&E fee, c* is such that the expected present value of loss is zero. i.e.

$$L_0(c^*) = 0 \quad (13)$$

DATA USED

Mortality tables, stock exchange market indices and returns, criteria for data extraction and source, calibrated diffusion (Black–Scholes) and jump-diffusion model parameters values are used as inputs in our study.

The selection of the stock markets and VA markets is based on the following criteria:

- Popularity of the VA (most popular countries, less popular countries and least popular countries).
- National and individual income (High income countries, middle income countries, low income countries).
- Stock market performance (highest to lowest stock market performance).

Individual countries in the same categories exhibit similar behaviours in terms popularity of the VA products, income and stock market performance, mortality and health risks. USA and Canada are most popular in VA products, Malaysia is medium, while Tanzania has less popularity of the products.

With respect to National and individual income, USA and Canada have the highest income. They are from the highly developed or first world countries. Malaysia is from the second world countries with middle income, while Tanzania is from the third world countries with low income.

In terms of stock market performance, USA has the highest performance followed by Canada, then Malaysia and the least performance is Tanzania stock market, refer Table 1.

Table 1. Stock market mean returns and standard deviations

Index market	Mean return (%)	Standard deviation (%)
DARSDSEI (Tanzania)	3.2	1.8
FBMKLCI (Malaysia)	9.2	9.3
SPTXS (Canada)	17.6	12.7
S&P500 (USA)	17.6	12.1

Table 1 shows the mean returns and standard deviations in the four stock markets of USA, Canada, Malaysia and Tanzania.

The criteria for selecting the dates of the daily asset prices:

- To include the recent global financial crisis of 2007–2008.
- To match the accumulation period of 10 years used in the study.

The time for the daily asset price is from 1 January 2005 to 1 October 2016. The criteria for using market index prices rather than individual asset prices:

- In order to calculate market risk.
- The index reflects the overall movements of a group of securities (tracks the portfolio performance, many different stocks).
- An index consists of many capitalized shares at the exchange market.
- Indices are excellent instruments for reducing risk exposure or increasing yields.

The study used Malaysian mortality table (M9903), Canadian Mortality table (CIA9704) and Tanzanian mortality table (T2000), refer Tables 2–4, respectively.

The study used stock exchange market index prices from Malaysia, Tanzania, Canada and USA, refer Figures 1–4.

The data were collected from Bloomberg database for the dates 1 January 2005 to 1 October 2016. The daily asset return, R_i distribution, refer Figures 5–8 is obtained from the daily closing price, St of the four market indices using the following equation:

$$R_i = \ln S_t - \ln S_t \tag{14}$$

where ln S_t is the natural logarithm of the asset price S_t at time t

The Black–Scholes model is calibrated using Gibbs Sampling procedures (Chan & Wong, 2015) to four stock exchange market indices and the resulting μ and σ parameters are tabulated in Table 5 Since g < r and μ is taken as the risk free rate, r then the study takes g as 50% of the mean rate in each of the four stock markets except DARSDSEI which uses 86%.

The Jump-diffusion model is calibrated to four stock exchange market indices using Gibbs Sampling procedures (Chan & Wong, 2015) and the resulting μ, σ, λ, μJ, σJ and g parameters are tabulated in Table 6. Since g < r and μ is taken as the risk free rate, r then the study takes g as 50% of the mean rate in each of the four stock markets except DARSDSEI which uses 86%.

Table 2. Malaysian mortality table

Attained age	Mortality rates		Attained age	Mortality rates	
	Male	Female		Male	Female
45	0.00240	0.00126	63	0.01580	0.00829
46	0.00268	0.00140	64	0.01757	0.00926
47	0.00299	0.00155	65	0.01951	0.01033
48	0.00334	0.00172	66	0.02163	0.01151
49	0.00372	0.00191	67	0.02394	0.01280
50	0.00413	0.00212	68	0.02645	0.01421
51	0.00458	0.00235	69	0.02919	0.01575
52	0.00507	0.00260	70	0.03217	0.01743
53	0.00559	0.00288	71	0.03540	0.01927
54	0.00614	0.00319	72	0.03890	0.02126
55	0.00674	0.00355	73	0.04269	0.02344
56	0.00737	0.00397	74	0.04680	0.02580
57	0.00804	0.00444	75	0.05123	0.02836
58	0.00875	0.00499	76	0.05602	0.03114
59	0.01015	0.00522	77	0.06118	0.03414
60	0.01137	0.00588	78	0.06675	0.03740
61	0.01271	0.00660	79	0.07274	0.04091
62	0.01419	0.00741	80	0.07918	0.04471

Table 3. Canadian mortality table

Attained age	Mortality rates		Attained age	Mortality rates	
	Male	Female		Male	Female
45	0.00148	0.00105	63	0.01028	0.00929
46	0.00162	0.00117	64	0.01160	0.01040
47	0.00178	0.00130	65	0.01304	0.01152
48	0.00196	0.00144	66	0.01462	0.01265
49	0.00216	0.00162	67	0.01632	0.01377
50	0.00238	0.00181	68	0.01816	0.01489
51	0.00263	0.00204	69	0.02013	0.01602
52	0.00291	0.00229	70	0.02225	0.01715
53	0.00323	0.00259	71	0.02450	0.01827
54	0.00359	0.00292	72	0.02691	0.01940
55	0.00400	0.00329	73	0.02946	0.02053
56	0.00445	0.00371	74	0.03216	0.02166
57	0.00498	0.00419	75	0.03501	0.02278
58	0.00558	0.00472	76	0.03802	0.02391
59	0.00626	0.00530	77	0.04119	0.02504
60	0.00705	0.00595	78	0.04453	0.02617
61	0.00801	0.00706	79	0.04802	0.02729
62	0.00908	0.00817	80	0.05169	0.02842

Table 4. Tanzania mortality table

Attained age	Mortality rates		Attained age	Mortality rates	
	Male	Female		Male	Female
0–1	0.089699	0.080493	50–54	0.015640	0.013854
1–4	0.013899	0.013899	55–59	0.021106	0.014641
5–9	0.005855	0.005504	60–64	0.028290	0.020588
10–14	0.003131	0.003070	65–69	0.038868	0.032480
15–19	0.003682	0.003392	70–74	0.060049	0.052626
20–24	0.006069	0.006159	75–79	0.095084	0.084197
25–29	0.010988	0.015461	80–84	0.164111	0.147358
30–34	0.014279	0.016065	85–89	0.250214	0.223177
35–39	0.017673	0.016012	90–94	0.365264	0.320709
40–44	0.016207	0.012830	95–99	0.511382	0.441152
45–49	0.015416	0.013037	100+	0.660961	0.582228

Table 5. Calibrated diffusion model parameters

Index	μ	σ	g
DARSDSEI (Tanzania)	0.032045	0.018269	0.027550
FRMKLCI (Malaysia)	0.092161	0.093089	0.046061
SPTXS (Canada)	0.176643	0.126744	0.088322
S&P500 (USA)	0.175795	0.121167	0.087898

Table 6. Calibrated jump-diffusion model parameters

Index	μ	σ	λ	μJ	σJ	g
DARSDSEI (Tanzania)	0.032231	0.018257	66.208454	0.000905	0.019716	0.027718
FBMKLCI (Malaysia)	0.092041	0.093049	11.114209	−0.003253	0.032613	0.046020
SPTXS (Canada)	0.177217	0.126589	19.595428	−0.005650	0.034590	0.088609
S&P500 (USA)	0.175513	0.121152	30.078617	−0.003544	0.031558	0.087757

Experiments

We evaluate the combined guarantee minimum death and accumulation benefits M&E fee using Black–Scholes model and jump-diffusion model for male and female annuitants aged 45–69. We consider a life insurance company based in Malaysia writing the VA to Malaysian annuitants and invests the premiums in the Malaysian stock exchange market. We also carry out a case study to determine the impact of stock market performance and regional mortality rates on the combined guarantee M&E fee. We consider the following cases:

- A local life insurance company issuing the guarantee to local and or foreign annuitants and investing into the local stock market. A Malaysian-based life insurance Company issuing the guarantee to annuitants of Malaysia, Tanzania and Canada.

- A local life insurance company issuing the guarantee to local annuitants and investing into different stock markets.
  - A Malaysian-based life insurance Company issuing the guarantee to Malaysian annuitants and invests the contributions into Malaysia, Tanzania, Canada and USA stock markets.
  - A Canadian-based life insurance company issuing the guarantee to Canadian annuitants and invests the contributions into Malaysia, Tanzania, Canada and USA stock markets.

RESULTS AND DISCUSSION

Jump-diffusion model vs. Black–Scholes model

Figures 9 and 10 show M&E fee for Malaysian male and female annuitants, respectively, where the FPVA premiums are invested in Malaysia stock market. The fees are calculated using Black–Scholes model and jump-diffusion model.

Black–Scholes model ignores the jumps in the asset prices caused by over or under reaction due to good or bad news coming from the market or an individual company. Hence it doesn't consider the impact of information arrival on the asset price changes. It only considers normal asset price changes but not abnormal. By assuming jumps in asset prices, the risks modelled in the jump-diffusion models are higher compared to the risks modelled in the Black–Scholes model. This makes the prices obtained using jump-diffusion model to be higher than the prices obtained using Black–Scholes model, refer Figures 9 and 10.

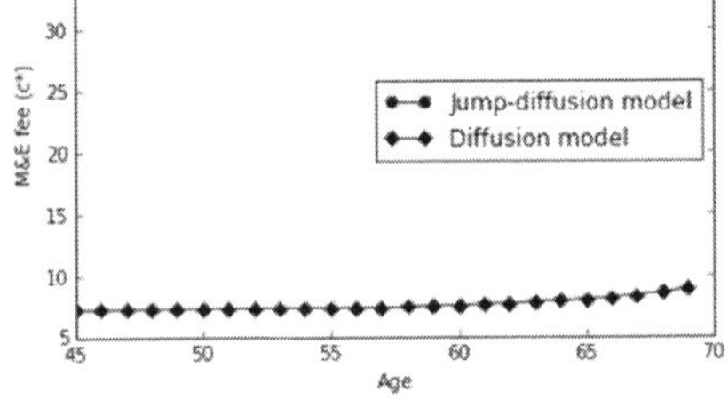

Figure 9. M&E fee for Malaysian male annuitants for different volatility models.

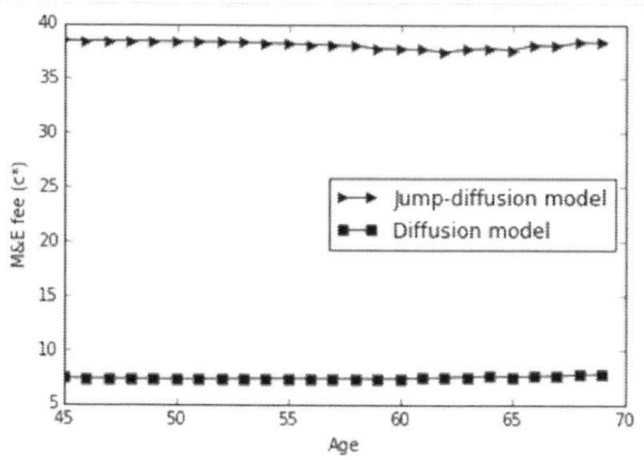

Figure 10. M&E fee for Malaysian female annuitants for different volatility models.

The nature of the changes of the asset prices determines the various features of the financial asset returns (returns distribution, asymmetry, volatility smile). There are many sources which determine the nature of the change of the asset prices. Some of these include: normal asset price changes due to disequilibrium in supply and demand on the market; abnormal asset price changes due to infrequent events caused by large-scale imbalance in the national or international market economy. VA pricing models strive to capture the various important features of the asset returns. Indirectly pricing models quantify the risks presented by the various sources.

Some models consider only one category of risk, while other models consider a combination of many sources of risk which an investor is exposed to in the stock market. In our study, we consider two models: the Black–Scholes model considers the risk of loss to the insurers due to normal changes of asset price; and the jump-diffusion model considers the risk of losses from normal and abnormal changes in the asset price.

Many sources of risks expose the investor to a significant risk of loss compared to few sources of risks. This in turn will compel the life insurance company issuing the VA with guarantees to charge high guarantee fees when using a pricing framework which accounts for many sources of risk of loss compared to a pricing model which takes into account few sources of risks of loss. Black–Scholes model produces lower guarantee fees than the jump-diffusion model because it accounts for only sources of risk of loss due to normal changes of the asset price, while jump-diffusion model accounts for risks of loss due to normal and abnormal changes in the asset price, refer Figures 9 and 10.

In reality there are many sources of risks of losses to the investor operating in the stock market. The modelling framework which considers many sources of risk will place the investor in a good position of reducing the risk of losses but at the expense of high fees! This is a challenge to the competitiveness of the company in attracting new annuitants.

Issuing VA to annuitants of different countries

Figures 11 and 12 show M&E fee for male and female annuitants, respectively, of Malaysia, Tanzania and Canada when the investment of the contributions is done in Malaysia stock exchange market. Mortality rates explain the level of mortality risk which individual annuitants bring into the group. To be fair, the life insurance company should charge the annuitants according to the risk they bring into the group. When the life insurance company charges same price to all annuitants of the same age regardless the region or country they come from, good risks (annuitants with lower mortality rate at same age) will feel they are overcharged and hence terminate the contract and surrender it while bad risks (annuitants with high mortality rates at same age) will feel that they are undercharged. In addition the company will be attracting bad risks and chasing away good risks. This will lead to an adverse selection problem which puts the company in a high risk of losses.

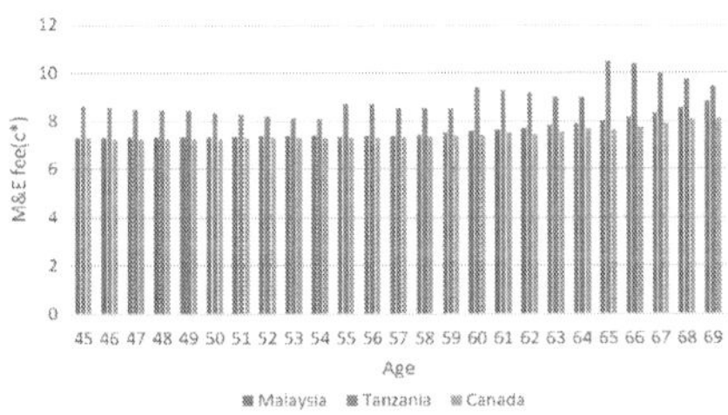

Figure 11. M&E fee for male annuitants in Malaysian stock market.

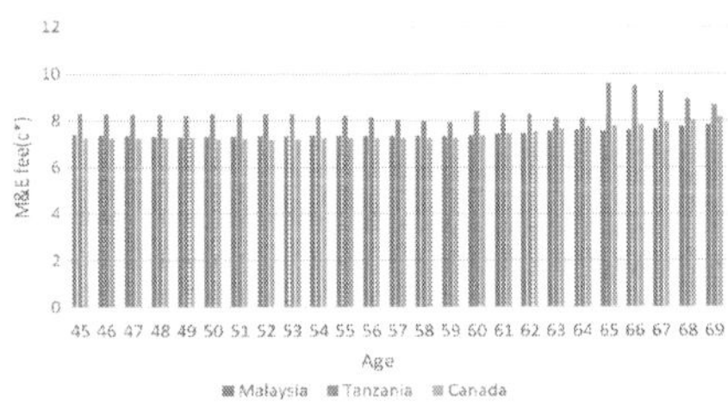

Figure 12. M&E fee for female annuitants in Malaysian stock market.

To avoid this, the M\&E fees should depend on the mortality rates of an individual person, refer Figures 11 and 12 and explained in detail in Juma, Lee, Goh, Chin, and Liew (2016) and Juma and Lee (2017). The higher the mortality rate of an individual, the higher the fees irrespective of the region although the mortality table of the region is used in the calculation of the fees.

Investment of the premiums into different stock markets

Figures 13 and 14 show M&E fee for male and female annuitants, respectively, of Malaysia when the investments are made in Malaysia, Tanzania, Canada and USA.

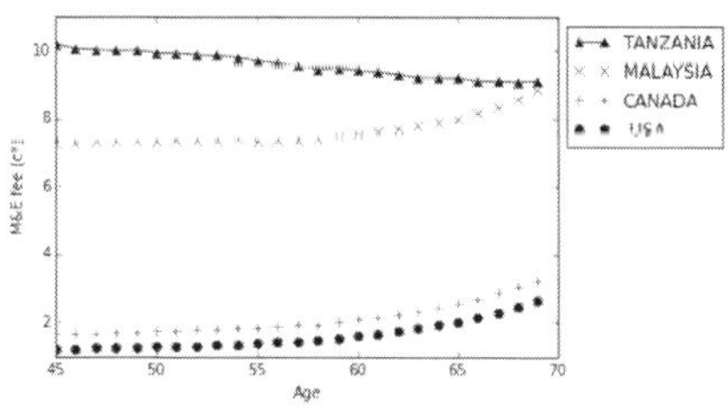

Figure 13. M&E fee for Malaysian male annuitants in various stock market.

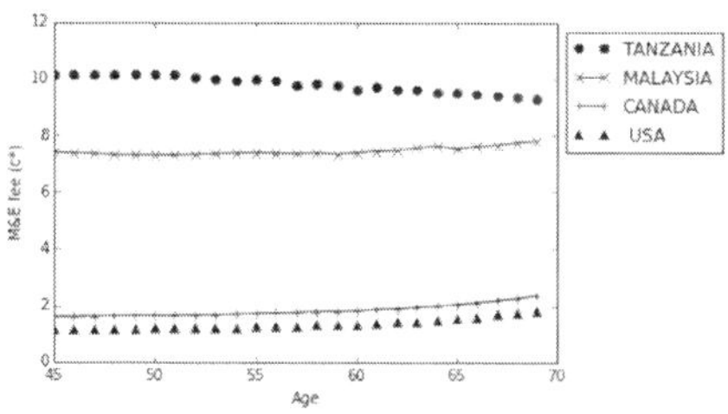

Figure 14. M&E fee for Malaysian female annuitants in various stock market.

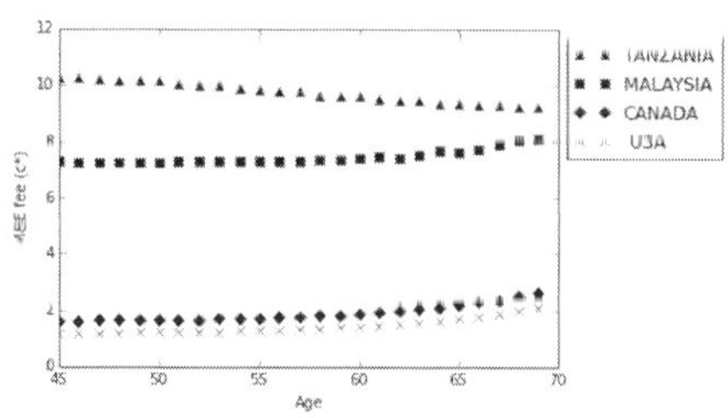

Figure 15. M&E fee for Canadian male annuitants in various stock market.

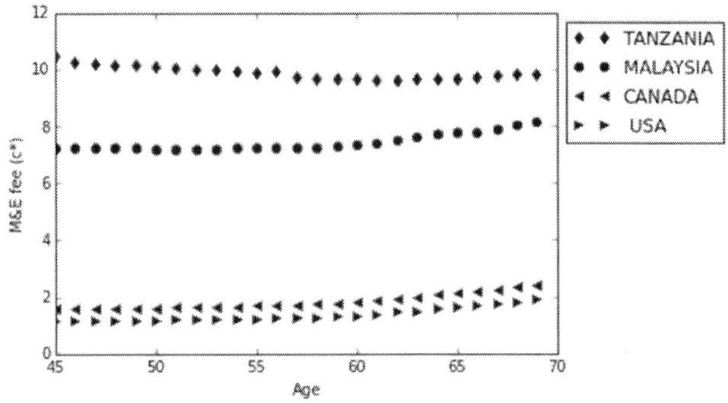

Figure 16. M&E fee for Canadian female annuitants in various stock market.

Figures 15 and 16 show M&E fee for male and female annuitants, respectively, of Canada when the investments are made in Malaysia, Tanzania, Canada and USA.

Mean return from investments and its dispersion are the two variables used to characterize the stock market performance. They are also risk factors used in the pricing of guarantees. High performing stock markets are characterized by high investment mean returns, while low performing stock markets are characterized by low investment mean returns. Since returns and volatility are negatively related, high-performing stock markets have lower volatility and vice versa. Volatility is a proxy measure of market risk; therefore high-performing stock markets have lower risk of loss to the investor. Also lower risk of the sub accounts falling below the guaranteed amount. In other words, it has lower risk to the life insurance company of paying the guarantee before it has accumulated enough. This will lower the guarantees fees, a situation which makes the company attractive, competitive and leads to the increase in sales and profitability.

Charging the same guarantees fees in every stock market the company invests the premiums, exposes the company to high risks of losses because low performing stock markets have high market risks which might outweigh the low market risks in high-performing stock markets. This also will reduce the incentives of expanding into foreign stock markets as well as reduction of guarantees fees in order to attract more annuitants, sales and profits. In order to fulfil the mission of expanding into international business, life insurance companies should charge lower guarantee fees when investing in high-performing stock markets and high guarantee fees when investing into low performing stock markets, refer Figures 13–16.

CONCLUSION

Following the financial crisis periods of 2007–2008 where asset returns are highly negative, this study sets to investigate the appropriateness of the Black–Scholes model in pricing the guaranteed minimum death and accumulation benefits embedded in FPVA. The study also conducted a case to examine the influence of stock market performance and regional mortality rates on the combined guarantee fees. The results showed that the Black–Scholes model ignores jumps in the asset prices, in other words it doesn't take into account the good or bad market information arrivals at discrete times which are random. This makes the jump-diffusion model appropriate in capturing the short-term smiles and leptokurtic distribution of the asset returns during the time of financial crisis where there are big downward jumps. Because the Black–Scholes model considers only normal asset price changes, it fails to model market risks brought by abnormal asset price changes and therefore under-prices the combined guarantee. This framework can also be extended to incorporate different volatility models such as those mentioned by Cheong, Cherng, and Yap (2016) and Chin, Lee, and Yap (2016a, 2016b).For the case when the life insurance company issues the VA to annuitants of different countries, the study found that the life insurance company has to charge different guarantee fees in order to be fair and to avoid adverse selection problem which results from charging same guarantee fees regardless of the differences in regional mortality rates. The adverse selection problem increases the risk of attracting bad risks (annuitants with high mortality rates at the same age) and chasing away good risks (annuitants with low mortality rates at the same age). This phenomenon increases the possibility of losses to the life insurance company.With respect to investment in different stock markets, the study found that the life insurance company has to charge lower guarantee fees when it invests the annuitants' premiums in highperforming stock markets and charge higher guarantee fees when investing into a low performing stock market. Charging same guarantee fees regardless of the stock market performance leads to over-pricing or under-pricing the guarantee. This increases the risk of losses to the company and reduces the competitiveness in attracting new annuitants. It also reduces the strategic motives of expanding into an international market. Specifically, Life Insurance Company should look to invest in stock markets which reduce the guarantee fees so that its strategy of competitiveness and attracting new annuitants is achieved.

ACKNOWLEDGEMENTS

We would like to thank in advance the editor and anonymous reviewers for their supportive comments and suggestions.

REFERENCES

1. Amin, K. I., & Ng, V. K. (1993). Option valuation with systematic stochastic volatility. The Journal of Finance, 48, 881–910. https://doi.org/10.1111/j.1540-6261.1993.tb04023.x
2. Bacinello, A. R., Millossovich, P., Olivieri, A., & Pitacco, E. (2011). Variable annuities: A unifying valuation approach. Insurance: Mathematics and Economics, 49, 285–297.
3. Bacinello, A. R., Millossovich, P., Olivieri, A., & Pitacco, E. (2012). Variable annuities as life insurance packages: A unifying approach to the valuation of guarantees. In M. Vanmaele, G. Deelstra, A. De Schepper, J. Dhaene, W. Schoutens, S. Vanduffel, & D. Vyncke (Eds.), Actuarial and financial mathematics conference. Interplay between finance and insurance (pp. 3–15). Wetteren: Universa Press.
4. Bates, D. S. (1996). Jumps and stochastic volatility: Exchange rate processes implicit in deutsche mark options. Review of Financial Studies, 9, 69–107. https://doi.org/10.1093/rfs/9.1.69 Bauer, D., Kling, A., & Russ, J. (2008). A universal pricing framework for guaranteed minimum benefits in variable annuities. Astin Bulletin, 38, 621–651. https://doi.org/10.1017/S0515036100015312
5. Bernard, C., Cui, Z., & Vanduffel, S. (2016). Impact of flexible periodic premiums on variable annuity guarantees. Retrieved from SSRN: 2739145
6. Chan, N. H., & Wong, H. Y. (2015). Simulation techniques in financial risk management. Hoboken, NJ: John Wiley & Sons.
7. Cheong, C. W., Cherng, L. M., & Yap, G. L. C. (2016). Heterogeneous market hypothesis evaluations using various jump-robust realized volatility. Journal for Economic Forecasting, 4, 50–64.
8. Chi, Y., & Lin, X. S. (2012). Are flexible premium variable annuities under-priced? Astin Bulletin, 42, 559–574.
9. Chin, W. C., Lee, M. C., & Yap, G. L. C. (2016a). Heterogeneous autoregressive model with structural break using nearest neighbor truncation volatility estimators for DAX. SpringerPlus, 5, 885. https://doi.org/10.1186/s40064-016-3465-x
10. Chin, W. C., Lee, M. C., & Yap, G. L. C. (2016b). Modelling financial market volatility using asymmetric-skewedarfimax and-harx models. Engineering Economics, 27, 373–381.
11. Costabile, M. (2013). Analytical valuation of periodical premiums

for equity linked policies with minimum guarantee. Insurance: Mathematics and Economics, 53, 597–600.

12. Cox, J. C., & Ross, S. A. (1976). The valuation of options for alternative stochastic processes. Journal of Financial Economics, 3, 145–166.
13. Hanson, F. B, & Westman, J. J. (2002). Jump-diffusion stock return models in finance: Stochastic process density with uniform-jump amplitude. In Proceedings of 15th International Symposium on Mathematical Theory of Networks and Systems (Vol. 7).
14. Hardy, M. (2003). Investment guarantees: Modeling and risk management for equity-linked life insurance (Vol. 215). Hoboken, NJ: John Wiley & Sons.
15. Heston, S. L. (1993). A closed-form solution for options with stochastic volatility with applications to bond and currency options. Review of Financial Studies, 6, 327–343. https://doi.org/10.1093/rfs/6.2.327
16. Hull, J., & White, A. (1987). The pricing of options on assets with stochastic volatilities. The Journal of Finance, 42, 281–300.
17. Juma, M., & Lee, M. C. (2017). A empirical study on annuity pricing with minimum guarantees. Applied Mathematical Sciences, 11, 59–75. https://doi.org/10.12988/ams.2017.610261
18. Juma, M., Lee, M. C., Goh, Y. K., Chin, S. T., & Liew, K. W. (2016). A study of impact of stochastic volatility on variable annuity pricing. Applied Mathematical Sciences, 10, 2953–2970. https://doi.org/10.12988/ams.2016.68228
19. Kou, S. (2007). Jump-diffusion models for asset pricing in financial engineering. Handbooks in Operations Research and Management Science, 15, 73–116. https://doi.org/10.1016/S0927-0507(07)15002-7
20. Kou, S. G. (2002). A jump-diffusion model for option pricing. Management Science, 48, 1086–1101. https://doi.org/10.1287/mnsc.48.8.1086.166
21. Merton, R. C. (1976). Option pricing when underlying stock returns are discontinuous. Journal of Financial Economics, 3, 125–144
22. Milevsky, M. A., & Posner, S. E. (2001). The titanic option: Valuation of the guaranteed minimum death benefit in variable annuities and mutual funds. The Journal of Risk and Insurance, 93–128. https://doi.org/10.2307/2678133
23. Scott, L. O. (1997). Pricing stock options in a jump-diffusion model with stochastic volatility and interest rates: Applications of Fourier

inversion methods. Mathematical Finance, 7, 413–426. https://doi.org/10.1111/mafi.1997.7.issue-4

24. Stein, E. M., & Stein, J. C. (1991). Stock price distributions with stochastic volatility: An analytic approach. Review of Financial Studies, 4, 727–752. https://doi.org/10.1093/rfs/4.4.727
25. Stone, R. (2003). Variable annuity guarantees-more than just acronyms. Society of Actuaries, the Actuary, 37, 3–5.
26. Yan, G., & Hanson, F. B. (2006). Option pricing for a stochasticvolatility jump-diffusion model with log-uniform jumpamplitudes. In American Control Conference (p. 6). IEEE.

SECTION 2
BONDS

CHAPTER

5

The Analysis of Corporate Bond Valuation Under an Infinite Dimensional Compound Poisson Framework

Sheng Fan

School of Economic Mathematics, Southwestern University of Finance and Economics, Chengdu 611130, China

INTRODUCTION

Corporate bonds trade at higher yield than the comparable government bonds due to default risk and embedded American options such as call, put, and convertibility. However, even accounts for the effect of embedded options, it has been difficult to reconcile the observed spread with credit risk and reasonable parameters and is thus frequently referred to as the "credit spread puzzle" [1]. Wide range models typically explain only 20%–30% of

Citation: Sheng Fan, "The Analysis of Corporate Bond Valuation under an Infinite Dimensional Compound Poisson Framework" Abstract and Applied Analysis Volume 2014, Article ID 282185, 10 pages http://dx.doi.org/10.1155/2014/282185

the observed credit spreads. The credit quality of bond, downward jump risk of the issuers' value, and the bond maturity have some explaining power to the observed spreads. For junk bonds, the credit risk accounts for much larger fraction of observed spreads than investment grade bonds do [2]. For long-maturity bonds, Collin-Dufresne et al. [3] have argued that firms with good credit quality are likely to issue more debt, which leads to higher yield spreads. The authors in [4–7] have shown that models with incorporating jump risk explain a significant part of observed credit spreads. Liquidity risk is an important element of spreads, if the debt market is of lower liquidity. It is shown in [8] by Longstaff et al. that liquidity risk may not construct the majority of corporate spreads as in the frictionless credit-default swaps market.

Since the structural approach is one of the most widely employed frameworks of credit risk valuation, by incorporating jump risk into the structural framework, this paper analyzes the valuation of callable corporate bond and embedded options. Allowing the economic variables driven by its own risk resource as in [9–12], we emphasize that the state nature of different maturities is an important determinant factor to the assets pricing. For the instantaneous forward interest rates, the risk resource depends not only on time t but also on the time-to-maturity. Therefore, we expand the dynamics of the instantaneous forward rates as an infinite dimensional jump-diffusion process.

This expansion has three features: (1) the model could flexibly capture the covariation and the term structure of correlation between different firm bonds and government bonds in a parsimonious way than their multifactor counterparts. (2) Any observed data set of bond prices is fully compatible with the expanding random process. (3) It is unnecessary to add an error term in the econometric model to calculate the parameters.

For the corporate bond valuation, based on different assumptions about the bankruptcy rules, exogenous and endogenous models have different implication and lead to different risk management behaviours. In the exogenous model, bankruptcy liquidation is made by the outside investors when the firm value reaches the trigger value such as bond par value [13–16]. For some exogenous bankruptcy models, there is arbitrage opportunity due to the negative spreads. In the endogenous model, the bankruptcy decision depends on the issuers who endogenously optimally exercise the embedded options [17–20]. We conduct our analysis within the endogenous bankruptcy framework.

To examine the corporate bond valuation and the interaction with embedded options, we introduce a benchmark bond which is a risk-free comparable government bond as in [21]. Treating the corporate bond as the benchmark bond minus the embedded options, we could easily clarify the interaction between call provisions and default risk.

In this paper, we first analyze the impact of benchmark bond price and firm value to the embedded option as well as the corporate bond valuation. It shows that credit spread is positively related with the benchmark bond price and negatively related with firm value. For the callable corporate bond, the embedded options value is higher than that of the pure default bond. The increment can be decomposed as the "American premium" part and the "call option" part. Secondly, we consider the embedded American options exercise boundaries and show that, for the callable corporate bond, the relationship between the firm values and the boundaries is more complicated than that of pure default bond. When the firm value is lower than the call price, the exercise boundaries are higher for higher firm value. When the firm value is higher than the call price, the strike boundary decreases with the firm value. Similar results are obtained in [21]. But our results are more accurate. Finally, we emphasize the effect of the exogenous shock and firm procyclicality to the credit spread. If the exogenous shock is negative to the benchmark bond and positive to the firm value, the credit spread depends on the total effect.

The remainder of this paper is organized as follows. Section 2 presents the basic valuation framework for the corporate bond. Section 3 derives the valuation model for the pure default corporate bond and analyzes the effecting factors. Section 4 examines the impact of benchmark bond price, firm value, and their covariation to the credit spread for the default and callable corporate bond. The conclusion follows in Section 5.

THE PRELIMINARY OF VALUATION MODEL

Let (Ω, Ψ, P) be a probability space with a stochastic process (t, x) and let $N_1(t), \ldots, N_M(t)$ be independent Poisson processes, where Ψ is a filtration. Assume that the instantaneous forward rates follow the dynamics

$$d_t f(t,x) = \alpha(t,x)\,dt + \beta(t,x)\,d_1 B(t,x) + \gamma(t,x)\,dJ(t), \tag{1}$$

where the stochastic process $B(t, x)$ is a generalization of the one-dimensional Brownian motion, which depends on both time t and maturity x. Given time-to-maturity x, the stochastic perturbation (t, x) is a standard

Brownian motion. (t, x) describes at time t the different magnitude's stochastic perturbation to the forward rates with different time-to-maturity. The instantaneous forward rates constitute a continuum. Such generalization is natural and acceptable. The generalized Brownian motion exists and can be obtained by solving SPDEs as in [12]. This model is different from that of [21] in three aspects. (1) We model the instantaneous forward rates rather than the interest rates. (2) The continuous stochastic risk source is infinite dimension. (3) We incorporate the compound jump process in the model. (t, x) satisfies the following requirements.

(1) (t, x) is continuous on t and x.

(2) $(t, x)$is a martingale in time t; that is, $Et[B(t, x)] = 0$, for all x.

(3) The variance of the increments with respect to time is equal to the time change; namely, $\text{Var}[d_1B(t, x)] = dt$, for all x. (4) The correlation of the increments, $\text{corr}[d_1B(t, x), d_1B(t, y)] = \rho(x, y)$, does not depend on time t.

At the same time, the correlation coefficient (x, y) satisfies the following requirements.

(1) Symmetric, $(x, y) = \rho(y, x)$.

(2) (x, y) takes values between −1 and 1.

(3) $(x, x) = 1$.

(4) The correlation coefficient matrix is positive semidefinite.

The notation $d_1(t, x)$ means that the increment is taken with respect to the first element time t. (t) is a compound Poisson process; $J(t) = \sum N(t)\ j=1$ Y_j. Poisson process (t) counts the number of jumps that occur at or before time t with the intensity λ. Y_j $(j = 1, 2, \ldots)$ are independent identically distributed discrete random variables with finite values a_1, a_2,...a_M; the probability $P(Y_j = a_i)=p_i$, $\sum M\ i=1\ p_i = 1$. Here, we allow that the impacts of exogenous shock to the economic basis are time-variable and assume that the jump size is finite. (t, x), $\beta(t, x)$, and $\gamma(t, x)$ are nonrandom functions and satisfy Lipschitz linear growth conditions. That is, for some constant L, α, β, and γ satisfy

$$|\alpha(t,x) - \alpha(t,y)| + |\beta(t,x) - \beta(t,y)| + |\gamma(t,x) - \gamma(t,y)| \le L|x-y|,$$

$$|\alpha(t,x)| + |\beta(t,x)| + |\gamma(t,x)| \le L(1+|x|), \tag{2}$$

for all $x, y, t \in \mathrm{R}^{+.}$ In the filtration Ψ_t, the compound jump is independent of the stochastic string process (t, x).

The decomposition theorem of a compound Poisson process (see [22]) shows that $J(t) = \sum_{m=1}^{N(t)} Y_m$ could be rewritten as $J(t) = \sum_{i=1}^{M} a_i N_i(t)$. So the dynamics of instantaneous forward rates can be written in the integral form

$$f(t,x) = f(0,x) + \int_0^t \alpha(u,x)\,du + \int_0^t \beta(u,x)\,d_1 B(u,x) + \sum_{i=1}^{M} \int_0^t \gamma(u,x)\, a_i dN_i(u). \tag{3}$$

The price at time t of a zero coupon bond with maturity s is expressed as

$$P(t,s) = \exp\left[-\int_0^{s-t} f(t,x)\,dx\right]. \tag{4}$$

Now, we consider a firm with one single bond to be outstanding. The bond has a series coupons paid c_j at each time t_j, $0< t_2 < \cdots < t_n = T$, and maturity at time T. In order to analyze the effect of the embedded default and call options on the corporate bond, we introduce a benchmark bond which is noncallable nondefault with the same coupon and maturity as the corporate bond. Corporate bond can be viewed as the benchmark bond minus a default and/or call option. The benchmark bond with maturity ($= t_n$) can be expressed as

$$P(t) = \sum_{j}^{n} c_j P(t,t_j) = \sum_{j}^{n} c_j \exp\left[-\int_0^{t_j - t} f(t,x)\,dx\right], \quad \text{for } t_{j-1} \le t < t_j, \quad j = 1,2,\ldots,n. \tag{5}$$

Assume that the market is complete. By the second fundamental theorem of asset pricing (see [22]), there exists an unique risk-neutral measure $\tilde{P}$. Under the risk-neutral measure $\tilde{P}$, for each fixed maturity x, define $\int_0^t \int_0^\infty \theta(u,y)c(x,y)\,dy\,du$, where $\theta(u, y)$ is the market price of risk. $\widetilde{B}(t,x)$ is the counterpart of $B(t, x)$ under the actual probability measure P. For each maturity x, $\widetilde{B}(t,x)$ is still a Brownian motion. $(t) = \sum N(t)\tilde{}\ j-1\ Yj$ is a compound Poisson process with intensity $\tilde{\lambda}$, where $\widetilde{N}(t)$ is the number of jumps till time t under the risk-neutral measure $\tilde{P}$. The jump sizes $Y1$, $Y2$ ... are still I.I.D. random variables, while the jump intensity satisfies $\widetilde{P}(Y_j = a_i) = p(a_i) = \widetilde{\lambda}_i/\widetilde{\lambda}, (i = 1,2,\ldots,M)$ for arbitrary j, where $\sum_{i=1}^{M} \widetilde{\lambda}_i = \widetilde{\lambda}$ and each $\widetilde{\lambda}_i\,(i = 1,2,\ldots,M)$ is a positive constant. Notice that, under the measure $\tilde{P}$, $\widetilde{B}(t,x)$ is still independent of $\tilde{N}(t)$ $(\iota = 1, 2, \ldots, M)$. Under the risk-neutral measure $\tilde{P}$, the price at time t of a zero coupon bond with maturity s can be expressed as (see [23])

$$\begin{aligned} &P(t,s) \\ &= P(0,s)\exp\Big[\Big[\int_0^t r(u)\,du \\ &\quad - \int_0^t \int_0^{s-u} \sigma(u,x)\,dx\,d_1\widetilde{B}(u,x) \\ &\quad - \frac{1}{2}\int_0^t du \int_0^{s-u} \sigma(u,x) \\ &\qquad \times \int_0^{s-u} \sigma(u,y) \\ &\qquad \times \rho(x,y)\,dy\,dx \\ &\quad - \sum_{i=1}^{M} \int_0^t [\exp\Big[\Big(-a_i \int_0^{s-u} \gamma(u,x)\,dx\Big) \\ &\qquad -1\Big]\widetilde{\lambda}_i du\Big] \\ &\times \exp\Big[-\sum_{i=1}^{M}\int_0^t \int_0^{s-u} \gamma(u,x)\,dx a_i\,d\widetilde{N}_i(u)\Big], \end{aligned} \tag{6}$$

where $\sigma(t, x)$ is a nonrandom function and satisfies $\widetilde{E}\int_0^t\int_0^{\infty}|\sigma(u,x)|\int_0^{\infty}|\sigma(u,y)|\rho(x,y)dy\,dx\,du < \infty$ for all $t \geq 0$. (t, x) is a left-continuous nonrandom function and satisfies $\widetilde{E}\int_0^t\int_0^{\infty}|\gamma(u,x)|\int_0^{\infty}|\gamma(u,y)|\rho(x,y)dy\,dx\,du < \infty$, for all $t \geq 0$. The discounted zero bond is a martingale. Notice that $(s, s) = 1$. We rewrite the value of zero coupon bond as

$$\begin{aligned} P(t,s) = \exp\Big\{ &-\int_t^s r(u)\,du \\ &+ \int_t^s \int_0^{s-u} \sigma(u,x)\,dx\,d_1\widetilde{B}(u,x) \\ &+ \frac{1}{2}\sigma_f^2(t,s) \\ &+ \sum_{i=1}^{M}\int_t^s \Big[\exp\Big(-a_i\int_0^{s-u}\gamma(u,x)\,dx\Big) - 1\Big] \\ &\qquad \times \widetilde{\lambda}_i du \\ &+ \sum_{i=1}^{M}\int_t^s\int_0^{s-u}\gamma(u,x)\,dx a_i d\widetilde{N}_i(u)\Big\}, \end{aligned} \tag{7}$$

Where

$$\sigma_f^2(t,s) = \int_t^s du \int_0^{s-u}\sigma(u,x)\int_0^{s-u}\sigma(u,y)\,\rho(x,y)\,dy\,dx. \tag{8}$$

Assume that the value of the firm S is independent of its capital structure and is a jump diffusion given by

$$
\begin{aligned}
dS(t) &= (a(t) - h(t))\, S(t)\, dt \\
&\quad + S(t) \int_0^{\overline{T}} b(t, x)\, d_1 B(t, x)\, dx \\
&\quad + S(t^-)\, c(t)\, dJ(t),
\end{aligned} \tag{9}
$$

where $\overline{T}$ $(0 < \overline{T} < \infty)$ is the survival period of the firm, $E \int_0^{\overline{T}} \int_0^{\infty} |b(t,x)| \int_0^{\infty} |b(t,y)| \rho(x,y)\, dy\, dx\, dt < \infty$, $a(t)$, $b(t, x)$, and $c(t)$ satisfy Lipschitz conditions, $c(t)$ is a left continuous nonrandom function and satisfies, for all jump size a_i, $c(t) > \max_i\{-1/a_i\}$ a.s., and $h(t) \geq 0$ is the firm's payout rate.

Under the risk-neutral measure $\tilde{P}$, we have

$$
\begin{aligned}
dJ(t) &= \sum_{i=1}^{M} a_i d\tilde{N}_i(t), \\
d_1 B(t, x) &= d_1 \tilde{B}(t, x) - \int_0^{\infty} \theta(t, y)\, \rho(x, y)\, dy\, dt.
\end{aligned} \tag{10}
$$

By the market price of risk equation

$$
\begin{aligned}
a(t) - r(t) &= \int_0^{\overline{T}} b(t, x) \int_0^{\infty} \theta(t, y)\, \rho(x, y)\, dy\, dx \\
&\quad - c(t) \sum_{i=1}^{M} a_i \tilde{\lambda}_i,
\end{aligned}
$$

the dynamics of the firm value is rewritten as

$$
\begin{aligned}
dS(t) &= [r(t) - h(t)]\, S(t)\, dt \\
&\quad + S(t) \int_0^{\overline{T}} b(t, x)\, d_1 \tilde{B}(t, x)\, dx \\
&\quad + S(t^-) \sum_{i=1}^{M} a_i c(t)\, d\left(\tilde{N}_i(t) - \tilde{\lambda}_i t\right),
\end{aligned} \tag{11}
$$

where $\tilde{N}_i(t) - \tilde{\lambda}_i t$ is a compensated Poisson process.

Theorem 1. Under the risk-neutral measure $\tilde{P}$, the firm value is

$$
\begin{aligned}
S(t) = s(0) \exp &\left\{ \int_0^t \left[r(u) - h(u) \right.\right. \\
&\quad - \frac{1}{2} \int_0^{\overline{T}} b(u, x)\, dx \\
&\quad \times \int_0^{\overline{T}} b(u, y)\, \rho(x, y)\, dy \\
&\quad \left. - \sum_{i=1}^{M} c(u)\, a_i \tilde{\lambda}_i \right] du \\
&\quad \left. + \int_0^t \int_0^{\overline{T}} b(u, x)\, d_1 \tilde{B}(u, x)\, dx \right\} \\
\times \exp &\left\{ \sum_{i=1}^{M} \int_0^t \ln\left(a_i c(u) + 1\right) d\tilde{N}_i(u) \right\}.
\end{aligned} \tag{13}
$$

Proof. We prove that (13) is the solution of (12). Let $(t) = S(0) \exp[X(t)]$. Denote

$$\begin{aligned} X^c(t) = \int_0^t \Bigg[r(u) - h(u) & - \frac{1}{2} \int_0^{\overline{T}} b(u,x)\,dx \\ & \times \int_0^{\overline{T}} b(u,y)\,\rho(x,y)\,dy \\ & - \sum_{i=1}^{M} c(u)\,a_i \widetilde{\lambda}_i \Bigg] du \\ & + \int_0^t \int_0^{\overline{T}} b(u,x)\,d_1 \bar{B}(u,x)\,dx \\ X^J(t) = \sum_{i=1}^{M} \int_0^t & \ln(a_i c(u) + 1)\,d\widetilde{N}_i(u). \end{aligned} \tag{14}$$

By Itô formula for jump process, we have

$$\begin{aligned} dS(t) &= S(t)\,dX^c(t) + \frac{1}{2} S(t)\,dX^c(t)\,dX^c(t) \\ &\quad + S(t) - S(t^-) \\ &= S(t) \Bigg\{ \Bigg[r(t) - h(t) - \sum_{i=1}^{M} c(t)\,a_i \widetilde{\lambda}_i \\ &\quad - \frac{1}{2} \int_0^{\overline{T}} b(t,x)\,dx \int_0^{\overline{T}} b(t,y)\,\rho(x,y)\,dy \Bigg] dt \\ &\quad + \int_0^{\overline{T}} b(t,x)\,d_1 \bar{B}(t,x)\,dx \Bigg\} \\ &\quad + \frac{1}{2} S(t) \int_0^{\overline{T}} b(t,x)\,dx \int_0^{\overline{T}} b(t,y)\,\rho(x,y)\,dy\,dt \\ &\quad + S(t^-) \left\{ \exp\left[\sum_{i=1}^{M} \ln(a_i c(t) + 1)\,d\widetilde{N}_i(t) \right] - 1 \right\}. \end{aligned} \tag{15}$$

Rearranging it, we obtain

$$\begin{aligned} dS(t) &= S(t) \Bigg\{ \Bigg[r(t) - h(t) - \sum_{i=1}^{M} c(t)\,a_i \widetilde{\lambda}_i \Bigg] dt \\ &\quad + \int_0^{\overline{T}} b(t,x)\,d_1 \bar{B}(t,x)\,dx \Bigg\} \\ &\quad + S(t^-) \sum_{i=1}^{M} c(t)\,a_i\,d\widetilde{N}_i(t). \end{aligned} \tag{16}$$

It completes the proof.

For $t \leq \tau \leq T$, the value of corporate bond can be rewritten as

$$\begin{aligned} S(\tau) = s(t) \exp \Bigg\{ & \int_t^\tau \Bigg[r(u) - h(u) - \sum_{i=1}^{M} c(u)\,a_i \widetilde{\lambda}_i \Bigg] du \\ & - \frac{1}{2} \sigma_S^2(t,\tau) + \int_t^\tau \int_0^{\overline{T}} b(u,x)\,d_1 \bar{B}(u,x)\,dx \\ & + \sum_{i=1}^{M} \int_t^\tau \ln(a_i c(u) + 1)\,d\widetilde{N}_i(u) \Bigg\}, \end{aligned} \tag{17}$$

where

$$\sigma_S^2(t,\tau) = \int_t^\tau du \int_0^{\overline{T}} b(u,x)\,dx \int_0^{\overline{T}} b(u,y)\,\rho(x,y)\,dy. \tag{18}$$

VALUING DEFAULTABLE BOND

In the endogenous bankruptcy model, the bankruptcy decision is optimally done by the bond issuers' themselves. For the pure default noncallable bond, when the embedded default option arrives at its maximum, the issuer would give up the firm to exempt its debt. The value of the pure defaultable bond with embedded default option is

$$P_D(t) = P(t) - f_D(t), \tag{19}$$

where $f_D(t)$ is the embedded default option value at time t. Consider

$$f_D(t) \equiv \sup_{t \le \tau \le T} \tilde{E}\left\{ e^{-\int_t^\tau r(u)du} [P(\tau) - S(\tau)]^+ \mid \Psi_t \right\}, \tag{20}$$

where Ψ_t, $0 \le t \le T$ is the filtration; $f_D(t)$ satisfies

$$f_D(t) \ge (P(t) - S(t))^+. \tag{21}$$

The yield spread of a given bond over its benchmark bond is a straightforward transformation of the bond's embedded option value. Define

$$\tau^* = \inf\{t \ge 0 : f_D(t) = (P(t) - S(t))^+\}, \tag{22}$$

as the optimal stopping time based on the state $S(t) = s^*$ at time t, which is the first time the default option value $fD(t)$ reaches the level $(P(t) - S(t))^+$.

Theorem 2. For different initial firm values and benchmark bond prices, the following properties hold

(1) *If* $S^x(0) > S^y(0)$, *then*

$$-\tilde{p}^s [S^x(t) - S^y(t)] \le f_D^x(t) - f_D^y(t) \le 0.$$

(2) *If* $S^x(0) > S^y(0)$, *then* $P_D^x(t) \le P_D^y(t)$.

(3) *If* $P^x(0) > P^y(0)$, *then*

$$0 \le f_D^x(t) - f_D^y(t) \le p^p [P^x(t) - P^y(t)].$$

(4) *If* $P^x(0) > P^y(0)$, *then* $P_D^x(t) \ge P_D^y(t)$. (23)

Here $\tilde{p}^s = P\{P(\tilde{\tau}^y) > S^y(\tau^y) \mid S(t) = s^y\}$ is the risk-neutral default probability under the condition that the firm value at time t is $S(t) = s$ and $\tilde{p}^p = P\{P\ x(\tau^x) > S(\tau^x) \mid P(t) = p^x\}$ is the risk-neutral default probability under the state in which the benchmark bond price at time t is $P(t) = \mathrm{p}$.

Remark 3. These results are consistent with those of [21], but our results are more accurate by providing closer embedded options spread

Proof. Due to $(t)a_i > -1$, $i = 1, 2, \ldots, M$, by the no-crossing properties for jump-diffusion process, we have

$$S^x(0) > S^y(0) \Longrightarrow S^x(t) \geq S^y(t),$$

$$\tilde{P} - \text{a.s.} \quad \forall 0 < t < \infty. \qquad (24)$$

(1) Given that the state at time t is $(t) = p$ and $S(t) = s^x$, assume that its optimal stopping time is τ^x and its associated embedded default options value is $f_D^x(t)$. Thus

$f_D^x(t) = \tilde{E}\{e^{-\int_t^{\tau^x} r(u)du}[P(\tau^x) - S^x(\tau^x)]^+ \mid \Psi_t\}$. Similarly, $f_D^y(t)$ is the embedded options value when $(t) = s^y$ at time t. Thus

$$\begin{aligned} & f_D^y(t) - f_D^x(t) \\ & = \sup_{t \leq \tau \leq T} \tilde{E}\left\{e^{-\int_t^{\tau} r(u)du}[P(\tau) - S^y(\tau)]^+ \mid \Psi_t\right\} \\ & \quad - \tilde{E}\left\{e^{-\int_t^{\tau^x} r(u)du}[P(\tau^x) - S^x(\tau^x)]^+ \mid \Psi_t\right\} \\ & \geq \tilde{E}\left\{e^{-\int_t^{\tau^x} r(u)du}[P(\tau^x) - S^y(\tau^x)]^+ \right. \\ & \quad \left. - e^{-\int_t^{\tau^x} r(u)du}[P(\tau^x) - S^x(\tau^x)]^+ \mid \Psi_t\right\} \\ & \geq 0. \end{aligned} \qquad (25)$$

For $(t) = s^y$, assume that its optimal stopping time is τ^y. We have

$$\begin{aligned} & f_D^y(t) - f_D^x(t) \\ & \leq \tilde{E}\left\{e^{-\int_t^{\tau^y} r(u)du}(P(\tau^y) - S^y(\tau^y))^+ \right. \\ & \quad \left. - e^{-\int_t^{\tau^y} r(u)du}(P(\tau^y) - S^x(\tau^y))^+ \mid \Psi_t\right\} \\ & = \tilde{E}\left\{\left[e^{-\int_t^{\tau^y} r(u)du}(P(\tau^y) - S^y(\tau^y)) \right.\right. \\ & \quad \left. -e^{-\int_t^{\tau^y} r(u)du}(P(\tau^y) - S^x(\tau^y))^+\right] \\ & \quad \left. \times I_{\{P(\tau^y) > S^y(\tau^y)\}} \mid \Psi_t\right\} \\ & \leq \tilde{E}\left\{\left[e^{-\int_t^{\tau^y} r(u)du}(P(\tau^y) - S^y(\tau^y)) \right.\right. \\ & \quad \left. -e^{-\int_t^{\tau^y} r(u)du}(P(\tau^y) - S^x(\tau^y))^+\right] \\ & \quad \left. \times I_{\{P(\tau^y) > S^y(\tau^y)\}} \mid \Psi_t\right\} \\ & = \tilde{E}\left\{e^{-\int_t^{\tau^y} r(u)du}[S^x(\tau^y) - S^y(\tau^y)] \mid \Psi_t\right\} \tilde{P} \\ & \quad \times \{P(\tau^y) > S^y(\tau^y)\} \\ & = [S^x(t) - S^y(t)]\, \tilde{p}^s, \end{aligned} \qquad (26)$$

where I_A denotes the indicator function of the set A. In the second last equation, we use the fact that $\tilde{E}\{XI_A \mid \Psi_t\} = \tilde{E}(X \mid \Psi_t)\tilde{P}(A)$ (by the same token, (30), (40), and (47) also hold).

(2) Note that $P(t) = P(t) - fD(t)$. For firm value $S(t) > S^y(t)$, we have

$$P_D^x(t) = P(t) - f_D^x(t) \le P(t) - f_D^y(t) = P_D^y(t). \tag{27}$$

$$\begin{aligned}
& P^x(0) > P^y(0) \\
& \Longrightarrow r^x(0) \le r^y(0) \\
& \Longrightarrow r^x(t) \le r^y(t) \\
& \Longrightarrow e^{-\int_0^t r^x(u)du} \ge e^{-\int_0^t r^y(u)du}, \\
& \tilde{P} - \text{a.s.} \quad \forall 0 < t < \infty, \\
& P^x(0) > P^y(0) \Longrightarrow P^x(t) \ge P^y(t), \\
& \tilde{P} - \text{a.s.} \quad \forall 0 < t < \infty.
\end{aligned} \tag{28}$$

Given that the state at time t is $(t) = p^y$ with the optimal stopping time τ^y, we have

$$\begin{aligned}
& f_D^x(t) - f_D^y(t) \\
& \ge \tilde{E}\left\{e^{-\int_t^{\tau^y} r^x(u)du}[P^x(\tau^y) - S(\tau^y)]^+ \mid \Psi_t\right\} \\
& \quad - \tilde{E}\left\{e^{-\int_t^{\tau^y} r^y(u)du}[P^y(\tau^y) - S(\tau^y)]^+ \mid \Psi_t\right\} \\
& = \tilde{E}\left\{e^{-\int_t^{\tau^y} r^x(u)du}[P^x(\tau^y) - S(\tau^y)]^+ \right. \\
& \qquad \left. - e^{-\int_t^{\tau^y} r^y(u)du}[P^y(\tau^y) - S(\tau^y)]^+ \mid \Psi_t\right\} \\
& \ge 0.
\end{aligned} \tag{29}$$

For the state of the value $(t) = p^x$ with the optimal stopping time τ^x, we have

$$\begin{aligned}
& f_D^x(t) - f_D^y(t) \\
& \le \tilde{E}\left\{e^{-\int_t^{\tau^x} r^x(u)du}[P^x(\tau^x) - S(\tau^x)]^+ \right. \\
& \qquad \left. - e^{-\int_t^{\tau^x} r^y(u)du}[P^y(\tau^x) - S(\tau^x)]^+ \mid \Psi_t\right\} \\
& = \tilde{E}\left\{\left[e^{-\int_t^{\tau^x} r^x(u)du}[P^x(\tau^x) - S(\tau^x)] \right.\right. \\
& \qquad \left. - e^{\int_t^{\tau^x} r^y(u)du}(P^y(\tau^u) - S(\tau^u))^+ \right] \\
& \qquad \left. \cdot I_{\{P^x(\tau^x) > S(\tau^x)\}} \mid \Psi_t\right\} \\
& \le \tilde{E}\left\{\left[e^{-\int_t^{\tau^x} r^x(u)du}[P^x(\tau^x) - S(\tau^x)] \right.\right. \\
& \qquad \left. - e^{-\int_t^{\tau^x} r^y(u)du}(P^y(\tau^x) - S(\tau^x))\right]
\end{aligned}$$

$$
\begin{aligned}
&\cdot I_{\{P^x(\tau^x)>S(\tau^x)\}} \mid \Psi_t \Big\} \\
&\le \tilde{E}\Big\{ e^{-\int_t^{\tau^x} r^x(u)du} \left[P^x(\tau^x) - S(\tau^x)\right] \\
&\quad - e^{-\int_t^{\tau^x} r^y(u)du} \left[P^y(\tau^x) - S(\tau^x)\right] \mid \Psi_t \Big\} \\
&\quad \times \tilde{P}\{P^x(\tau^x) > S(\tau^x)\} \\
&\le \tilde{E}\Big\{ e^{-\int_t^{\tau^x} r^x(u)du} P^x(\tau^x) \\
&\quad - e^{-\int_t^{\tau^x} r^y(u)du} P^y(\tau^x) \mid \Psi_t \Big\} \tilde{p}^P \\
&= \left[P^x(t) - P^y(t)\right] \tilde{p}^P.
\end{aligned}
\tag{30}
$$

(4) Note that $P(t) = P(t) - f_D(t)$. For $P(t) \ge P^y(t)$, we have

$$P_D^x(t) = P^x(t) - f_D^x(t) \ge P^y(t) - f_D^y(t) = P_D^y(t). \tag{31}$$

This completes the proof.

VALUING CALLABLE DEFAULTABLE BOND

For the callable defaultable bond, when the firm value falls far below its obligation such that the default option arrives at its maximum, the issuers would exempt its obligation by giving up the firm. When the benchmark bond exceeds the preset provisional call price schedule Kt, which is the strike price to call the debt, the issuers intend to exercise the call option to buy back the bond or reissue a bond with lower interest rates. The value of the callable corporate bond is

$$P_{CD}(t) = P(t) - f_{CD}(t), \tag{32}$$

where $f_{CD}(t)$ is the embedded option value at time t,

$$
\begin{aligned}
&f_{CD}(t) \\
&= \sup_{t\le\tau\le T} \tilde{E}\left\{ \exp\left[-\int_t^{\tau} r(u)\,du\right] \left[P(\tau) - K_\tau \wedge S(\tau)\right]^+ \mid \Psi_t \right\}.
\end{aligned}
\tag{33}
$$

The embedded option $f_C(t)$ satisfies

$$f_{CD}(t) \ge \left(P(t) - K_t \wedge S(t)\right)^+. \tag{34}$$

The optimal stopping time is

$$\tau = \inf\left\{t \ge 0 : f_{CD}(t) = \left(P(t) - K_t \wedge S(t)\right)^+\right\}. \tag{35}$$

Similarly, we have the following theorem.

Theorem 4. For the different initial value of firm value and benchmark bond, the following properties hold.

$$\begin{aligned}
&(1)\ If\ S^{x}(0) > S^{y}(0),\ then \\
&\quad -\left[S^{x}(t) - S^{y}(t)\right]\bar{p}^{s} \le f_{CD}^{x}(t) - f_{CD}^{y}(t) \le 0. \\
&(2)\ If\ S^{x}(0) > S^{y}(0),\ then\ P_{CD}^{x} \le P_{CD}^{y}(t). \\
&(3)\ If\ P^{x}(0) > P^{y}(0),\ then \\
&\quad -\left[S^{x}(t) - S^{y}(t)\right]\bar{p}^{P} \le f_{CD}^{x}(t) - f_{CD}^{y}(t) \le 0. \\
&(4)\ If\ P^{x}(0) > P^{y}(0),\ then\ P_{CD}^{x} \ge P_{CD}^{y}(t).
\end{aligned} \tag{36}$$

Proof. It is similar with the case of the pure default corporate bond. By the no-crossing properties for jump-diffusion process, we have

$$\begin{aligned}
&S^{x}(0) > S^{y}(0) \\
&\Longrightarrow S^{x}(t) \ge S^{y}(t) \\
&\Longrightarrow K_t \wedge S^{x}(t) \ge K_t \wedge S^{y}(t), \\
&\widetilde{P}-\text{a.s.} \quad \forall 0 < t < \infty.
\end{aligned} \tag{37}$$

So, replacing the strike price $S(t)$ by $K_t \wedge S^x(t)$ and $S^y(t)$ by $K_t \wedge S^y(t)$ in the proof of Theorem 2, we obtain the results of Theorem 4.

Theorem 5. $f_C(t)$ satisfies

$$\begin{aligned}
&(1)\ f_{CD}(t) - f_D(t) \\
&\quad = AP(t) + \widetilde{E}\left\{e^{-\int_t^{\tau_d} r(u)du}\left[P(\tau_d) \wedge S(\tau_d) - K_{\tau_d}\right]^{+} \mid \Psi_t\right\} \\
&(2)\ f_{CD}(t) - f_D(t) \le f_C(t),
\end{aligned} \tag{38}$$

where

$$\begin{aligned}
AP(t) = \widetilde{E}\Big\{&e^{-\int_t^{\tau_{cd}} r(u)du}\left[P(\tau_{cd}) - K_{\tau_{cd}} \wedge S(\tau_{cd})\right]^{+} \\
&- e^{-\int_t^{\tau_d} r(u)du}\left[P(\tau_d) - K_{\tau_d} \wedge S(\tau_d)\right]^{+} \mid \Psi_t\Big\}, \\
f_C(t) = \sup_{t \le \tau \le T} \widetilde{E}&\left\{e^{-\int_t^{\tau} r(u)du}\left[P(\tau) - K_\tau\right]^{+} \mid \Psi_t\right\},
\end{aligned} \tag{39}$$

where τ_d is the optimal stopping time for the pure default corporate bond and τcd is the optimal stopping time for the callable default corporate bond.

Remark 6. Due to the embedded call option, $f_C(t)$ is worth more than $f_D(t)$. The increment of $f_C(t)$ over $f_D(t)$ can be decomposed into two parts.

(1) The part of the "American premium" ($AP(t)$). That is, the firm has the right to exercise the embedded option at the optimal stopping time.

(2) The part of the "call option" $\widetilde{E}\{\exp[-\int_t^{\tau_d} r(u)du][P(\tau_d) \wedge S(\tau_d) - K_{\tau_d}]^{+} \mid \Psi_t\}$. Due to the call provision, the firm has the option either to buy back the debt when the firm works well ($K\tau_d \le (\tau_d) < S(\tau_d)$) or to keep the firm when the firm value is less than the benchmark bond but before bankruptcy ($K\tau_d \le S(\tau_d) \le P(\tau_d)$).

From the second part of Theorem 4, the increment of $f_C(t)$ over $f_D(t)$ is less than that of embedded pure call option $f_c(t)$ due to the interaction between the default option and call option.

Proof. (1) Notice that $(t) \geq K_t \wedge S(t)$. Let τ_{cd} be the optimal stopping time of $f_C(t)$, and let τ_d be the optimal stopping time of $f_D(t)$,

$$\begin{aligned} & f_{CD}(t) - f_D(t) \\ & = \tilde{E}\left\{e^{-\int_t^{\tau_{cd}} r(u)du}\left[P(\tau_{cd}) - K_{\tau_{cd}} \wedge S(\tau_{cd})\right]^+ \mid \Psi_t\right\} \\ & \quad - \tilde{E}\left\{e^{-\int_t^{\tau_d} r(u)du}\left[P(\tau_d) - S(\tau_d)\right]^+ \mid \Psi_t\right\} \\ & = \tilde{E}\left\{e^{-\int_t^{\tau_{cd}} r(u)du}\left[P(\tau_{cd}) - K_{\tau_{cd}} \wedge S(\tau_{cd})\right]^+\right. \\ & \quad \left. - e^{-\int_t^{\tau_d} r(u)du}\left[P(\tau_d) - K_{\tau_d} \wedge S(\tau_d)\right]^+ \mid \Psi_t\right\} \\ & + \tilde{E}\left\{e^{-\int_t^{\tau_d} r(u)du}\left(P(\tau_d) - K_{\tau_d}\right)^+ \mid \Psi_t\right\} \tilde{P} \\ & \times \left[K_{\tau_d} \leq P(\tau_d) < S(\tau_d)\right] \\ & + \tilde{E}\left\{e^{-\int_t^{\tau_d} r(u)du}\left(S(\tau_d) - K_{\tau_d}\right)^+ \mid \Psi_t\right\} \\ & 1 \times \tilde{P}\left[K_{\tau_d} \leq S(\tau_d) \leq P(\tau_d)\right] \\ & = \mathrm{AP}(t) + \tilde{E}\left\{e^{-\int_t^{\tau_d} r(u)du}\right. \\ & \qquad \left. \times \left[P(\tau_d) \wedge S(\tau_d) - K_{\tau_d}\right]^+ \mid \Psi_t\right\}. \end{aligned} \tag{40}$$

(2) On the other hand, we have

$$\begin{aligned} & f_{CD}(t) - f_D(t) \\ & = \sup_{t \leq \tau \leq T} \tilde{E}\left\{e^{-\int_t^{\tau} r(u)du}\left[P(\tau) - K_\tau \wedge S(\tau)\right]^+ \mid \Psi_t\right\} \\ & \quad - \sup_{t \leq \tau \leq T} \tilde{E}\left\{e^{-\int_t^{\tau} r(u)du}\left[P(\tau) - S(\tau)\right]^+ \mid \Psi_t\right\} \\ & \leq \sup_{t \leq \tau \leq T} \tilde{E}\left\{e^{-\int_t^{\tau} r(u)du}\left[P(\tau) - K_\tau\right]^+\right. \\ & \qquad \left. + \left[P(\tau) - S(\tau)\right]^+ \mid \Psi_t\right\} \\ & \quad - \sup_{t \leq \tau \leq T} \tilde{E}\left\{e^{-\int_t^{\tau} r(u)du}\left[P(\tau) - S(\tau)\right]^+ \mid \Psi_t\right\} \\ & \leq \sup_{t \leq \tau \leq T} \tilde{E}\left\{e^{-\int_t^{\tau} r(u)du}\left[P(\tau) - K_\tau\right]^+ \mid \Psi_t\right\}. \end{aligned} \tag{41}$$

This completes the proof.

For the embedded options exercise boundaries, denote P^*_{DB} as the exercise boundary of the pure defaultable corporate bond, when $S(t) = s^*$ at time t, which means that as soon as the benchmark bond price reaches P^*_{DB}, the firm should exercise options; that is, the firm should default. Denote P^*_{CDB} as the exercise boundary of the callable defaultable corporate bond, when $S(t) = s*$ at time t, which means that, in the case $Kt > S(t)$, the firm would default as soon as the benchmark bond price reaches P^*_{CDB}; in the case $Kt \leq S(t)$, the firm would call back its debt as soon as the benchmark bond price reaches P^*_{CDB}. In terms of the benchmark bond price, we have the following theorem.

Theorem 7. For each $t \in [0, T)$, it holds that

$$\begin{aligned}
&(1)\ if\ S^x(t) > S^y(t),\ then\ P^x_{DB}(t) \geq P^y_{DB}(t),\\
&(2)\ when\ K_t > S(t),\ if\ S^x(t) > S^y(t),\ then\\
&\quad P^x_{CDB}(t) \geq P^y_{CDB}(t),\\
&(3)\ when\ K_t \leq S(t),\ if\ S^x(t) > S^y(t),\ then\\
&\quad P^x_{CDB}(t) \leq P^y_{CDB}(t),\\
&(4)\ if\ S(t) \leq K_t,\ then\ P_{CDB}(t) \geq P_{DB}(t).
\end{aligned} \tag{42}$$

Proof. (1) When the benchmark bond price $P(t) < P^y_{DB}(t)$, the firm with $Sy(t)$ would not exercise the embedded default option. By part 1 of Theorem 2, we have

$$\begin{aligned}
f^x_D(t) &\geq f^y_D(t) - (S^x(t) - S^y(t))\\
&> P(t) - S^y(t) - (S^x(t) - S^y(t))\\
&= P(t) - S^x(t).
\end{aligned} \tag{43}$$

So the firm with $S(t)$ would also not exercise the embedded default option, which means that the default boundary is no lower than that of the firm value $Sy(t)$. Thus, $P^x_{DB}(t) \geq P^y_{DB}(t)$.

(2) When $Kt > (t)$, the firm would consider to exercise the default option. If $S^x(t) > S^y(t)$, for $P(t) < P^y_{CDB}(t)$, we have

$$\begin{aligned}
f^x_{CD}(t) &\geq f^y_{CD}(t) - (S^x(t) - S^y(t))\\
&\geq P(t) - S^y(t) - (S^x(t) - S^y(t))\\
&= P(t) - S^x(t).
\end{aligned} \tag{44}$$

So

$P^x_{CDB}(t) \geq P^y_{CDB}(t)$.

(3) When $K_t < (t)$, the firm would consider to exercise the call option. If $S(t) > S^y(t)$, which means $P(t) - K_t \wedge S^y(t) \geq P(t) - K_t \wedge S^x(t)$, for $P(t) < P^x_{CDB}(t)$, we have

$$f^y_{CD}(t) > f^x_{CD}(t) > (P(t) - K_t)^+ \tag{45}$$

So $P(t) < P^y_{CDB}(t)$, which means $P^x_{CDB}(t) \leq P^y_{CDB}(t)$. (4) When $S(t) \leq K_t$, for $P(t) < P_{DB}(t)$, we have

$$f_{CD}(t) \geq f_D(t) \geq (P(t) - S(t))^+ > P(t) - K_t \wedge S(t). \tag{46}$$

So $P(t) < P_{CDB}(t)$, which means $P_{CDB}(t) \geq P_{DB}(t)$. This completes the proof

For the value of the embedded option, it can be re-expressed as

$$
\begin{aligned}
&f_{CD}(t) \\
&= \sup_{t \le \tau \le T} \widetilde{E}\left\{e^{-\int_t^\tau r(u)du}\left[P(\tau) - K_\tau \wedge S(\tau)\right]^+ \right. \\
&\qquad \left. \times I_{P(\tau) > K_\tau \wedge S(\tau)} \mid \Psi_t\right\} \\
&= \sup_{t \le \tau \le T} \widetilde{E}\left\{e^{-\int_t^\tau r(u)du}\left[(P(\tau) - K_\tau)^+ \mid K_\tau \right.\right. \\
&\qquad\qquad \left. < S(\tau), \Psi_t\right] \times \widetilde{P}(K_\tau < S(\tau)) \\
&\qquad + e^{-\int_t^\tau r(u)du}\left[(P(\tau) - S(\tau))^+ \mid K_\tau \ge S(\tau), \Psi_t\right] \\
&\qquad \left. \times \widetilde{P}(K_\tau \ge S(\tau))\right\} \\
&= \sup_{t \le \tau \le T} \widetilde{E}\left\{e^{-\int_t^\tau r(u)du}\right. \\
&\qquad \times \left[(P(\tau) - K_\tau) \mid K_\tau < S(\tau),\right. \\
&\qquad\quad \left. K_\tau < P(\tau), \Psi_t\right] \widetilde{P}(K_\tau \ge S(\tau)) \\
&\qquad\qquad\qquad \times \widetilde{P}(K_\tau < S(\tau))\, \widetilde{P} \\
&\qquad + e^{-\int_t^\tau r(u)du}\left[(P(\tau) - S(\tau)) \mid K_\tau \ge S(\tau),\right. \\
&\qquad\qquad\qquad \left. P(\tau) \ge S(\tau), \Psi_t\right] \widetilde{P}(K_\tau \ge S(\tau)) \\
&\qquad\qquad\qquad \left. \times \widetilde{P}(P(\tau) \ge S(\tau))\right\}.
\end{aligned} \tag{47}
$$

For $K_\tau < (\tau)$, we have

$$
\begin{aligned}
&\widetilde{P}\left\{\ln K_\tau < \ln S(t) + \int_t^\tau [r(u) - h(u)]\, du \right. \\
&\qquad \left. + I_S(t,\tau) - \frac{1}{2}\sigma_S^2(t,\tau) + J_S(t,\tau) - C_{JS}(t,\tau)\right\} \\
&= \widetilde{P}\left\{\ln K_\tau - \ln S(t) - \int_t^\tau [r(u) - h(u)]\, du \right. \\
&\qquad \left. < I_S(t,\tau) - \frac{1}{2}\sigma_S^2(t,\tau) + J_S(t,\tau) - C_{JS}(t,\tau)\right\},
\end{aligned} \tag{48}
$$

where

$$
\begin{aligned}
I_S(t,\tau) &= \int_t^\tau \int_0^{\overline{T}} b(u,x)\, dx\, d_1\widetilde{B}(u,x), \\
J_S(t,\tau) &= \sum_{i=1}^{M} \int_t^\tau \ln(a_i c(u) + 1)\, d\widetilde{N}_i(u) \\
C_{JS}(t,\tau) &= \sum_{i=1}^{M} \int_t^\tau c(u)\, a_i \widetilde{\lambda}_i\, du.
\end{aligned} \tag{49}
$$

Notice that the price of benchmark bond satisfies

$$
P(t,\tau)\, P(\tau) = P(t) - \sum_{t \le t_j < \tau} c_j P(t, t_j). \tag{50}
$$

For $(\widetilde{K}_\tau < P(\tau))$, we have

$$\begin{aligned}
&\tilde{P}\left\{K_\tau P(t,\tau) < P(t) - \sum_{t \le t_j < \tau} c_j P(t,t_j)\right\} \\
&= \tilde{P}\left\{\ln K_\tau - \int_t^\tau r(u)\,du + I_f(t,\tau)\right. \\
&\qquad + \frac{1}{2}\sigma_f^2(t,\tau) + J_f(t,\tau) + C_{Jf}(t,\tau) \\
&\qquad \left. < \ln\left[p(t) - \sum_{t \le t_j < \tau} c_j P(t,t_j)\right]\right\} \\
&= \tilde{P}\left\{I_f(t,\tau) + \frac{1}{2}\sigma_f^2(t,\tau) + J_f(t,\tau) + C_{Jf}(t,\tau)\right. \\
&\qquad < \int_t^\tau r(u)\,du - \ln K_\tau \\
&\qquad \left. + \ln\left[p(t) - \sum_{t \le t_j < \tau} c_j P(t,t_j)\right]\right\},
\end{aligned} \tag{51}$$

where

$$\begin{aligned}
I_f(t,\tau) &= \int_t^\tau \int_0^{\tau-u} \sigma(u,x)\,dx\,d_1\tilde{B}(u,x), \\
J_f(t,\tau) &= \sum_{i=1}^{M} \int_t^\tau \int_0^{\tau-u} \gamma(u,x)\,dx\,a_i\,d\tilde{N}_i(u), \\
C_{Jf}(t,\tau) &= \sum_{i=1}^{M} \int_t^\tau \left[\exp\left(-a_i \int_0^{\tau-u} \gamma(u,x)\,dx\right) - 1\right] \tilde{\lambda}_i\,du.
\end{aligned} \tag{52}$$

For $\tilde{P}(P(\tau) \ge S(\tau))$, we have

$$\begin{aligned}
&\tilde{P}\{P(\tau) \ge S(\tau)\} \\
&= \tilde{P}\left\{P(t) - \sum_{t \le t_j < \tau} c_j P(t,t_j)\right. \\
&\qquad \ge S(t)\exp\left[I_f(t,\tau) + \frac{1}{2}\sigma_f^2(t,\tau) + J_f(t,\tau)\right. \\
&\qquad\qquad + C_{Jf}(t,\tau) - \int_t^\tau h(u)\,du + I_S(t,\tau) \\
&\qquad\qquad \left.\left. - \frac{1}{2}\sigma_S^2(t,\tau) + J_S(t,\tau) - C_{JS}(t,\tau)\right]\right\} \\
&= \tilde{P}\left\{I_{Sf}(t,\tau) + \sigma_f^2(t,\tau) - \frac{1}{2}\sigma_{Sf}^2(t,\tau) + cov\left(I_f, I_S\right)\right. \\
&\qquad + I_f(t,\tau) + C_{Jf}(t,\tau) + J_S(t,\tau) - C_{JS}(t,\tau) \\
&\qquad < \ln\left[P(t) - \sum_{t \le t_j < \tau} c_j P(t,t_j)\right] \\
&\qquad \left. + \int_t^\tau h(u)\,du - \ln S(t)\right\},
\end{aligned} \tag{53}$$

Where

$$I_{Sf}(t,\tau) = I_S(t,\tau) + I_f(t,\tau),$$

$$\sigma_{Sf}^2(t,\tau) = \sigma_S^2(t,\tau) + \sigma_f^2(t,\tau) + 2\operatorname{cov}\left(I_f, I_S\right),$$

$$\operatorname{cov}\left(I_f, I_S\right) = \int_t^{\tau} du \int_0^{\tau-u} \sigma(u,x)\,dx \int_0^{\bar{T}} b(u,y)\,\rho(x,y)\,dy. \tag{54}$$

From (47)–(54), we have the following theorem.

Theorem 8. For the exogenous shock,

(1) *if* $J_f < 0$, *then* $f_{CD} \uparrow$,

(2) *if* $J_S > 0$, *then*

$$\tilde{P}(K_\tau < S(\tau)) \uparrow, \tilde{P}(P(\tau) \geq S(\tau)) \downarrow, P(\tau) - S(\tau) \downarrow,$$

where f_{CD} *is dependent on the total impact,*

(3) *if* $J_S + J_f > 0$, *then*

$$(P(\tau) - S(\tau))^+ \downarrow, \tilde{P}(S(\tau) < P(\tau)) \downarrow, f_{CD} \downarrow,$$

(4) *if* $J_S + J_f < 0$, *then*

$$(P(\tau) - S(\tau))^+ \uparrow, \tilde{P}(S(\tau) < P(\tau)) \uparrow, f_{CD} \uparrow.$$

For the impact of covariation between I_S *and* I_f, *we have*

(1) *when* $S(\tau) > K_\tau, \operatorname{cov}\left(I_f, I_S\right) > 0$, (55)

if $S(\tau) \uparrow$, *then* $f_{CD} \downarrow$,

(2) *when* $S(\tau) \leq K_\tau, \operatorname{cov}\left(I_f, I_S\right) < 0$,

if $S(\tau) \downarrow$, *then* $f_{CD} \uparrow$.

For the impact of covariation between I_S and I_f, we have

(1) *when* $S(\tau) > K_\tau, \operatorname{cov}\left(I_f, I_S\right) > 0$,

if $S(\tau) \uparrow$, *then* $f_{CD} \downarrow$,

(2) *when* $S(\tau) \leq K_\tau, \operatorname{cov}\left(I_f, I_S\right) < 0$,

if $S(\tau) \downarrow$, *then* $f_{CD} \uparrow$. (56)

Remark 9. (1) The present default distance ln (t) − ln $P(t)$ is negatively related with the default probability $P_d(S, P)$. Therefore, it negatively relates with the credit spread.

(2) The covariation between the economic basis and firm value will affect the default probability and the credit spread. If the economy is boom and the covariation is negative, then the credit spread will narrow along with the increasing firm value. If the economy is recession and the covariation is positive, then the credit spread would widen along with the decreasing firm value.

(3) The shocks cause jump in benchmark bond price and the firm value. When the exogenous shock increases the firm value, the default probability would decrease. Therefore, the embedded options and the credit spread tend to decrease. However, the call provision would offset this effect to some degree. The effect of exogenous shock depends on the total impact of J_f and J_S. As Theorem 8 shows, if $J_S + J_f > 0$, then the credit spread would narrow. If $J_S + J_f < 0$, then the credit spread would widen.

(4) Due to the free boundary problems of American options and the firm's payout rate, we have not obtained the accurate pricing for the firm bond and the embedded American options.

CONCLUSION

We study the valuation of the firm bonds and their embedded options through an infinite dimensional jump-diffusion process. Some analytical results on corporate bond valuation, embedded options, and their exercise boundaries are obtained. By decomposing the corporate bond as benchmark bond minus or plus embedded options, analyzing the credit spread's effect factors, and dismantling the comprehensive effects of the exogenous shock on the embedded options, we find that the credit spread is affected by the firm value. Lower initial firm value implies higher credit spread. But the difference between the spreads is less than the difference of firm value times the risk-neutral default probability. A formal empirical test of the model's implications especially on the "credit spread puzzle" would be an interesting subject for future researches.

ACKNOWLEDGMENTS

Thanks are due to referees whose meaningful suggestions are very helpful to revise the paper. This work is supported by the Fundamental Research Funds for the Central Universities (JBK130401 and JBK131106).

REFERENCES

1. J. Amato and E. Remolona, "The credit spread puzzle," BIS Quarterly Review, pp. 51–63, 2003.
2. E. J. Elton, M. J. Gruber, D. Agrawal, and C. Mann, "Explaining the rate spread on corporate bonds," Journal of Finance, vol. 56, no. 1, pp. 247–277, 2001.
3. P. Collin-Dufresne, R. S. Goldstein, and J. S. Martin, "The determinants of credit spread changes," Journal of Finance, vol. 56, no. 6, pp. 2177–2207, 2001.
4. K. J. M. Cremers, J. Driessen, and P. Maenhout, "Explaining the level of credit spreads: option-implied jump risk premia in a firm value model," Review of Financial Studies, vol. 21, no. 5, pp. 2209–2242, 2008.
5. D. Duffie, J. Pan, and K. Singleton, "Transform analysis and asset pricing for affine jump-diffusions," Econometrica, vol. 68, no. 6, pp. 1343–1376, 2000.
6. S. G. Kou and H. Wang, "Option pricing under a double exponential jump diffusion model," Management Science, vol. 50, no. 9, pp. 1178–1192, 2004.
7. C. L. Yu, H. Li, and M. T. Wells, "Mcmc estimation of levy jump ´ models using stock and option prices," Mathematical Finance, vol. 21, no. 3, pp. 383–422, 2011.
8. F. A. Longstaff, S. Mithal, and E. Neis, "Corporate yield spreads: default risk or liquidity? New evidence from the credit default swap market," Journal of Finance, vol. 60, no. 5, pp. 2213–2253, 2005.
9. D. P. Kennedy, "The term structure of interest rates as a Gaussian random field," Mathematical Finance, vol. 4, no. 3, pp. 247–258, 1994.
10. D. P. Kennedy, "Characterizing Gaussian models of the term structure of interest rates," Mathemnatical Finance, vol. 7, no. 2, pp. 107–118, 1997.
11. R. S. Goldstein, "The term structure of interest rates as a random field," Review of Financial Studies, vol. 13, no. 2, pp. 365–384, 2000.
12. P. Santa-Clara and D. Sornette, "The dynamics of the forward interest rate curve with stochastic string shocks," Review of Financial Studies, vol. 14, no. 1, pp. 149–185, 2001.
13. R. C. Merton, "On the pricing of corporate debt: the risk structure of

interest rates," Journal of Finance, vol. 29, no. 2, pp. 449– 470, 1974.
14. F. Black and J. C. Cox, "Valuing corporate securities: some effects of bond indenture provisions," Journal of Finance, vol. 31, no. 2, pp. 351–367, 1976.
15. M. J. Brennan and E. S. Schwartz, "Analyzing convertible bonds," Journal of Financial and Quantitative Analysis, vol. 15, no. 4, pp. 907–929, 1980.
16. F. A. Longsta and E. S. Schwartz, "A simple approach to valuing risky fixed and oating rate debt," Journal of Finance, vol. 50, pp. 789–819, 1995.
17. R. Geske, "The valuation of corporate liabilities as compound options," Journal of Financial and Quantitative Economics, vol. 12, no. 4, pp. 541–552, 1977.
18. M. J. Brennan and E. S. Schwartz, "Convertible bonds: valuation and optimal strategies for call and conversion," Journal of Finance, vol. 32, no. 5, pp. 1699–1715, 1977.
19. E. O. Fischer, R. Heinkel, and J. Zechner, "Dynamic capital structure choice: theory and tests," Journal of Finance, vol. 44, no. 1, pp. 19–40, 1989.
20. H. Leland and K. Toft, "Optimal capital structure, endogenous bankruptcy, and the term structure of credit spreads," Journal of Finance, vol. 51, no. 3, pp. 987–1019, 1996.
21. V. V. Acharya and J. N. Carpenter, "Corporate Bond Valuation and Hedging with Stochastic Interest Rates and Endogenous Bankruptcy," Review of Financial Studies, vol. 15, no. 5, pp. 1355– 1383, 2002.
22. S. E. Shreve, Stochastic Calculus for Finance II: Continuous-Time Models, Springer Finance, 2004.
23. S. Fan, "A stochastic string with a compound Poisson process," Abstract and Applicated Analysis, vol. 2013, Article ID 857678, 8 pages, 2013.

CHAPTER

6

The Time Decay of Bond Premium and Discount—An Analysis of the Time Passage Effect on Bond Prices

Jorge Brusa[1], **Jenny Gu**[2], **and Grace Yaru Liu**[3]

[1]Division of International Banking & Finance Studies, Sanchez School of Business, Texas A&M International University, Laredo, USA

[2]College of Business, University of Dallas, Irving, USA

[3]Department of Supply Chain and Information Systems, Pennsylvania State University, University Park, USA

ABSTRACT

In this paper, we show that the price of a premium bond and the price of a discount bond will both move toward face value at an increasing rate as the bonds approach maturity. We present a mathematical proof to show that the

Citation: Brusa, J., Gu, J. and Liu, G. (2014) "The Time Decay of Bond Premium and Discount—An Analysis of the Time Passage Effect on Bond Prices". Theoretical Economics Letters, 4, 323-330. http://dx.doi.org/10.4236/tel.2014.45043

decline in premium and discount decline over time, to be referred to as time decay, accelerates as time passes by. We also provide numerical examples and graphical representations to illustrate the time passage effect on bond prices and discuss the implications of the findings to bond investor and asset managers in light of the quantitative easing policies taken by central banks after the 2008 financial crisis.

Keywords:-Time Decay, Premium Bond, Discount Bond, Time Passage Effect, Quantitative Easing

INTRODUCTION

Researchers in economics and finance have studied the relationships among bond prices, interest rates, and maturity for many years. One of the best known results in the study is that, if the market interest rate on a bond remains unchanged, the price of a bond will approach its face value as the bond approaches maturity. However, how exactly and in what way the bond price approaches its face value has not been well discussed in textbooks. For example, when a bond is selling at premium (i.e. when the bond price is above its face value), the only way

for the bond price to approach its face value is to move downward. Therefore the price path toward maturity for a premium bond must be downward-sloping. However, a downward-sloping path could be convex (decreasing at a decreasing rate), concave (decreasing at an increasing rate), or a straight line (decreasing at constant rate), to name just three. Similarly, when a bond is selling at discount (i.e. when the bond price is below its face value), the only way for the bond price to reach its face value is to move upward. Therefore the price path toward maturity for a discount bond must be upward-sloping. However, an upward-sloping path could also be convex (increasing at an increasing rate), concave (increasing at a decreasing rate), or a straight line (increasing at constant rate).

In this paper we formally present a mathematical proof to show that the price path for a premium bond is concave (decreasing toward the par at an increasing rate) and the price path for a discount bond is convex (increasing toward the par at an increasing rate) as the bond approaches maturity. In other words, we show that both the premium (defined as price minus the par) and the discount (defined as par minus the price) are diminishing at an increasing rate as the bond approaches maturity. This is the first contribution of our study because, to the best of our knowledge, no study in the literature has formally presented a mathematical proof to show that both the premium and the discount diminish at an increasing rate as bond approaches maturity[1]. Since both the premium and the discount will decrease over time when interest rate remains unchanged, we will refer to the decrease in premium and discount as time decay of premium and discount.

One of the reasons that most textbooks only briefly address the issue of bond price path toward maturity (e.g. Brigham and Houston, 2012) [1] and that some books do not even mention the issue (e.g. Booth, Cleary, and Drake, 2014; Gitman, Joehnk, Smart, 2011; Keown, Martin, and Petty, 2014) [2]-[4] may be that, in order to examine the time passage effect on bond price, the condition that the interest rate remains unchanged must hold. Some authors consider that condition too strong and therefore practically impossible[2]. It is true that interest rate may be constantly changing. However, the change in interest rate should not the reason to restrain us from examining the price change due to time passage. As Fabozzi (2007) [5] indicates, the change in bond price can always be decomposed into two components: one component due to time passage toward maturity without any change in interest rate, and one component due to the change in interest rate. In other words, regardless of whether interest rate changes or remains constant, there is always a component of change in bond prices due to time passage and that component should be thoroughly examined and fully understood[3]. This is the second reason that we thoroughly examine how time passage affects bond prices.

The results of this study should be interesting not only to scholars studying fixed income securities, but also to asset managers with clients holding bonds. As we show in Sections 2 and 3 in the paper, for premium bonds, their value decreases toward the par at an increasing rate over time. The quantitative easing policies initiated by the Federal Reserve, Bank of England, European Central Bank, and Bank of Japan as a response to the 2008 financial crisis have driven the global interest rates to historical lows. As a result, most bonds issued prior to 2008 (which usually carry coupon rates higher than the current interest rates), have become premium bonds in the past few years. As these bonds approach maturity, their prices will move lower over time even as the central banks continue to maintain the interest rate at the current level. Furthermore, since the prices of these premium bonds decrease at an increasing rate, the decrease in value for these bonds will accelerate as they approach maturity in the next few years. As many individual bond investors do not understand the time passage effect on bond prices, asset managers with clients investing in individual bonds or bond funds are likely to be questioned by their client as the clients see the value of their bonds declining at an increasing rate. The analyses of the bond price path will provide to asset managers a tool to explain to their clients why the value of these bonds declines over time. This is the third reason for the study.

The rest of the paper is organized as follows. In Section 2, we present a mathematical proof to show how both the premium and the discount will decrease at an increasing rate as the bond approaches maturity. In Section 3, we present numerical examples to illustrate our findings. In Section 4 we discuss the implications of the findings in the paper. We conclude the paper in Section 5.

MATHEMATICAL MODEL

In this section, we will present a mathematical proof to show that the price of a premium bond will decline toward face value at an increasing rate over time, and the price of a discount bond will rise toward face value, also at an increasing rate as it approaches maturity. The way we prove the arguments is as follows: we will first show that the price of a premium bond increases at a decreasing rate as the bond's term to maturity increases. However, in practice, a bond's term to maturity decreases over time. Consequently, as a bond approaches maturity, the price of a premium decreases at an increasing rate. Similarly, for a discount bond we will show that as term to maturity increases, the price decreases at a decreasing rate. Therefore, as the bond approaches the maturity and the term to maturity decreases, the price of a discount bond increases at an increasing rate.

We now formally present the mathematical proof:

Let T be the term to maturity of the bond (and thus T decreases when a bond moves toward maturity as time passes by), cr be the coupon rate, i be the market rate of interest, and F be the face (par) value of the bond. Then the price of the bond, P, is determined by the equation:

$$P = cr \cdot F\left[\frac{1-\frac{1}{(1+i)^T}}{i}\right] + F\left[\frac{1}{(1+i)^T}\right]$$

$$P = \frac{cr \cdot F}{i}\left[1-\frac{1}{(1+i)^T}\right] + F\left[\frac{1}{(1+i)^T}\right] \qquad (1)$$

$$\frac{\partial P}{\partial T} = \frac{cr \cdot F}{i}\left[(1+i)^{-T}\ln(1+i)\right] - F(1+i)^{-T}\ln(1+i)$$

$$\frac{\partial P}{\partial T} = \left[\frac{cr}{i}-1\right]F(1+i)^{-T}\ln(1+i) \qquad (2)$$

$$\frac{\partial P}{\partial T} = 0 \text{ if } \left(\frac{cr}{i} - 1\right) = 0 \text{ (for a par bond where } cr = i) \tag{3}$$

$$\frac{\partial P}{\partial T} > 0 \text{ if } \left(\frac{cr}{i} - 1\right) > 0 \text{ (for a premium bond where } cr > i) \tag{4}$$

$$\frac{\partial P}{\partial T} < 0 \text{ if } \left(\frac{cr}{i} - 1\right) < 0 \text{ (for a discount bond where } cr < i) \tag{5}$$

Equation (3) shows that for a par bond, the price is a constant (equal to the par value), and the bond price path toward maturity for a par bond is a horizontal straight line.

Equation (4) shows that for a premium bond, the price is an increasing function of its term to maturity. As the bond approaches maturity, the term to maturity T decreases. Consequently, the price for a premium bond decreases as it approaches maturity.

Equation (5) shows that for a discount bond, the price is a decreasing function of term to maturity. As a bond approaches maturity, the term to maturity T decreases. As a result, the price for a discount bond increases as it approaches maturity.

To find out whether the bond price path for a premium bond is decreasing at an increasing rate or at a decreasing rate, and whether the bond price path for a discount bond is increasing at an increasing rate or at a decreasing rate, we differentiate Equation (2) with respect to T again and obtain:

$$\frac{\partial^2 P}{\partial T^2} = \left[\frac{cr}{i} - 1\right] F \ln(1+i) \frac{\partial}{\partial T}\left[(1+i)^{-T}\right]$$

$$\frac{\partial^2 P}{\partial T^2} = \left[\frac{cr}{i} - 1\right] F \ln(1+i)\left[-(1+i)^{-T} \ln(1+i)\right]$$

$$\frac{\partial^2 P}{\partial T^2} = -\left[\frac{cr}{i} - 1\right] F \left[\ln(1+i)\right]^2 (1+i)^{-T}$$

$$\frac{\partial^2 P}{\partial T^2} = 0 \text{ if } \left(\frac{cr}{i} - 1\right) = 0 \text{ (for a par bond)} \tag{6}$$

$$\frac{\partial^2 P}{\partial T^2} < 0 \text{ if } \left(\frac{cr}{i} - 1\right) > 0 \text{ (for a premium bond)} \tag{7}$$

$$\frac{\partial^2 P}{\partial T^2} > 0 \text{ if } \left(\frac{cr}{i} - 1\right) < 0 \text{ (for a discount bond)} \tag{8}$$

Combining Equations (4) and (7), the results show that the price for a premium bond is an increasing function with respect to T at a decreasing rate. However, as a bond approaches maturity, the term to maturity T decreases. Therefore, the price path for a premium bond toward maturity is decreasing at an increasing rate.

Equation (5) suggests that the price of a discount bond is a decreasing function of T. However, Equation (8) suggests that the downward-sloping curve becomes less negative as T increases. Therefore Equations (5) and (8) suggest that the price for a discount bond is decreasing with respect to T at a decreasing rate. However, as a bond approaches maturity, the term to maturity T decreases. Therefore, the price path for a discount bond toward maturity is increasing at an increasing rate.

In the next section, we provide a few numerical examples and graphical representations to illustrate the mathematical proofs.

NUMERICAL ILLUSTRATIONS

In Table 1 we present the prices for bonds with a face value of $1000, five different coupon rates: 12%, 10%, 8%, 6%, and 4%; and 21 different terms to maturity from 0 to 20 years. We assume the market rate of interest to be 8%, and coupon payments are made annually.

In the table, we can observe the prices for premium bonds (bonds with coupon rates of 12% and 10%) are increasing functions of term to maturity at a decreasing rate. For example, for the 12% bond the price increases by $37.04 when term to maturity increases from 0 to 1 year, but the price increases by only $8.59 when the term to maturity increases from 19 to 20 years. Since the price is increasing at a decreasing rate, it would not increase indefinitely as term to maturity approaches infinity. The price is equal to $P=\frac{cr\cdot F}{i}$ as T approaches infinity (known as a perpetual bond) according to Equation (1), or $1500 for a 12% bond when the market rate of interest is 8%. Figure 1 is the graphical representation of Table 1, showing bond prices as a function of term to maturity.

Since the price for a premium bond increases at a decreasing rate as term to maturity increases, the price path toward maturity date, which is the price movement with a decrease in term to maturity, is decreasing with an increasing rate.

Similarly, the prices for discount bonds (bonds with coupon rates of 6% and 4%) are decreasing functions of term to maturity at a decreasing rate. For example, for the 4% bond the price decreases by $37.04 when term to maturity increases from 0 to 1 year, but the price decreases by only $8.59 when the term to maturity increases from 19 to 20 years.

Since the price is decreasing at a decreasing rate, it does not decrease indefinitely as term to maturity approaches infinity. The price for a perpetual discount bond (with term to maturity approaches to infinity) is $P = \frac{cr \cdot F}{i}$, according to Equation (1), or \$500 for a 4% bond when the market rate of interest is 8%.

Table 1. The prices for bonds with a face value of \$1000, coupon rates of 12%, 10%, 8%, 6%, and 4% respectively (assuming annual payment), and terms to maturity from 0 to 20 years when the market rate of interest is 8%.

Term to Maturity	Coupon Rate				
	12%	10%	8%	6%	4%
0	$1000.00	$1000.00	$1000	$1000.00	$1000.00
1	$1037.04	$1018.52	$1000	$981.48	$962.96
2	$1071.33	$1035.67	$1000	$964.33	$928.67
3	$1103.08	$1051.54	$1000	$948.46	$896.92
4	$1132.49	$1066.24	$1000	$933.76	$867.51
5	$1159.71	$1079.85	$1000	$920.15	$840.29
6	$1184.92	$1092.46	$1000	$907.54	$815.08
7	$1208.25	$1104.13	$1000	$895.87	$791.75
8	$1229.87	$1114.93	$1000	$885.07	$770.13
9	$1249.88	$1124.94	$1000	$875.06	$750.12
10	$1268.40	$1134.20	$1000	$865.80	$731.60
11	$1285.56	$1142.78	$1000	$857.22	$714.44
12	$1301.44	$1150.72	$1000	$849.28	$698.56
13	$1316.15	$1158.08	$1000	$841.92	$683.85
14	$1329.77	$1164.88	$1000	$835.12	$670.23
15	$1342.38	$1171.19	$1000	$828.81	$657.62
16	$1354.05	$1177.03	$1000	$822.97	$645.95
17	$1364.87	$1182.43	$1000	$817.57	$635.13
18	$1374.88	$1187.44	$1000	$812.56	$625.12
19	$1384.14	$1,192.07	$1000	$807.93	$615.86
20	$1392.73	$1,196.36	$1000	$803.64	$607.27

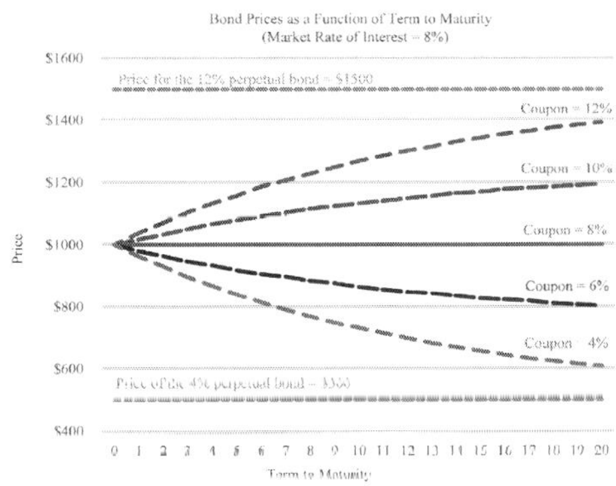

Figure 1. Bond prices as a function of term to maturity: the face value is \$1000, coupon rates are 12%, 10%, 8%, 6%, and 4% respectively (assuming annual payment), and terms to maturity are from 0 to 20 years when the market rate of interest is 8%.

Since the price for a discount bond decreases at a decreasing rate as term to maturity increases, the price path toward maturity date, which is the price movement with a decrease in term to maturity, is increasing with an increasing rate.

To enhance the understanding of the effect of time passage on bond prices, we construct Table 2 to illustrate how the prices of 20-year bonds with different coupon rates converge toward face value as time approaches maturity date.

The results in Table 2 show that the price for the 10% premium bond declines by $4.29 from $1196.36 to $1192.07 in the first year after it is issued (from 20 years remaining until maturity to 19 years remaining until maturity). However, during the very last year immediately before maturity (from one year remaining until maturity to the maturity date), the price declines by $18.52 from $1018.52 to $1000. The results show the price path toward maturity for a premium bond decreases at an increasing rate.

As the bond approaches maturity, the price approaches the face value and the premium (defined as the dollar amount above the face value) declines purely due to time passage without any change in interest rate, we refer to the decline as time decay in bond premium.

Table 2. The convergence of bond prices toward face value as they approaches maturity date. The bonds have coupon rates of 12%, 10%, 8%, 6%, and 4% respectively and a face value of $1000. We assume all the bonds start with an initial term to maturity of 20 years, the market rate of interest is 8%, and interests are paid annually.

Term Remaining Until Maturity	Coupon Rate				
	12%	10%	8%	6%	4%
20	$1392.73	$1196.36	$1000	$803.64	$607.27
19	$1384.14	$1192.07	$1000	$807.93	$615.86
18	$1374.88	$1187.44	$1000	$812.56	$625.12
17	$1364.87	$1182.43	$1000	$817.57	$635.13
16	$1354.05	$1177.03	$1000	$822.97	$645.95
15	$1342.38	$1171.19	$1000	$828.81	$657.62
14	$1329.77	$1164.88	$1000	$835.12	$670.23
13	$1316.15	$1158.08	$1000	$841.92	$683.85
12	$1301.44	$1150.72	$1000	$849.28	$698.56
11	$1285.56	$1142.78	$1000	$857.22	$714.44
10	$1268.40	$1134.20	$1000	$865.80	$731.60
9	$1249.88	$1124.94	$1000	$875.06	$750.12
8	$1229.87	$1114.93	$1000	$885.07	$770.13
7	$1208.25	$1104.13	$1000	$895.87	$791.75
6	$1184.92	$1092.46	$1000	$907.54	$815.08
5	$1159.71	$1079.85	$1000	$920.15	$840.29
4	$1132.49	$1066.24	$1000	$933.76	$867.51
3	$1103.08	$1051.54	$1000	$948.46	$896.92
2	$1071.33	$1035.67	$1000	$964.33	$928.67
1	$1037.04	$1018.52	$1000	$981.48	$962.96
0	$1000.00	$1000.00	$1000	$1000.00	$1000.00

Similarly, the results in Table 2 show that, the price for the 6% premium bond rises by $4.29 from $803.64 to $807.93 in the first year after it is issued (from 20 years remaining until maturity to 19 years remaining until maturity). However, during the very last year right before maturity (from one year remaining until maturity to the maturity date), the price rises by $18.52 from $981.48 to $1000. The results show the price path toward maturity for a discount bond rises at an increasing rate.

As the bond approaches maturity, the price approaches the face value and the discount (defined as the dollar amount below the face value) declines purely due to time passage without any change in interest rate, we refer to the decline as time decay in bond discount. Figure 2 is the graphical representation of Table 2 showing the price path of bonds with different coupon rates as these bonds approach maturity when all the bonds start with an initial term to maturity of 20 years. Note that the bond prices in Figure 1 and the bond prices in Figure 2 are symmetric with respect to the y-axis as the x-axis in Figure 1 is the term to maturity (number of years in ascending order) while the x-axis in Figure 2 is the term remaining until maturity (number of years in descending order).

IMPLICATIONS OF THE FINDINGS

The findings in the paper in and by itself are interesting and important. As we discussed, the change in bond prices can always be decomposed into at least two components: one component due to time passage without taking into account the change in interest rate, and one component due to the change in interest rate. To fully understand bond pricing, it is necessary to understand the driving forces behind both components. The findings in this paper contribute to the understanding of the first component. Moreover, the findings in the paper can also enhance the management of fixed income securities and the communication between asset managers and their clients. As we have shown in the paper, the value of premium bonds declines at an increasing rate as they approach maturity. To respond to the 2008 financial crisis, central banks worldwide have taken programs in the past few years. These programs have driven interest rates to historical lows and many bonds issued before 2008 (which carry coupon rates higher than the current interest rates) have become premium bonds. As these bonds approach maturity in the next few years, their value will decline.

Moreover, because the price of premium bonds declines at an increasing rate, the decline of these bonds' value will accelerate even central banks maintain the current policy unchanged. As most investors do not fully understand the time decay of bond premium, they may unfairly blame asset managers for the decline in bond value even the decline is purely due to the passage of time and the central banks' monetary policies. A better understanding of the price path effect on bond prices will help asset managers manage bond portfolios more effectively and enhance the communications between asset manager and their clients.

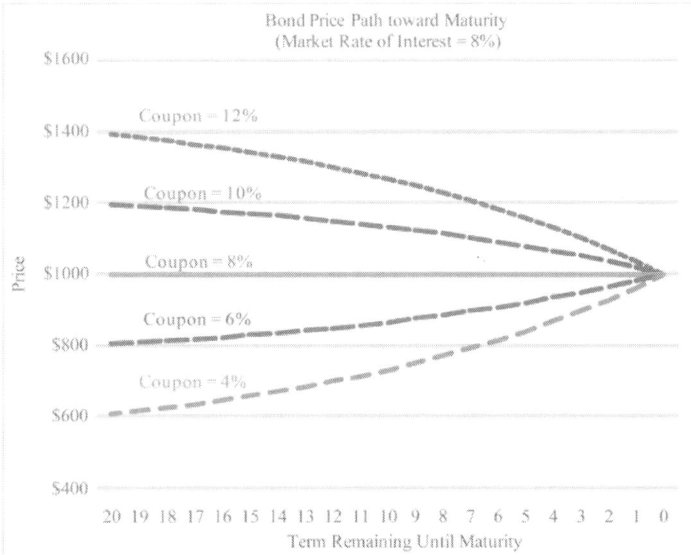

Figure 2. Bond prices as a function of term remaining until maturity: the face value is \$1000, coupon rates are 12%, 10%, 8%, 6%, and 4% respectively (assuming annual payment), and all the bonds start with an initial term to maturity of 20 years when the market rate of interest is 8%.

CONCLUSION

In this paper, we show that as bonds approach maturity, the price of a premium bond will decline to the face value at an increasing rate, and the price of a discount bond will rise to the face value at an increasing rate. As both the prices of premium and discount bonds move toward face value, the premium and the discount will decrease over time even when interest rate remains unchanged. We refer to the decrease purely caused by time passage as time decay in premium and discount. We present a mathematical proof to show that the time decay accelerates over time. We also provide numerical examples and graphical representations to illustrate the time passage effect on bond prices. Finally we discuss the implications of the findings to bond investors and asset managers.

REFERENCES

1. Brigham, E. and Houston, J. (2012) Fundamentals of Financial Management. 7th Edition, South-Western Cengage, Boston.
2. Booth, L., Cleary, S. and Drake, P. (2014) Corporate Finance. John Wiley & Sons, Hoboken.
3. Gitman, L., Joehnk, M. and Smart, S. (2011) Fundamentals of Investing. 11th Edition, Prentice Hall, Upper Saddle River.
4. Keown, A., Martin, J. and Petty, W. (2014) Foundation of Finance. 8th Edition, Pearson, London.
5. Fabozzi, F. (2007) Fixed Income Analysis. 2nd Edition, Wiley, Hoboken.
6. Sharpe, W., Alexander, G. and Bailey, J. (1998) Investments. 6th Edition, Prentice Hall, Upper Saddle River.
7. Lasher, W. (2014) Practical Financial Management. 7th Edition, Cengage Learning, Boston.
8. Merton, R. (1973) Theory of Rational Option Pricing. The Bell Journal of Economics and Management Science, 4, 141-183. http://dx.doi.org/10.2307/3003143
9. Smith, C. (1976) Option Pricing: A Review. Journal of Financial Economics, 3, 3-51. http://dx.doi.org/10.1016/0304-405X(76)90019-2

CHAPTER

7

On a New Corporate Bond Pricing Model with Potential Credit Rating Change and Stochastic Interest Rate

Hong-Ming Yin[1], Jin Liang [2], and Yuan Wu[3]

[1]Department of Mathematics and Statistics, Washington State University, Pullman, WA 99164, USA; hyin@wsu.edu

[2]Department of Mathematics, Tongji University, Shanghai 200092, China

3 School of Statistics and Mathematics, Shanghai Lixin University of Accounting and Finance,

ABSTRACT

In this paper, we consider a new corporate bond-pricing model with credit-rating migration risks and a stochastic interest rate. In the new model, the criterion for rating change is based on a predetermined ratio of the

Citation: Yin, H.-M.; Liang, J.; Wu, Y. "On a New Corporate Bond Pricing Model with Potential Credit Rating Change and Stochastic Interest Rate". J. Risk Financial Manag. 2018, 11, 87. https://doi.org/10.3390/jrfm11040087

corporation's total asset and debt. Moreover, the rating changes are allowed to happen a finite number of times during the life-span of the bond. The volatility of a corporate bond price may have a jump when a credit rating for the bond is changed. Moreover, the volatility of the bond is also assumed to depend on the interest rate. This new model improves the previous existing bond models in which the rating change is only allowed to occur once with an interest-dependent volatility or multi-ratings with constant interest rate. By using a Feynman-Kac formula, we obtain a free boundary problem. Global existence and uniqueness are established when the interest rate follows a Vasicek's stochastic process. Calibration of the model parameters and some numerical calculations are shown.

Keywords:-Corporate bond-pricing model; multi credit rating migration; jump volatility; stochastic interest rate

INTRODUCTION

The study for bond-pricing models is always an attractive topic in financial research fields, as bonds and their derivatives are the largest capital markets in the financial industry Richelson and Richelson (2011). For a for-profit corporation, the price of a bond plays an essential role in determining the valuation of the corporation. This is particularly true for a corporation that requires a large capital such as corporations in mining industry. One of those examples is a corporation conducting crude-oil exploration and production. A corporation's financial stress very much depends on the bond price of the corporation. It is a challenge in the financial markets to develop a comprehensive and dynamic bond-pricing model when an interest rate is uncertain.

There are two important factors that affect a bond price for a corporation. The first is the credit rating given by various rating agencies, which apply on their own measures. The second is the interest rate. For a long time, credit risks were mainly considered default risks, while credit-rating migration risks were usually neglected. After the financial crisis in 2008 and late the European sovereign debt crisis, researchers and bond traders realized the importance of credit-rating migration. One of the main factors for a dramatic change of a bond price is the credit-rating migration including default (see, for examples, Elton et al. (2001); Bessembinder et al. (2018)). The inflection and the liquidity risks in a corporate bond model are the other important factors (see, for examples, Kariya et al. (2016) and Kang and Pflueger (2015)). We would like to point out that the tax is another important factor in bond modelling (see, for example, Kenneth et al. (2018)). However, we will not consider this factor in this paper.

Traditional bond models with potential defaults have two broad categories: structural and reduced-form (or intensity-based). In a structural model, the default event occurs when the firm's value falls below some insolvency threshold. A reduced-form approach concerns exogenous reasons rather than the firm's value itself, where the default time is modelled by introducing a hazard rate (default intensity). The structural model, which will be applied in this paper, was developed first by Merton (1974). He used a predesigned structure to price a corporate bond with a fixed interest rate. The default event may only occur at the maturity of the claim (similar to a European option). The values of such bonds are considered as contingent claims. The value of a corporation, denoted by S(t), follows a geometric Brownian motion:

$$\frac{dS(t)}{S(t)} = \mu_0 dt + \sigma_0 dW_t.$$

where the constants μ_0 and σ_0 represent, respectively, the drift rate and the volatility. In 1976, Black and Cox (1976) extended Merton's model by introducing safety covenants. This extended model allows bondholders to recognize the remaining value of a corporation if its assets fall below a certain given threshold (default ratio). In the modified model Black and Cox (1976), the default of a bond is allowed to occur any time up to the bond maturity (similar to a barrier option). Since those fundamental works by Merton and Black-Cox, various extensions have been considered by many researchers, see, for examples, Jarrow and Turnbull (1995) and Longstaff and Schwartz (1995), Leland and Toft (1996), Briys and De Varenne (1997), Lando (1998), Tsiveriotis and Fernandes (1998). The reader may find more references in Monograph by Dixit and Pindyck (1994). Recently, the authors of Li et al. (2018) used a risk-adjusted inside debt metric and corporate conservatism to study the credit risks and found some interesting results which closely matched data from the financial markets. However, those well-known models do not reflect a potential risk of credit-rating change and the interest-rate change. From the real financial market one can clearly observe that either the rating change or the interest-rate change will dramatically affect the bond price for a corporation.

In the study of credit-rating migration, a major tool is to introduce a Markov chain process. Most researchers mainly adopt a transfer intensity matrix. Through this approach, the reduced form framework is naturally developed for the dynamic process of credit rating migrations, see Das and Tufano (1996); Jarrow et al. (1997); Lando (2000); Thomas et al. (2002); Duffie and Singleton (1999); Hurd and Kuznetsov (2007) and so forth. The transfer intensity matrix usually comes from general statistical data, which do not include any particular firm's assets. However, a firm's asset value plays a key role in credit rating by rating agencies. In this situation, the Markov chain alone cannot fully capture the credit-rating mitigation for the bond price of a corporation. In 2015, Liang and Zeng (2015) used a structural model for pricing a bond where a credit rating may change. They gave a predetermined migration threshold where a firm's asset value is divided into high and low rating regions, under the assumption that the value of a firm follows different stochastic processes in different regions. By using the Feynman-Kac theorem, the model can be reduced to a boundary value problem of a partial differential equation. This PDE method is very different from the traditional approaches in the research field for corporate bond modelling. However, a crucial assumption in Liang-Zeng's model is that the rating migration boundary is predetermined. This assumption is inappropriate in the real financial world. It often depends on the proportion of a firm's debt to the value of the firm. From this point of view, the migration boundary should be unknown. Under a fixed-interest rate, Hu-Liang-Wu in 2016 (Hu et al. 2015; Liang et al. 2016a, 2016b) proposed a new model to reflect this factor. They obtained a free boundary problem for the bond model. The existence and uniqueness for the free boundary problem are established.

As we stated before, when pricing a corporate bond in the real financial market, the interest rate is a very sensitive factor. On the other hand, the interest rate itself is a stochastic process. However, when a stochastic interest rate is considered in the bond pricing model, the corresponding mathematical model becomes extremely difficult to analyze. In 2017, Liang et al. (2017) used Vasicek's model for the stochastic interest rate in the bond pricing model with a potential credit migration. This is a considerable improvement in comparison to existing bond models such as Hu et al. (2015); Liang et al. (2016b). They established the well-posedness for the mathematical model in Liang et al. (2017). Some properties of the free boundary such as the monotonicity of free boundary are also derived. However, there are two shortcomings in the model developed in Liang et al. (2017). The first is that the rating change is only allowed to occur one time during the life-span of the bond, which is not practical in the financial market. The second is that the current rating (known information) is not fully reflected in the model.

In the present paper, we extend the model considered in Liang et al. (2017) to overcome those shortcomings. The new bond model will allow the rating change to happen a finite number of times before the maturity of the bond. The rating change is determined solely by the debt-to-asset ratio. Moreover, by taking the current asset of a corporation and its bond rating into consideration, we assume that the assets of a corporation follows a general geometric Brownian motion. For the new model, we apply the Feynman-Kac theory to obtain a partial differential equation with free boundary. Particularly, we employ some PDE techniques used in Liang et al. (2017) to obtain the global existence and uniqueness when the interest rate follows a Vasicek's process. We would like to mention that in Wu and Liang (2018) recently extended the model in Liang et al. (2017) to the multi-rating case under the assumption of a constant interest rate.

The paper is organized as follows: In Section 2, we propose some basic assumptions for the model with credit-rating migration and stochastic interest rate. In Section 3, by employing the Feynman-Kac theorem, we convert the new model to a free boundary problem. In Section 4, we consider a special case where the interest rate is given by Vasicek's process. The global existence and uniqueness are established. Some concluding remarks are given in Section 5.

THE DERIVATION OF THE NEW MODEL

We begin with some basic assumptions for our bond model.

Let (Ω,F,P) be a complete probability space. We assume that a firm issues only a corporate bond, which is a contingent claim of its value on the space FF.

There are several models to describe the stochastic interest rate. We use a general model which includes Vasicek's model as a special case.

The Interest Rate Model

Suppose that an interest rate r=r(t) follows a general stochastic process:

$$dr(t) = \beta(t,r(t))dt + \sigma_r(t)dW_t^r, \qquad (1)$$

where β(t,s) is a known function, which is typically bounded by US current interest rate and 30-year treasury bond. W_t^r is the standard Brownian motion which generates the filtration $\{F_t\}$ and $\sigma_r(t)$ is the volatility which can also be determined by historical data.

An important example for (1) is Vasicek's model where

$$\beta(t,r(t)) = a(t)(\vartheta(t) - r(t)),$$

where a(t), ϑ (t) and σr(t) are given positive functions. We would like to point out that Vasicek's model is widely accepted in the financial market when parameters a, ϑ and σ_r are chosen properly.

Further Value of a Corporation

Let S(t) denote a firm's value in a risk-neutral world. Suppose that there is a jump for the volatility when the bond rating changes from one rating region to another rating region. The firm's asset value, denoted by S(t), follows the standard geometric Brownian motion:

$$\frac{dS(t)}{S(t)} = \mu dt + \sigma dW_t, \qquad (2)$$

where μ is expected the growth or decay rate of the asset and σ represents the volatility, Wt is the Brownian motion which generates the filtration {Ft}, and the correlation between the interest rate and firm's value is given by

$$Cov(dW_t, dW_t^r) = \rho dt,$$

where $-1<\rho<1$ is a constant.

From the financial market, the expected growth of a corporation with a large debt load strongly depends on r(t). Therefore, a natural assumption is that the expected growth is a function of time, interest rate and the value of the corporation:

$$\mu = \mu(t, r(t), S(t)),$$

where μ could be negative when the cash flow of a corporation becomes negative.

Moreover, the volatility σ is a function of time, interest rate and the value of corporation:

$$\sigma = \sigma(t, r(t), S(t)).$$

We would like to point out that μ and σ are assumed to be deterministic functions, which are determined by using the empirical data in the past. They also carry the up-to-date information for the corporation. One of important features for σ(t,r(t),S(t)) and μ(t,r(t),S(t)) is that there may have a jump where the rating for a bond changes.

For a rating agency, assume that a rating for a corporate's bond is purely determined by the proportion, denoted by $\gamma(t)=\frac{D(t)}{S(t)}$, of the firm's debt D(t) to the asset S(t).

Suppose that there are at most a finite number of possible rating over the finite time horizon [0,T]. Let γ_i,i=1,2,⋯,n, be denoted the critical values of the debt-to-asset ratio with

$$0<\gamma_1<\gamma_2<\cdots<\gamma_n\leqslant 1.$$

Those values $\{\gamma_i\}_{i=1}^{n}$ can be found from the empirical data for all corporations in every industrial section.

We can also use the empirical data to determine the approximate expected return and volatility in each rating range:

$$\begin{aligned}\mu(t,r,S) &= \mu_i(t,r,s), \qquad \gamma_i<\frac{D(t)}{S(t)}<\gamma_{i+1}, i=1,\cdots,n-1,\\ \sigma(t,r,s) &= \sigma_i(t,r,s), \qquad \gamma_i<\frac{D(t)}{S(t)}<\gamma_{i+1}, i=1,\cdots,n-1,\end{aligned}$$

where

$$0<\mu_n\leqslant\cdots\leqslant\mu_1,$$
$$0<\sigma_1\leqslant\cdots\leqslant\sigma_n.$$

Although there is no difference for the order of μ_i and σ_i in the model, one can observe a fact in the financial market that the volatility is larger and growth of the asset is slower when debt-to-asset is increasing.

Please note that there are *n* interfaces, denoted by Γ_t^i, where

$$\frac{D(t)}{S(t)}=\gamma_i, i=1,\cdots,n$$

which are unknown.

We assume that the firm issues only one corporate zero-coupon bond with face value *F*. Denote by Φ_t the discount value of the bond at time *t*, which is the same as D(t) (debt obligation by Merton's assumption). Therefore, at the maturity time *T*, an investor can get Φ_T=min{S(T),F}. Here we assume that the bondholder has the priority to claim the firm's asset than the equity holders.

It is now clear that Φ(s,r,t) is the conditional expectation of present value of payoff value at the maturity subject to s=S(t) and r=r(t):

$$\Phi(s,r,t) = E_{s,t}\left[e^{-\int_t^T r(\tau)d\tau}\min(S_T,F)\Big|s=S(t),r=r(t)\right], \qquad (3)$$

where r(t) and S(t) satisfy the stochastic processes (1) and (2), respectively.

THE MATHEMATICAL MODEL

Due to the potential rating change, the firm's asset may display a jump. However, for a large corporation, we may assume that the bond price is continuous throughout the life-span of the bond. Moreover, we may also assume that Φ(s,y,t) is differentiable in each rating interval.

Let

$$Q_T = \{(s,r,t) : 0 < s < \infty, -\infty < r < \infty, 0 < t \leqslant T\}.$$

Suppose there exists a risk-neutral measure in the probability space (Ω,F,P). By using the Feynman-Kac formula, we see that Φ satisfies the following partial differential equation in Q_T (see Liang et al. (2017) detailed derivation):

$$\frac{\partial \Phi}{\partial t} + \frac{1}{2}\sigma^2 s^2 \frac{\partial^2 \Phi}{\partial s^2} + \sigma_r \sigma \rho s \frac{\partial^2 \Phi}{\partial s \partial r} + \frac{1}{2}\sigma_r^2 \frac{\partial^2 \Phi}{\partial r^2} + rs\frac{\partial \Phi}{\partial s}$$
$$+\beta\frac{\partial \Phi}{\partial r} - r(t)\Phi = 0, \tag{4}$$

where the functions σ will take the value in the corresponding *i*th-rate region.

We would like to point out that the expected rate of return μ does not occur in the model since we used a risk-neutral measure in the derivation.

At the maturity date *T*, the bond value is known:

$$\Phi(s,r,T) = \min\{s,F\}. \tag{5}$$

Also, at the boundary s=0, the price of the bond is also equal to 0:

$$\Phi(0,r,T) = 0. \tag{6}$$

Let

$$\Gamma_T^i = \{(s,r,t) : \Phi(s,r,t) = s\gamma_i\}, i = 1, \cdots, n,$$

be the interface between two adjacent regions with different bond rating. We also use Φ_i to be the value of Φ in *i*th-rating region.

To see the free boundary conditions for Φ on Γ_t^i, we note that the value of the bond is continuous when it passes a rating threshold, i.e.,

$$\Phi_i = \Phi_{i+1} = s\gamma_i, i = 1, 2, \cdots, n-1. \tag{7}$$

Also, as derived in Hu et al. (2015); Liang et al. (2016b), another condition on the free boundary ΓiT,

$$\frac{\partial \Phi_i}{\partial s} = \frac{\partial \Phi_{i+1}}{\partial s}, i = 1, 2, \cdots, n-1. \tag{8}$$

Equations (4)–(8) forms a complete mathematical model for a corporate bond under the assumptions given in Section 2. We would like to point out that it is a mathematical challenge to establish the global existence and uniqueness for the problem (4)–(8) due to discontinuous coefficients in Equation (4) and free boundary ΓiT. However, one can find a numerical solution by using a finite difference method. Due to the scope and page limitation of the paper, we will carry out the numerical simulation for the problem (4)–(8) in a separate paper.

STOCHASTIC INTEREST RATE IN VASICEK MODEL

In this section, we consider a special case where the interest rate is given by Vasicek's process:

$$dr(t) = a(\vartheta - r(t))dt + \sigma_r dW_t. \tag{9}$$

Let P(t,r) be the value of a guaranteed zero-coupon bond with face value 1 at the maturity t=T. From the Feynman-Kac's formula, we see that P(t,r) satisfies the following PDE Hall (1989):

$$\frac{\partial P}{\partial t} + \frac{1}{2} r\sigma_r^2 \frac{\partial^2 P}{\partial r^2} + a(\vartheta - r(t))\frac{\partial P}{\partial r} - rP = 0, \quad r > 0, 0 < t < T, \tag{10}$$

$$P(T,r) = 1, \tag{11}$$

With the Vasicek's model, there is an explicit solution for P(t,r).

Lemma 1.

For the problem (10), there exists a unique solution $P(t,r) \in C^{1,2}([0,T]\times(0,\infty))$:

$$P(t,r) = A(t)e^{B(t)},$$

where

$$A(t) = exp\{\frac{1}{a^2}[B(t)^2 - (T-t)](a^2\vartheta - \frac{\sigma_r^2}{2} - \frac{\sigma_r^2}{4a}B(t)^2)\},$$

$$B(t) = \frac{1}{a}[1 - e^{-a(T-t)}].$$

The proof can be found in Jiang (2005).

Similar to Liang et al. (2017), we introduce a transform by

$$y = \frac{s}{P(r,t)}, \quad V_i(y,t) = \frac{\Phi_i(s,r,t)}{P(r,t)},$$

where

$$\Phi_i(s,r,t) = \Phi(s,r,t), \qquad \frac{\Phi_{i+1}}{\gamma_{i+1}} < s < \frac{\Phi_i}{\gamma_i}.$$

Then after some elementary calculations, we see that $V_i(y,t)$ satisfies

$$\frac{\partial V_i}{\partial t} + \frac{1}{2}\hat{\sigma}_i^2 y^2 \frac{\partial^2 V_i}{\partial y^2} = 0, \quad \frac{V_i}{\gamma_{i+1}} < y < \frac{V_i}{\gamma_i}, \ 0 < t < T, \tag{12}$$

$$V_i(0,t) = 0, \qquad t > 0, \tag{13}$$

$$V_i(y,T) = \min\{y, F\}, \tag{14}$$

where

$$\hat{\sigma}_i = \sqrt{\sigma_i^2 + 2\rho\sigma_i\sigma_r B(t) + \sigma_r^2 B^2(t)}, \quad i = 1, 2, \cdots, n. \tag{15}$$

The interface Γ_T^i, which is the credit rating migration boundary, is defined by

$$\Gamma_T^i = \{(t,y) : y = \frac{V_i}{\gamma_i}\}.$$

For the conditions on free boundary, we use the same assumptions as in Liang et al. (2017): for $i=1,2,\cdots,n-1$ and $(y,t)\in\Gamma_T$,

$$V_i(y,t) = V_{i+1}(y,t) = \gamma_i y, \tag{16}$$

$$\frac{\partial V_i(,t)}{\partial y} = \frac{\partial V_{i+1}(y,t)}{\partial y}. \tag{17}$$

By using the standard change of variables $x=\log y$ and renaming $T-t$ as t, and defining

$\phi(x,t) = V_i(e^x, T-t)$ in ith rating region,

we obtain the following equation:

$$\frac{\partial \phi}{\partial t} - \frac{1}{2}\sigma^2\left(\frac{\partial^2 \phi}{\partial x^2} - \frac{\partial \phi}{\partial x}\right) = 0, \qquad (x,t) \in Q_T \backslash \Gamma_T^i, \tag{18}$$

where

$$Q_T = (-\infty, +\infty) \times (0, T], \quad \Gamma_T^i = \{(x,t) | \phi(x,t) = \gamma_i e^x\} \tag{19}$$

and σ is a function of φ and (x,t), i.e.,

$$\sigma = \hat{\sigma}_i(x,t), \quad \text{if } \gamma_i e^x < \phi < \gamma_{i+1} e^x \tag{20}$$

The constant γi is defined in Section 3.

Without loss of generality, we assume that the payoff face value $F=1$. Now the initial condition is as follows:

$$\phi(x,0) = \phi_0(x) = \min\{e^x, 1\}, \qquad -\infty < x < \infty. \tag{21}$$

The domain will be divided into the different rating region $Q_T(i)$.

We shall prove that these exist n subdomains which are separated by free boundary $x=s_i(t)$, $i=1,\cdots,n$, and

$$Q_T(i) = \{(x,t) : s_i(t) < x < s_{i+1}(t), 0 < t < T\}, \tag{22}$$

Hypothesis 1.

Suppose that $\sigma i(t,r,s)$ *is differentiable on* $[0,T]\times R^2$. *Moreover,*

$$0 < a_1 \leqslant \sigma_i(t,r,s) \leqslant a_2, i = 1,2,\cdots,n.$$

Theorem 1.

Under the assumption Hypothesis 1, for the problem (12)–(17) there exists a unique solution $\phi(x,t) \in C^{2,1}(Q_T \setminus (\cup_{i=1}^{n} \Gamma_T^i)$. *Moreover, there exist n-free boundaries* $x = s_i(t) \in C^{\infty}(0,T], i = 1,2,\cdot,n,$, *such that*

$$\Gamma_T^i = \{(x,t) : \phi(s_i(t),t) = \gamma_i e^{s_i(t)}\}.$$

The mathematical proof for Theorem 1 is rather technical, but can be done by using the same idea as in Liang et al. (2017). We will skip the details here. The interested reader can find a special case when there exists only one free boundary in Liang et al. (2017). We will provide a complete proof in a separate paper.

SIMULATION

Calibration

For a model implementation, the parameter calibration is important. In this model, the parameters are: $r,\gamma_i,\sigma_i, i=1,\cdots,n$.

1. The risk free interest rate r can be directly estimated from the market data. In practice, it can be taken by the rate of long term government bonds
2. $\gamma_i, i=1,\cdots,n$, can be estimated from the credit rating data and the firm's balance sheet, where the proportion of the debt to the firm's value is shown in different rating regions.
3. The volatility of a firm's asset value in different rating regions $\sigma_i, i=1,\cdots,n$, can be calibrated through the market quotes of the firm's stock prices in different rating respectively. The details are shown as follows:

1. Find equity volatilities from stock prices in the market

If the firm is in *i*th rating, S_{ik} is defined as the *k*th day stock price. Then we denote by σ_{ik} as the daily volatility, or the standard variance, of the stock price from *N* observations and an unbiased estimate of the variance rate per day are as follows (i=1,⋯,n):

$$u_i = \frac{1}{N}\sum_{k=1}^{N}\frac{S_{ik}}{S_{ik}-S_{i(k-1)}}, \quad \sigma_{iN}^2 = \frac{1}{N-1}\sum_{k=1}^{N}\left(\frac{S_{ik}}{S_{ik}-S_{i(k-1)}}-u_i\right)^2$$

Then the volatility per annum is $\sigma_{iN}\sqrt{M_i}$, where M_i is the trading days in *i*th rating per annum.

2. Estimate asset volatility from the equity volatility.

Once obtained the equity values and volatilities, we can calibrate the firm's volatility by using Merton's model Merton (1974), where the relationship between firm's volatility σ_V and equity one σ_E is given by the Black-Scholes formula. Then σ_V and V_0 can be estimated through the following equations:

$$E_0 = V_0 N(d_1) - De^{-rT} N(d_2), \quad \sigma_E E_0 = N(d_1)\sigma_V V_0,$$

where

$$d_1 = \frac{\log(V_0/D) + (r + \sigma_V^2)T}{\sigma_V\sqrt{T}}, d_2 = d_1 - \sigma_V\sqrt{T},$$

and D is the amount of debt to be paid at time T, V_0 and E_0 are the firm's value and equity value at the initial time.

A Real Example

We choose MTR Corporation Ltd. as an example, which is listed in Hong Kong and the rating is assigned by Moodys (see Wu and Liang (2018)). The firm's data are quoted in three different ratings. The equity value, credit ratings data, debt, and calculated corresponding volatilities, firm's value are shown in Table 1 (the prices and values are in billions of Hong Kong dollars, data Source: Wind).

Table 1. The values, credit ratings, volatilities and debt of MTR.

Date	Close Price	Rating	σ_E	σ_V	D	E_0	V_0
3 January 2001 ... 30 December 2002	9.14 ... 5.8	Low (A3)	0.25	0.18	18.64	45.35	10.66
2 January 2004 ... 29 December 2005	7.65 ... 12.26	Middle (A1)	0.21	0.15	18.79	40.46	8.88
3 January 2011 ... 31 December 2012	25.33 ... 228.03	High (A1a)	0.17	0.13	41.32	146.22	16.05

The migration coefficients of γ_1=0.37 and γ_2=0.43, are referred from the balance sheet. The risk free rate r=0.035, which is fixed, is used by the rate of long-term treasury bonds that time, and T=6. By numerical calculation of our model, the bond-value function graph $\Phi(x,t)$ is shown in Figure 1.

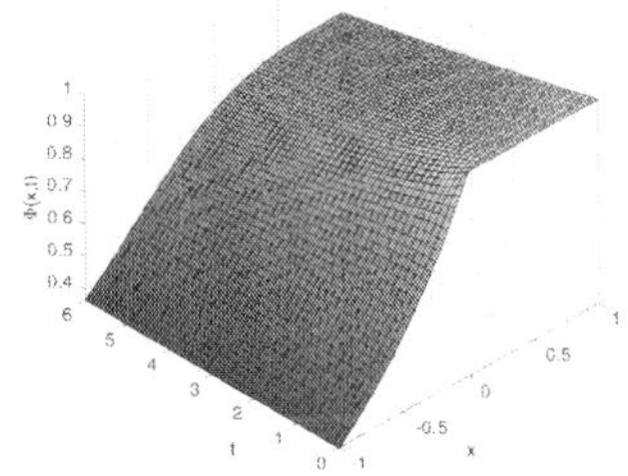

Figure 1. Bond-Value Function Φ vs. firm's log value x and time t when fixed interest rate r.

Migration Free Interfaces

Take interest rate to be stochastic, then the bond value function in a three-dimension region (S,r,t), and the migration interfaces become two-dimension free surfaces. Choose the example which has three rating regions: High, Middle and Low and an interest rate process (9). The free boundaries actually are two free surfaces. Take the parameters with credit rating above are shown:

$$a = 1,\ \vartheta = 0.03,\ \sigma_r = 0.15,\ F = 1, \rho = 0.5,\ T = 6,$$

$$\sigma_L = 0.18,\ \sigma_M = 0.15,\ \sigma_H = 0.13,\ \gamma_1 = 0.37,\ \gamma_2 = 0.43.$$

The numerical calculation for our model shows that there are two free surfaces in Figure 2:

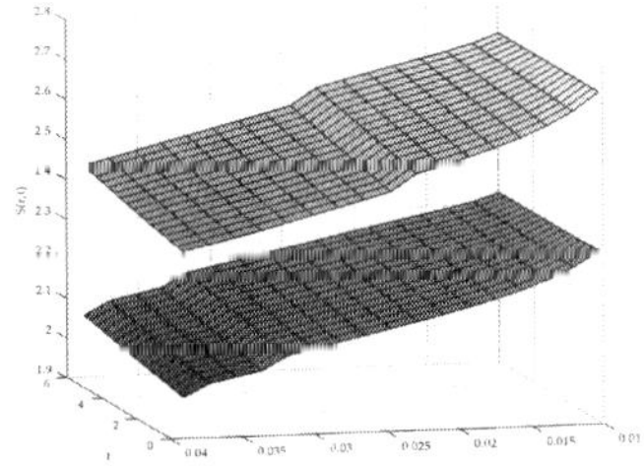

Figure 2. Multi Migration Free Surfaces.

Figure 1 shows that the value of the bond price function in which the domain of the solution has been divided by free boundary surfaces into three regions: high, middle and low rating regions. From **Figure 2**, the value changes quite significantly across the free boundary surfaces. Since s(t) has an opposite tendency with v(τ), where τ=T−t, the free boundary v(t) is increasing as expected.

CONCLUSIONS

In this paper we derived a new model for pricing a corporate bond with potential credit-rating migration risks and a stochastic interest rate. The new model allows a finite number of potential rating changes during the life-span of the bond. The firm's value is assumed to follow a general geometric Brownian motion with jumps for expected growth (or decay) and volatility. When the interest rate follows a Vasicek's process, we used a dimension-deduction technique to establish the well-posedness of the free boundary problem for the new model. Calibration to the model parameters is also discussed with an example. The bond-value function's numerical graph and their migration interfaces are also shown.

This new model extends the previous bond models in two aspects. The first is that a rating change may occur multiple times before the maturity of a bond. The second is that the interest rate itself is uncertain and is assumed to follow a stochastic process. This is particularly important when pricing a long-term corporate bond for corporations due to certain economic cycles. Moreover, our method is based on PDE analysis which is very different from traditional ones in the literature. The new model provides a new insight for evaluating the values of corporations with large debt in different macroeconomic and microeconomic environments. It also provides some additional qualitative information for bond investors when investing in corporate bonds.

ACKNOWLEDGMENTS

The authors would like to thank Song-Ping Zhu from University of Wollongon, Australia, for some discussions during the preparation of the paper. They also would like to express their gratitude to Charles Moore and Brian Yin for their proofreading of the paper. Many thanks further go to the anonymous referees for their comments and suggestions which improve a lot the original version of the paper.

REFERENCES

1. Black, Fischer, and John C. Cox. 1976. Some Effects of Bond Indenture Provisions. Journal of Finance 31: 351–67.
2. Bessembinder, Hendrik, Stacey Jacobsen, William Maxwell, and Kumar Venkataraman. 2018. Venkataraman, Capital Commitment and Illiquidity in Corporate Bonds. The Journal of Finance 73: 1665–61. .
3. Briys, Eric, and François De Varenne. 1997. Valuing Risky Fixed Rate Debt: An Extension. Journal of Financial and Quantitative Analysis 32: 239–49. .
4. Das, Sanjiv, and Peter Tufano. 1996. Pricing credit-sensitive debt when interest rates, credit ratings, and credit spreads are stochastic. Journal of Financial Engineering 5: 161–98
5. Dixit, Avinash K., and Robert S. Pindyck. 1994. Investment under Uncertainty. Princeton: Princeton University Press.
6. Duffie, Darrell, and Kenneth J. Singleton. 1999. Modeling Term Structures of Defaultable Bonds. The Review of Financial Studies 12: 687–720. .
7. Elton, Edwin J., Martin J. Gruber, Deepak Agrawal, and Christopher Mann. 2001. Explaining the Rate Spread on Corporate Bonds. The Journal of Finance 56: 247–77. .
8. Hall, John. 1989. Options, Futures, and Other Derivatives. Upper Saddle River: Prentice-Hall, Inc.
9. Hu, Bei, Jin Liang, and Yuan Wu. 2015. A Free Boundary Problem for Corporate Bond with Credit Rating Migration. Journal of Mathematical Analysis and Applications 428: 896–909. .
10. Hurd, Tom, and Alexey Kuznetsov. 2007. Affine Markov chain models of multifirm credit migration. Journal of Credit Risk 3: 3–29. .
11. Jarrow, Robert, and Stuart Turnbull. 1995. Pricing Derivatives on Financial Securities Subject to Credit Risk. Journal of Finance 50: 53–86. .
12. Jarrow, Robert A., David Lando, and Stuart M. Turnbull. 1997. A Markov model for the term structure of credit risk spreads. Review of Financial Studies 10: 481–523.
13. Jiang, Lishang. 2005. Mathematical Modeling and Methods for Option Pricing. Singapore: World Scientific.
14. Kang, Johnny, and Carolin E. Pflueger. 2015. Inflation Risk in Corporate

Bonds. The Journal of Finance 70: 115–62. .

15. Kariya, Takeaki, Yoko Tanokura, Hideyuki Takada, and Yoshiro Yamamura. 2016. Measuring Credit Risk of Individual Corporate Bonds in US Energy Sector. Asia-Pacific Finan Markets 23: 229–62. .
16. Kenneth, J. Kopecky, Zhichuan Li, Timothy F. Sugrue, and Alan L. Tucker. 2018. Revising M&M with Taxes: An alternative equilibriusm process. International Journal of Financial Studies 6: 1–12.
17. Lando, David. 1998. On Cox Processes and Credit-risky Securities. Review of Derivatives Research 2: 99–120. .
18. Lando, David. 2000. Some elements of rating based credit risk modeling. In Advanced Fixed-Income Valuation Tools. New York: John Wiley & Sons, Inc., pp. 193–215.
19. Leland, Hayne, and Klaus Bjerre Toft. 1996. Optimal capital structure, endogenous bankruptcy, and the term strcuture of credit spreads. Journal of Finance 51: 987–1019. .
20. Li, Zhichuan Frank, Shannon Lin, Shuna Sun, and Alan Tucker. 2018. Risk-adjusted inside debt. Global Finance Journal 35: 12–42. .
21. Liang, Jin, and Chukun Zeng. 2015. Corporate bonds pricing under credit rating migration and structure framework. A Journal of Chinese University (A) 30: 61–70.
22. Liang, Jin, Yuejuan Zhao, and Xudan Zhang. 2016a. Utility Indifference Valuation of Corporate Bond with Credit Rating Migration by Structure Approach. Economic Modelling 54: 339–46. .
23. Liang, Jin, Yuan Wu, and Bei Hu. 2016b. Asymptotic Traveling Wave Solution for a Credit Rating Migration Problem. Journal of Differential Equations 261: 1017–45. .
24. Liang, Jin, Hong-Ming Yin, Xinfu Chen, and Yuan Wu. 2017. On a corporate bond pricing model with credit rating migration risks and stochastic interest rate. Qualitative Finance and Economics 1: 300–19. .
25. Longstaff, Francis, and Eduardo Schwartz. 1995. A Simple Approach to Valuing Risky Fixed and Floating Rate Debt. Journal of Finance 50: 789–819.
26. Merton, Robert C. 1974. On the Pricing of Corporate Debt: The Risk Structure of Interest Rates. Journal of Finance 29: 449–70.
27. Richelson, Hildy, and Stan Richelson. 2011. Bonds: The Unbeaten Path to Secure Investment Growth, 2nd ed. Hoboken: John Wiley &

Sons, Inc.

28. Thomas, Lyn, David Allen, and Nigel Morkel-Kingsbury. 2002. A hidden Markov chain model for the term structure of bond credit risk spreads. International Review of Financial Analysis 11: 311–29. .
29. Tsiveriotis, Kostas, and Chris Fernandes. 1998. Valuing convertible bonds with credit risk. The Journal of Fixed Income 8: 95–102. .
30. Wu, Yuan, and Jin Liang. 2018. A new model and its numerical method to identify multi credit migration boundaries. International Journal of Computer Mathematics 95: 1688–702

CHAPTER

8

Forecasting Term Structure of Interest Rates in Japan

Hokuto Ishii

Graduate School of Economics, Nagoya University, Furo-cho, Chikusa-ku, Nagoya 464-8601, Japan;

ABSTRACT

In this paper, we examined and compared the forecast performances of the dynamic Nelson–Siegel (DNS), dynamic Nelson–Siegel–Svensson (DNSS), and arbitrage-free Nelson–Siegel (AFNS) models after the financial crisis period. The best model for the forecast performance is the DNSS model in the middle and long periods. The AFNS is inferior to the DNS model for long-period forecasting. In U.S. bond markets, AFNS is shown to be

Citation: Ishii, H. "Forecasting Term Structure of Interest Rates in Japan". Int. J. Financial Stud. 2019, 7, 39. https://doi.org/10.3390/ijfs7030039

superior to DNS in the U.S. However, for Japanese data, there is no evidence that the AFNS is superior to the DNS model in the long forecast horizon.

Keywords:- Dynamic Nelson–Siegel; arbitrage-free Nelson–Siegel; affine term structure; forecasting

INTRODUCTION

After the financial crisis, the Japanese economy faced an economic environment unlike it had ever experienced before. In January 2013, under the new governor of the central bank, the Bank of Japan began carrying out a new monetary policy framework. In general, this is called the "unconventional monetary policy".

In this paper, we examine and compare forecast performances for Japanese Government Bonds (JGB) from the financial crisis onwards. To analyze the dynamics of JGB yield, we adopt the Dynamic Nelson–Siegel (DNS), Dynamic Nelson–Siegel–Svensson (DNSS), and Arbitrage-free Nelson–Siegel (AFNS) models. The former two models are curve-fitting models, which do not impose theoretical fundamentals such as arbitrage pricing theory. On the other hand, the last model imposes the arbitrage pricing theory and, thus, it is theoretically more rigorous than the former two models.

To forecast the future dynamics of the yield curves is a key concern of economists, policymakers, and those in the private sector, as the dynamics of the yields are considered to include the expectations of the market.

If we can forecast the dynamics of the yield curve or term structure of interest rates, we may capture future fluctuations of the economic fundamentals. Therefore, forecasting the term structure of interest rates is a major step towards discerning the economic activity of a country.

This study provides one main result: The DNSS model is the best, among the three models, to forecast future yields; it outperforms the random walk model (a benchmark model) for 6- and 12-month forecasts. Moreover, the AFNS model is inferior to the DNS model for long forecast horizons, with evidence suggesting that the arbitrage-free assumption is not appropriate for long-period forecasting; although, Christensen et al. (2011) showed that the AFNS model is superior to the DNS in terms of forecasting performance. The next sub-section reviews the literature. The organization of the paper is at end of the section.

Literature Review

The literature of the curve-fitting model started with the pioneering work of Nelson and Siegel (1987). Diebold and Li (2006) extended the cross-sectional Nelson–Siegel model to the time series. We refer to the Diebold–Li extension of Nelson–Siegel model as the dynamic Nelson-Siegel (DNS) model. The DNS model is a three-factor model. On the other hand, Diebold and Diebold and Rudebusch (2013) introduced a four-factor1 model as an extension of the DNS. We refer to the four-factor model as the dynamic Nelson–Siegel–Svensson (DNSS) model.

A fundamental characteristic of curve-fitting models, such as DNS and DNSS, is that theoretically rigorous restrictions (like the arbitrage-free theory) are not imposed. Although this point has attracted criticism, Gasha et al. (2010) showed that these models have empirically better performances2 than arbitrage-free models; in particular, the affine term structure model, which was pioneered by Vasicek (1977). Cox et al. (1985) and Hull and White (1990) presented representative models that extend upon the Vasicek model. These arbitrage-free models impose the rigorous arbitrage-free theory. After Duffie and Kan (1996) derived the analytical solution of the affine term structure model, arbitrage-free models have rapidly developed (Ang and Piazzesi 2003). Since then, a lot of studies (Rudebusch and Wu 2008; Hördahl et al. 2005; Wu 2006; and Moench, Emanuel 2008. Forecasting the yield curve in a data-rich environment: a no-arbitrage factor augmented VAR approach 2008) have been written on arbitrage-free macro-finance equilibrium models.

Fama (2006) discussed the forecasting of future spot rates by using the forward interest rate. Al-Al-Zoubi (2009) examined a wide range of the interest rate models and reported the model parameter by using the generalized method of moment (GMM). De Rezende and Ferreira (2013) reported the improvement of the out-of-sample forecasts of the Nelson–Siegel class of models by using the quantile autoregression approach. There are a lot of the applied studies for the Nelson—Siegel class of models. Chen and Tsang (2013) applied the DNS model to domestic and foreign interest rate differential. Ishii (2018) extended the original Chen and Tsang (2013) model to the four- and five-factor models.

Christensen et al. (2009, 2011) set the Nelson–Siegel model into an affine term structure framework. Their model is called the arbitrage-free Nelson–Siegel (AFNS) model, which is considered to be both an empirically and theoretically good model. In fact, Christensen et al. (2011) and Diebold and Rudebusch (2013) showed that the AFNS was superior to the DNS for forecasting the performance of U.S. bond markets. Hence, they concluded that the imposition of the arbitrage-free theory improves the DNS.

There are many varieties of AFNS models. As Christensen et al. (2009, 2011) used, some are continuous time models[3] . Others (e.g., Alfaro 2011 and Niu 2012) proposed discrete-time AFNS models. The advantage of the discrete model is that matching the data with the model is easier, such that we can estimate the model easily and simply

A Kalman filter, through maximum likelihood estimation, is often used when estimating the parameters of the DNS and AFNS models. However, Hamilton and Wu (2012) pointed out that the maximum likelihood estimator might be a local estimator and, so, they suggested another method, such as the ordinary least squares (OLS) method. Diebold and Li (2006) and Niu (2012) used the multiple regression method. Following these studies, our study employs the multiple regression method. In the first step, using the non-linear least-squares (NLS) method, the scale parameters of DNS, DNSS, and AFNS are estimated. In the second step, using OLS, the state variables are estimated.

Gurkaynak and Wright (2012) surveyed affine term structure models, including the Nelson–Siegel frameworks. Diebold and Rudebusch (2013) explained the framework of the DNS and the AFNS in detail.

Few studies have investigated JGB yields. One of the contributions of this study is to examine JGB yields and to compare the forecast performance of AFNS with those of DNS and DNSS. The second contribution is to provide evidence that the arbitrage-free assumption is not appropriate for the forecasting of JGB yields, unlike the forecasting of U.S. government bond yields.

The rest of this paper is organized as follows. In Section 2, we show the main theoretical models: DNS, DNSS, and AFNS. Section 3 characterizes the empirical models derived from the theoretical models. In Section 4, we show the data used in the analyses. Section 5 describes the empirical results. Our conclusions are drawn in Section 6.

THEORETICAL MODEL

This section explains three yield curve models: DNS, DNSS, and the discrete- and continuous-time versions of the AFNS model.

The Framework of the DNS and the DNSS Model

In the continuous-time model, the bond price P(t, T) of the maturity (T − t) and the yield y(t, T) at time t are represented as:

$$P(t,T) = e^{-(T-t)y(t,T)}. \tag{1}$$

The instantaneous forward rate is represented as:

$$f(t,T) = -\frac{\partial P(t,T)/\partial(T-t)}{P(t,T)}. \tag{2}$$

Then, the yield is written as:

$$y(t,T) = \frac{1}{T-t}\int_0^{T-t} f(s,T)ds. \tag{3}$$

Nelson and Siegel (1987) proposed the following forward rate:

$$f(t,T) = \beta_1 + \beta_2 e^{-\lambda(T-t)} + \beta_3(T-t)\lambda e^{-\lambda(T-t)}, \tag{4}$$

where β and λ are parameters. This model is a three-factor model. Substituting Equation (4) into Equation (3), we obtain the Nelson–Siegel yield function:

$$y(t,T) = \beta_1 + \left(\frac{1-e^{-\lambda(T-t)}}{\lambda(T-t)}\right)\beta_2 + \left(\frac{1-e^{-\lambda(T-t)}}{\lambda(T-t)} - e^{-\lambda(T-t)}\right)\beta_3. \tag{5}$$

Svensson (1994) extended the three-factor Nelson-Siegel model to a four-factor model. In general, the extended version of the Nelson–Siegel model, which was proposed by Svensson (1994), is called the generalized Nelson–Siegel model. Its forward rate is written as:

$$f(t,T) = \beta_1 + \beta_2 e^{-\lambda_1(T-t)} + \beta_3(T-t)\lambda_1 e^{-\lambda_1(T-t)} + \beta_4(T-t)\lambda_2 e^{-\lambda_2(T-t)}. \tag{6}$$

Substituting Equation (6) into Equation (3), we obtain the Nelson–Siegel–Svensson yield:

$$y(t,T) = \beta_1 + \left(\frac{1-e^{-\lambda_1(T-t)}}{\lambda_1(T-t)}\right)\beta_2 + \left(\frac{1-e^{-\lambda_1(T-t)}}{\lambda_1(T-t)} - e^{-\lambda_1(T-t)}\right)\beta_3 + \left(\frac{1-e^{-\lambda_2(T-t)}}{\lambda_2(T-t)} - e^{-\lambda_2(T-t)}\right)\beta_4. \tag{7}$$

Diebold and Li (2006) proposed the dynamic Nelson–Siegel model by allowing time-varying parameters. For convenience, we denote y(t, T) by $y_t(\tau)$ with τ = T − t. The DNS model is written as:

$$y_t(\tau) = \beta_{1t} + \beta_{2t}\left(\frac{1-e^{-\lambda_t\tau}}{\lambda_t\tau}\right) + \beta_{3t}\left(\frac{1-e^{-\lambda_t\tau}}{\lambda_t\tau} - e^{-\lambda_t\tau}\right), \tag{8}$$

where t denotes time.

Similarly, the DNSS is described as:

$$y_t(\tau) = \beta_{1t} + \beta_{2t}\left(\frac{1-e^{-\lambda_{1t}\tau}}{\lambda_{1t}\tau}\right) + \beta_{3t}\left(\frac{1-e^{-\lambda_{1t}\tau}}{\lambda_{1t}\tau} - e^{-\lambda_{1t}\tau}\right) + \beta_{4t}\left(\frac{1-e^{-\lambda_{2t}\tau}}{\lambda_{2t}\tau} - e^{-\lambda_{2t}\tau}\right). \quad (9)$$

In the next section, we show the empirical models of Equations (8) and (9).

Affine Term Structure Model

In this sub-section, following Duffie and Kan (1996), we show the properties of the affine term structure model. Suppose that N state variables follow a vector autoregression(VAR)(1) process under the physical measure denoted by P. The continuous representation of the VAR (1) process is:

$$dX_t = \kappa^{\mathbb{P}}\left(\theta^{\mathbb{P}} - X_t\right)dt + \Sigma dW_t^{\mathbb{P}}, \quad (10)$$

where κ^P is an N × N matrix, θ^P is an N × 1 matrix, Σ is an N × N matrix, and WP t is a Brownian motion under the physical measure P. This form is called the Ornstein–Uhlenbeck process, as used in Vasicek (1977).

Suppose that the risk premium Γ_t is defined in the essentially affine form, as proposed by Duffee (2002),

$$\Gamma_t = \Gamma_0 + \Gamma_1' X_t, \quad (11)$$

where the scalar Γ_0 is a constant term, and Γ1 is an N × 1 matrix. Under the risk-neutral measure Q, the process $dW_t^{\mathbb{Q}}$ is defined[4] as:

$$dW_t^{\mathbb{Q}} = dW_t^{\mathbb{P}} + \Gamma_t dt. \quad (12)$$

Substituting Equation (12) into Equation (10), the risk-neutral measure process of dX_t is derived as:

$$dX_t = \kappa^{\mathbb{Q}}\left(\theta^{\mathbb{Q}} - X_t\right)dt + \Sigma dW^{\mathbb{Q}}, \quad (13)$$

where κ^Q is an N × N matrix, θ^Q is an N ×1 matrix, and W^Q is a Brownian motion under the risk-neutral measure Q.

Suppose that the instantaneous risk-free rate (r) is also an affine function:

$$r = \rho_0 + \rho_1 X_t, \quad (14)$$

where ρ_0 is a scalar and ρ_1 is a row vector

The zero-coupon bond prices are an exponential affine function:

$$P(t,T) = \exp\left[B(t,T)' X_t + C(t,T)\right], \quad (15)$$

where B(t, T) and C(t, T) are the solutions of the ordinary differential equations:

$$\frac{dB(t,T)}{dt} = \rho_1 + \left(\kappa^{\mathbb{Q}}\right)' B(t,T), \text{ and} \tag{16}$$

$$\frac{dC(t,T)}{dt} = \rho_0 - B(t,T)'\kappa^{\mathbb{Q}}\theta^{\mathbb{Q}} - \frac{1}{2}\sum_{j=1}^{n}\left(\Sigma' B(t,T) B(t,T)\Sigma\right), \tag{17}$$

where n denote the number of the factor.

When B(T, T) = 0 and C(T, T) = 0, the zero-coupon bond price P(t, T) is equal to one. This implies that the bond price corresponds with the par value at maturity, so that the boundary conditions are B(T, T) = 0 and C(T, T) = 0. From the price function of the zero-coupon bond prices, the yield function is written as:

$$y(t,T) = -\frac{\log P(t,T)}{T-t} = -\frac{B(t,T)'}{T-t}X_t - \frac{C(t,T)}{T-t}. \tag{18}$$

This function is also the affine function.

Continuous-Time AFNS Framework

In the previous two sub-sections, we have shown the derivations of DNS models and the affine term structure model. In this sub-section, following Christensen et al. (2011), we derived the AFNS model by combining the DNS model with the affine term structure model. In the three-factor model, the j-th elements of B(t, T) in Equation (18), denoted by B^j , are as follows:

$$B^1(t,T) = -(T-t), \tag{19}$$

$$B^2(t,T) = -\frac{1-e^{-\lambda(T-t)}}{\lambda}, \tag{20}$$

$$B^3(t,T) = (T-t)e^{-\lambda(T-t)} - \frac{1-e^{-\lambda(T-t)}}{\lambda}. \tag{21}$$

Then, Equation (18) is written as:

$$y(t,T) = X_t^1 + \left[\frac{1-e^{-\lambda(T-t)}}{\lambda(T-t)}\right]X_t^2 + \left[\frac{1-e^{-\lambda(T-t)}}{\lambda(T-t)} - e^{-\lambda(T-t)}\right]X_t^3 - \frac{C(t,T)}{T-t}. \tag{22}$$

This equation matches DNS model, except for the last term—which is often called the yield adjustment term or Jensen's inequality term. This term arises from the convexity of the function and plays a role describing the non-linearity of the yield curve.

When we suppose the three-factor model, Equations (16) and (17) are rewritten as:

$$\frac{dB(t,T)}{dt} = \rho_1 + \left(\kappa^{\mathbb{Q}}\right)' B(t,T), \tag{23}$$

$$\frac{dC(t,T)}{dt} = \rho_0 - B(t,T)'\kappa^{\mathbb{Q}}\theta^{\mathbb{Q}} - \frac{1}{2}\sum_{j=1}^{3}(\Sigma' B(t,T)B(t,T)\Sigma). \tag{24}$$

From Equation (23) and the relationship

$$\frac{d}{dt}\left[e^{(\kappa^{\mathbb{Q}})'(T-t)}B(t,T)\right] = e^{(\kappa^{\mathbb{Q}})'(T-t)}\frac{dB(t,T)}{dt} - e^{(\kappa^{\mathbb{Q}})'(T-t)}\left(\kappa^{\mathbb{Q}}\right)'B(t,T), \tag{25}$$

we obtain

$$\frac{d}{dt}\left[e^{(\kappa^{\mathbb{Q}})'(T-t)}B(t,T)\right] = e^{(\kappa^{\mathbb{Q}})'(T-t)}\rho_1. \tag{26}$$

Integrating Equation (26), we get

$$\int_t^T \frac{d}{ds}\left[e^{(\kappa^{\mathbb{Q}})'(T-s)}B(s,T)\right]ds = \int_t^T e^{(\kappa^{\mathbb{Q}})'(T-s)}\rho_1 ds. \tag{27}$$

From the boundary condition B(T, T) = 0, the solution is:

$$B(t,T) = -e^{-(\kappa^{\mathbb{Q}})'(T-s)}\int_t^T e^{(\kappa^{\mathbb{Q}})'(T-s)}\rho_1 ds. \tag{28}$$

$$\kappa^{\mathbb{Q}} = \begin{bmatrix} 0 & 0 & 0 \\ 0 & \lambda & 0 \\ 0 & -\lambda & \lambda \end{bmatrix}, \tag{29}$$

and

$$\rho_0 = 0,\ \rho_1 = \begin{bmatrix} 1 \\ 1 \\ 0 \end{bmatrix}. \tag{30}$$

From the structures (30) and (31),

$$e^{-(\kappa^{\mathbb{Q}})'(T-t)} = \begin{bmatrix} 1 & 0 & 0 \\ 0 & e^{\lambda(T-t)} & 0 \\ 0 & \lambda(T-t)e^{-\lambda(T-t)} & e^{-\lambda(T-t)} \end{bmatrix}. \tag{31}$$

Substituting above results into the ordinary differential equation system (29), we obtain

$$B(t,T) = -\begin{bmatrix} 1 & 0 & 0 \\ 0 & e^{-\lambda(T-t)} & 0 \\ 0 & \lambda(T-t)e^{-\lambda(T-t)} & e^{-\lambda(T-t)} \end{bmatrix} \times \int_t^T \begin{bmatrix} 1 \\ e^{-\lambda(T-t)} \\ -\lambda(T-t)e^{-\lambda(T-t)} \end{bmatrix} ds, \tag{32}$$

and the solution is:

$$B(t,T) = \begin{bmatrix} -(T-t) \\ -\frac{1-e^{-\lambda(T-t)}}{\lambda} \\ (T-t)e^{-\lambda(T-t)} - \frac{1-e^{-\lambda(T-t)}}{\lambda} \end{bmatrix}. \tag{33}$$

Each element of Equation (33) corresponds to Equations (19)–(21), respectively, such that we can set the DNS model into the affine term structure model, except for the term C(t, T).

Next, we consider the yield adjustment term −C(t, T)/(T − t). From Equation (24), we can analytically derive the yield adjustment term. In the risk-neutral measure Q, θ^Q is zero. Therefore, the yield adjustment term is:

$$\frac{C(t,T)}{T-t} = \frac{1}{2(T-t)} \sum_{j=1}^{3} \int_t^T (\Sigma' B(t,T) B(t,T) \Sigma) ds. \tag{34}$$

As for the dynamics of the state transition equations, there are two patterns in the yield adjustment term, depending on whether the factor dynamics are independent or correlated. If the state factor has independent factor dynamics, the volatility matrix Σ is represented as:

$$\Sigma = \begin{bmatrix} \sigma_{11} & 0 & 0 \\ 0 & \sigma_{22} & 0 \\ 0 & 0 & \sigma_{33} \end{bmatrix}. \tag{35}$$

Then, the yield adjustment term is written as:

$$\frac{C(t,T)}{T-t} = \overline{A}\left(\frac{(T-t)^2}{6}\right) + \overline{B}\left(\frac{1}{2\lambda^2} - \frac{1}{\lambda^3}\frac{1-e^{-\lambda T-t}}{T-t} + \frac{1}{4\lambda^3}\frac{1-e^{-2\lambda T-t}}{T-t}\right) + \overline{C}\left(\frac{1}{2\lambda^2} + \frac{1}{\lambda^2}e^{-\lambda(T-t)} - \frac{1}{4\lambda}(T-t)e^{-2\lambda(T-t)} - \frac{3}{4\lambda^2}e^{-2\lambda(T-t)}\frac{2}{\lambda^3}\frac{1-e^{-\lambda(T-t)}}{T-t}\frac{2}{8\lambda^3}\frac{1-e^{-2\lambda(T-t)}}{T-t}\right), \tag{36}$$

where:

$$\overline{A} = \sigma_{11}^2 + \sigma_{12}^2 + \sigma_{13}^2,$$

$$\overline{B} = \sigma_{21}^2 + \sigma_{22}^2 + \sigma_{23}^2,$$

$$\overline{C} = \sigma_{31}^2 + \sigma_{32}^2 + \sigma_{33}^2.$$

On the other hand, if the factor dynamics are correlated, the volatility matrix Σ is represented as[6]:

$$\Sigma = \begin{bmatrix} \sigma_{11} & \sigma_{12} & \sigma_{13} \\ \sigma_{21} & \sigma_{22} & \sigma_{23} \\ \sigma_{31} & \sigma_{32} & \sigma_{33} \end{bmatrix}. \tag{37}$$

$$\frac{C(t,T)}{T-t} = \overline{A}\left(\frac{(T-t)^2}{6}\right) + \overline{B}\left(\frac{1}{2\lambda^2} - \frac{1}{\lambda^3}\frac{1-e^{-\lambda(T-t)}}{T-t} + \frac{1}{4\lambda^3}\frac{1-e^{-2\lambda(T-t)}}{T-t}\right) + \overline{C}\left(\frac{1}{2\lambda^2} + \frac{1}{\lambda^2}e^{-\lambda(T-t)} - \frac{1}{4\lambda}(T-t)e^{-2\lambda(T-t)} - \frac{3}{4\lambda^2}e^{-2\lambda(T-t)} - \frac{2}{\lambda^3}\frac{1-e^{-\lambda(T-t)}}{T-t} + \frac{5}{8\lambda^3}\frac{1-e^{-2\lambda(T-t)}}{T-t}\right) + \overline{D}\left(\frac{1}{2\lambda}(T-t) + \frac{1}{\lambda^2}e^{-\lambda(T-t)} - \frac{1}{\lambda^3}\frac{1-e^{-\lambda(T-t)}}{T-t}\right) + \overline{E}\left(\frac{3}{\lambda^2}e^{-\lambda(T-t)} + \frac{1}{2\lambda}(T-t) + \frac{1}{\lambda}(T-t)e^{-\lambda(T-t)} - \frac{3}{\lambda^3}\frac{1-e^{-\lambda(T-t)}}{T-t}\right) + \overline{F}\left(\frac{1}{\lambda^2} + \frac{1}{\lambda^2}e^{-\lambda(T-t)} - \frac{1}{2\lambda^2}e^{-2\lambda(T-t)} - \frac{3}{\lambda^3}\frac{1-e^{-\lambda(T-t)}}{T-t} + \frac{3}{4\lambda^3}\frac{1-e^{-\lambda(T-t)}}{T-t}\right), \tag{38}$$

where

$$\overline{D} = \sigma_{11}\sigma_{21} + \sigma_{12}\sigma_{22} + \sigma_{13}\sigma_{23},$$

$$\overline{E} = \sigma_{11}\sigma_{31} + \sigma_{12}\sigma_{32} + \sigma_{13}\sigma_{33},$$

$$\overline{F} = \sigma_{21}\sigma_{31} + \sigma_{22}\sigma_{32} + \sigma_{23}\sigma_{33}.$$

The yield adjustment term depends only on the time of maturity τ (= T – t). Therefore, it is a constant value with time t. Moreover, as the maturity becomes larger, the value of the yield adjustment term is larger.

Discrete-Time AFNS

As it is easier to match data with a discrete AFNS model, rather than with the continuous version, we derive the discrete-time AFNS model, following Niu (2012). Under the physical measure P, the dynamics of the state variables are written as:

$$X_t = \mu^{\mathbb{P}} + A^{\mathbb{P}} X_{t-1} + \eta_t^{\mathbb{P}}, \tag{39}$$

where $\eta_t^{\mathbb{P}} \overset{\mathbb{P}}{\sim} N(0,Q)$. Under the risk-neutral measure Q, the dynamics of the state variables are written as:

$$X_t = \mu^{\mathbb{Q}} + A^{\mathbb{Q}} X_{t-1} + \eta_t^{\mathbb{Q}}, \tag{40}$$

where $\eta_t^{\mathbb{Q}} \sim N(0,Q)$. The discrete versions7 of Equations (23) and (24) are represented as:

$$B(t,T)' = -\rho_1' \sum_{k=0}^{T-t-1} \left(\Phi^{\mathbb{Q}}\right)^k \text{ and} \tag{41}$$

$$C(t,T) = C(t,T-1) + C(t,T-1)'\theta^{\mathbb{Q}} + \frac{1}{2}C(t,T-1)QC(t,T-1), \tag{42}$$

$$\rho_1' = \begin{bmatrix} 1 \\ \frac{1-e^{-\lambda}}{\lambda} \\ \frac{1-e^{-\lambda}}{\lambda} - e^{-\lambda} \end{bmatrix}, \tag{43}$$

$$\left(\Phi^{\mathbb{Q}}\right)^k = \begin{bmatrix} 1 & 0 & 0 \\ 0 & e^{-\lambda} & \lambda e^{-\lambda} \\ 0 & 0 & e^{-\lambda} \end{bmatrix}^k = \begin{bmatrix} 1 & 0 & 0 \\ 0 & e^{-k\lambda} & k\lambda e^{-k\lambda} \\ 0 & 0 & e^{-k\lambda} \end{bmatrix}. \tag{44}$$

From Equations (41) and (44), we obtain:

$$B(t,T)' = -\rho_1' \sum_{k=0}^{T-t-1} \left(\Phi^{\mathbb{Q}}\right)^k$$
$$= \begin{bmatrix} -n \\ -\frac{1-e^{-\lambda n}}{\lambda} \\ ne^{-\lambda n} - \frac{1-e^{-\lambda n}}{\lambda} \end{bmatrix}. \tag{45}$$

We have that Equation (45) is the discrete version of Equation (33). Moreover, in the risk-neutral measure $\mathbb{Q}, \theta^{\mathbb{Q}} = 0$. Therefore, the yield adjustment term of the discrete version is written as:

$$C(t,T) = C(t,T-1) + \frac{1}{2}C(t,T-1)QC(t,T-1). \tag{46}$$

Substituting Equations (43) and (44) into Equation (19), we obtain the discrete-time AFNS model as:

$$y(t,T) = X_t^1 + \left[\frac{1-e^{-\lambda(T-t)}}{\lambda(T-t)}\right]X_t^2 + \left[\frac{1-e^{-\lambda(T-t)}}{\lambda(T-t)} - e^{-\lambda(T-t)}\right]X_t^3 - \frac{C(t,T)}{T-t}, \tag{47}$$

where C(t, T) is given by Equation (46).

Until now, we have shown theoretical models of DNS, DNSS, and AFNS (continuous- and discrete-time versions). In the next section, we show empirical models of Equations (8), (9), and (45). Moreover, we compare the forecast performances of the yields.

EMPIRICAL MODELS

For the purpose of evaluating forecasting performance of the three yield curve models, it is convenient to represent the empirical counterparts of the theoretical models explicitly.

As we cannot observe the state variables, we must estimate them first. As Piazzesi (2010) used, the most frequently used way of estimating state variables is to use the principal component method. As Litterman and Scheinkman (1991) first showed, it is usually considered that the yield curve is explained almost by the first principal components. Moreover, the first, second, and third components are usually identified as the level, slope, and curvature of the yield curve, respectively.

In this paper, we supposed that the transition equation of the state variables followed a VAR (1) process, and that the state variables were uncorrelated or correlated with each other. The uncorrelated transition equation of the state variables is represented as:

$$\begin{bmatrix} X_t^{1,DNS} \\ X_t^{2,DNS} \\ X_t^{3,DNS} \end{bmatrix} = (I-A)\mu + \begin{bmatrix} A_{11} & 0 & 0 \\ 0 & A_{22} & 0 \\ 0 & 0 & A_{33} \end{bmatrix} \begin{bmatrix} X_{t-1}^{1,DNS} \\ X_{t-1}^{2,DNS} \\ X_{t-1}^{3,DNS} \end{bmatrix} + \begin{bmatrix} \eta_t^1 \\ \eta_t^2 \\ \eta_t^3 \end{bmatrix}, \tag{48}$$

where μ is a vector of means. The variance–covariance matrix of the stochastic shocks is represented as:

$$Q^{DNS} = \begin{bmatrix} Q_{11} & 0 & 0 \\ 0 & Q_{22} & 0 \\ 0 & 0 & Q_{33} \end{bmatrix}. \tag{49}$$

From Equation (8), the empirical model of DNS is represented as:

$$\begin{bmatrix} y_t(\tau_1) \\ y_t(\tau_2) \\ \vdots \\ y_t(\tau_N) \end{bmatrix} = \begin{bmatrix} 1 & \frac{1-e^{-\lambda\tau_1}}{\lambda\tau_1} & \frac{1-e^{-\lambda\tau_1}}{\lambda\tau_1} - e^{-\lambda\tau_1} \\ 1 & \frac{1-e^{-\lambda\tau_2}}{\lambda\tau_2} & \frac{1-e^{-\lambda\tau_2}}{\lambda\tau_2} - e^{-\lambda\tau_2} \\ \vdots & \vdots & \vdots \\ 1 & \frac{1-e^{-\lambda\tau_N}}{\lambda\tau_N} & \frac{1-e^{-\lambda\tau_N}}{\lambda\tau_N} - e^{-\lambda\tau_N} \end{bmatrix} \begin{bmatrix} X_t^{1,DNS} \\ X_t^{2,DNS} \\ X_t^{3,DNS} \end{bmatrix} + \begin{bmatrix} \varepsilon_t(\tau_1) \\ \varepsilon_t(\tau_2) \\ \vdots \\ \varepsilon_t(\tau_N) \end{bmatrix}, \tag{50}$$

where $\varepsilon_t(\tau_i)$ is the measurement error and i.i.d. white noise and τ_N denotes a maturity N. Similarly, we consider the uncorrelated state dynamics for the DNSS,

$$\begin{bmatrix} X_t^{1,DNSS} \\ X_t^{2,DNSS} \\ X_t^{3,DNSS} \\ X_t^{4,DNSS} \end{bmatrix} = (I-A)u + \begin{bmatrix} A_{11} & 0 & 0 & 0 \\ 0 & A_{22} & 0 & 0 \\ 0 & 0 & A_{33} & 0 \\ 0 & 0 & 0 & A_{44} \end{bmatrix} \begin{bmatrix} X_{t-1}^{1,DNSS} \\ X_{t-1}^{2,DNSS} \\ X_{t-1}^{3,DNSS} \\ X_{t-1}^{4,DNSS} \end{bmatrix} + \begin{bmatrix} \eta_t^1 \\ \eta_t^2 \\ \eta_t^3 \\ \eta_t^4 \end{bmatrix}, \tag{51}$$

where the variance–covariance matrix of the stochastic shocks is represented as:

$$Q^{DNSS} = \begin{bmatrix} Q_{11} & 0 & 0 & 0 \\ 0 & Q_{22} & 0 & 0 \\ 0 & 0 & Q_{33} & 0 \\ 0 & 0 & 0 & Q_{44} \end{bmatrix}. \tag{52}$$

From Equation (9), the empirical model of DNSS is represented as:

$$\begin{bmatrix} y_t(\tau_1) \\ y_t(\tau_2) \\ \vdots \\ y_t(\tau_N) \end{bmatrix} = \begin{bmatrix} 1 & \frac{1-e^{-\lambda_1\tau_1}}{\lambda_1\tau_1} & \frac{1-e^{-\lambda_1\tau_1}}{\lambda\tau_1} - e^{-\lambda_1\tau_1} & \frac{1-e^{-\lambda_2\tau_1}}{\lambda_2\tau_1} - e^{-\lambda_2\tau_1} \\ 1 & \frac{1-e^{-\lambda_1\tau_2}}{\lambda_1\tau_2} & \frac{1-e^{-\lambda_1\tau_2}}{\lambda_1\tau_2} - e^{-\lambda_1\tau_2} & \frac{1-e^{-\lambda_2\tau_2}}{\lambda_2\tau_2} - e^{-\lambda_2\tau_2} \\ \vdots & \vdots & \vdots & \vdots \\ 1 & \frac{1-e^{-\lambda_1\tau_N}}{\lambda_1\tau_N} & \frac{1-e^{-\lambda_1\tau_N}}{\lambda_1\tau_N} - e^{-\lambda_1\tau_N} & \frac{1-e^{-\lambda_2\tau_N}}{\lambda_2\tau_N} - e^{-\lambda_2\tau_N} \end{bmatrix} \begin{bmatrix} X^{1,DNSS} \\ X^{2,DNSS} \\ X^{3,DNSS} \\ X^{4,DNSS} \end{bmatrix} + \begin{bmatrix} \varepsilon_t(\tau_1) \\ \varepsilon_t(\tau_2) \\ \vdots \\ \varepsilon_t(\tau_N) \end{bmatrix}. \tag{53}$$

From Equation (39), suppose that the state transition equation of AFNS also follows a VAR (1) process, as in DNS

$$\begin{bmatrix} X_t^{1,AFNS} \\ X_t^{2,AFNS} \\ X_t^{3,AFNS} \end{bmatrix} = (I-A)\mu + \begin{bmatrix} A_{11} & 0 & 0 \\ 0 & A_{22} & 0 \\ 0 & 0 & A_{33} \end{bmatrix} \begin{bmatrix} X_{t-1}^{1,AFNS} \\ X_{t-1}^{2,AFNS} \\ X_{t-1}^{3,AFNS} \end{bmatrix} + \begin{bmatrix} \eta_t^1 \\ \eta_t^2 \\ \eta_t^3 \end{bmatrix}, \quad (54)$$

where the stochastic shock variance–covariance matrix is represented as:

$$Q^{AFNS} = \begin{bmatrix} Q_{11} & 0 & 0 \\ 0 & Q_{22} & 0 \\ 0 & 0 & Q_{33} \end{bmatrix}. \quad (55)$$

From Equation (45), the empirical model of AFNS is represented as:

$$\begin{bmatrix} y_t(\tau_1) \\ y_t(\tau_2) \\ \vdots \\ y_t(\tau_N) \end{bmatrix} = \begin{bmatrix} 1 & \frac{1-e^{-\lambda\tau_1}}{\lambda\tau_1} & \frac{1-e^{-\lambda\tau_1}}{\lambda\tau_1} - e^{-\lambda\tau_1} \\ 1 & \frac{1-e^{-\lambda\tau_2}}{\lambda\tau_2} & \frac{1-e^{-\lambda\tau_2}}{\lambda\tau_2} - e^{-\lambda\tau_2} \\ \vdots & \vdots & \vdots \\ 1 & \frac{1-e^{-\lambda\tau_N}}{\lambda\tau_N} & \frac{1-e^{-\lambda\tau_N}}{\lambda\tau_N} - e^{-\lambda\tau_N} \end{bmatrix} \begin{bmatrix} X_t^{1,AFNS} \\ X_t^{2,AFNS} \\ X_t^{3,AFNS} \end{bmatrix} - \frac{C(t,T)}{T-t} + \begin{bmatrix} \varepsilon_t(\tau_1) \\ \varepsilon_t(\tau_2) \\ \vdots \\ \varepsilon_t(\tau_N) \end{bmatrix}. \quad (56)$$

There is a yield adjustment term C(t, T)/(T − t) in this equation, which generates a different forecast than the DNS model. Simply stated, the arbitrage-free condition imposes this term, which is a feature of the AFNS and affine term structure models that distinguishes them from curve-fitting models.

Next, we considered the correlated transition equation of the state variables. The representations of the state transition equation of the correlated state variables are as follows:

$$A^{DNS} = \begin{bmatrix} A_{11} & A_{12} & A_{13} \\ A_{21} & A_{22} & A_{23} \\ A_{31} & A_{32} & A_{33} \end{bmatrix}, \quad (57)$$

$$A^{DNSS} = \begin{bmatrix} A_{11} & A_{12} & A_{13} & A_{14} \\ A_{21} & A_{22} & A_{23} & A_{23} \\ A_{31} & A_{32} & A_{33} & A_{34} \\ A_{41} & A_{42} & A_{43} & A_{44} \end{bmatrix}, \quad (58)$$

$$A^{AFNS} = \begin{bmatrix} A_{11} & A_{12} & A_{13} \\ A_{21} & A_{22} & A_{23} \\ A_{31} & A_{32} & A_{33} \end{bmatrix}. \quad (59)$$

The variance-covariance matrixes of η_t are written as:

$$Q^{DNS} = \begin{bmatrix} Q_{11} & Q_{12} & Q_{13} \\ Q_{21} & Q_{22} & Q_{23} \\ Q_{31} & Q_{32} & Q_{33} \end{bmatrix}, \quad (60)$$

$$Q^{DNSS} = \begin{bmatrix} Q_{11} & Q_{12} & Q_{13} & Q_{14} \\ Q_{21} & Q_{22} & Q_{23} & Q_{23} \\ Q_{31} & Q_{32} & Q_{33} & Q_{34} \\ Q_{41} & Q_{42} & Q_{43} & Q_{44} \end{bmatrix}, \quad (61)$$

$$Q^{AFNS} = \begin{bmatrix} Q_{11} & Q_{12} & Q_{13} \\ Q_{21} & Q_{22} & Q_{23} \\ Q_{31} & Q_{32} & Q_{33} \end{bmatrix}. \quad (62)$$

Even though each state variable is correlated, the empirical DNS, DNSS, and AFNS models have the same representation. In summary, we can write the above equation simply, in matrix representation.

Each state transition equation of the DNS, DNSS, and AFNS models, in matrix form, is represented as:

$$X_t = (I - A)\mu + AX_{t-1} + \mathrm{H}_t, \quad (63)$$

where X_t is a 3 × 1 matrix (DNS and AFNS) or 4× 1 matrix (DNSS), A is a state transition matrix of 3 × 3 (DNS and AFNS) or 4 × 4 (DNSS), H_t is a N × 1 matrix such that Ht ~ N(0, Q), and Q is a variance–covariance matrix of 3 × 3 (DNS and AFNS) or 4 × 4 (DNSS). In matrix representation, the empirical DNS and DNSS models are represented as:

$$Y_t = BX_t +_t, \quad (64)$$

where Y_t is a N × 1 matrix, B_t is a matrix of 3 × N (DNS) or 4 × N (DNSS), and t is a N × 1 matrix such that t ~ N(0, I).

On the other hand, the empirical model of AFNS is represented as:

$$Y_t = BX_t + C +_t, \quad (65)$$

where C is the yield adjustment term, which is a N ×1 matrix corresponding to the term −C(t, T)/(T −t) in Equation (23) or (45)

ESTIMATION

In this section, we explained the details of the data used in the empirical analysis and report the estimated parameters and the state variables of the DNS, DNSS, and AFNS models. We employed monthly JGB zero-coupon yield from September 2009 to June 2015. The zero-yields were calculated8 using the treasury discount bill and the prevailing interest rates, published by the Ministry of Finance. The maturities are 3, 6, 12, 24, 36, 48, 60, 72, 84, 96, 108, 120, 180, 240, 300, 360, and 480 months. Figure 1 is the JGB zero-yield curve (September 2009 to June 2015).

In the first step, we estimated the scale parameters λ, λ1, and λ2 for the DNS and DNSS models using non-linear least-squares. In the second step, we estimated the state variables for the DNS and DNSS models using the estimated parameters λ, λ_1, and λ_2, and OLS. Moreover, we estimated the transition matrix A and the variance–covariance matrix Q for each independent and correlated state variable. The estimated state variables for DNS are invalid for using for the AFNS, since the constant term includes a bias. Using the estimated variance–covariance matrix, we could obtain the values of −C(t, T)/(T − t) from Equation (42). Subtracting the yield adjustment term −C(t, T)/(T − t) from the yields, we excluded the constant term bias, so that we could estimate the state variable for the AFNS model.

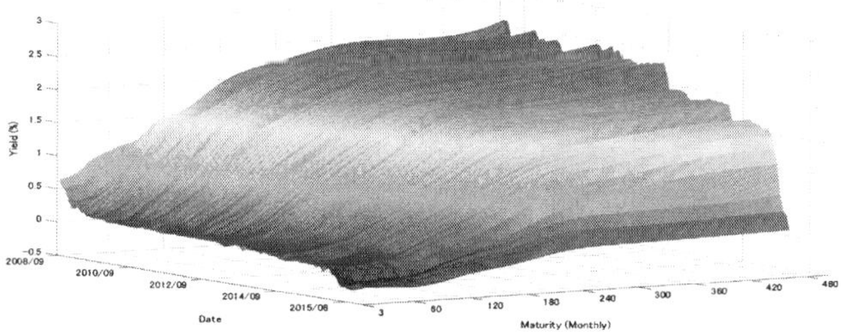

Figure 1. Japanese government bond Zero-Yield curve (2008/09–2015/06). Note: Figure 1 plotted the Japanese government bond Zero-Yield curve. We used the yields of maturities of 3, 6, 12, 24, 36, 48, 60, 72, 84, 96, 108, 120, 180, 240, 300, 360, and 480 months, and the term was from September 2009 to June 2015 monthly data. To calculate the zero-yield of the Japanese government bond, the treasury bill and the prevailing interest rates, which are published by the Japanese Ministry of Finance were used. To interpolate of the missing maturities, the cubic spline was used.

As the DNS and DNSS are time-series models, we approximated each parameter, using nonlinear least-squares, at each time point. The estimated DNS and DNSS models are written as:

$$\hat{y}_t(\tau) = \hat{\beta}_{1,t} + \hat{\beta}_{2,t}\left(\frac{1-e^{-\hat{\lambda}_t\tau}}{\hat{\lambda}_t\tau}\right) + \hat{\beta}_{3,t}\left(\frac{1-e^{-\hat{\lambda}_t\tau}}{\hat{\lambda}_t\tau} - e^{-\hat{\lambda}_t\tau}\right), \tag{66}$$

$$\hat{y}_t(\tau) = \hat{\beta}_{1,t} + \hat{\beta}_{2,t}\left(\frac{1-e^{-\hat{\lambda}_{1,t}\tau}}{\hat{\lambda}_{1,t}\tau}\right) + \hat{\beta}_{3,t}\left(\frac{1-e^{-\hat{\lambda}_{1,t}\tau}}{\hat{\lambda}_{1,t}\tau} - e^{-\hat{\lambda}_{1,t}\tau}\right) + \hat{\beta}_{4,t}\left(\frac{1-e^{-\hat{\lambda}_{2,t}\tau}}{\hat{\lambda}_{2,t}\tau} - e^{-\hat{\lambda}_{2,t}\tau}\right), \tag{67}$$

In the DNS and DNSS models, the parameters $\hat{\lambda}$, $\hat{\lambda}_1$, and $\hat{\lambda}_2$ changed over time. However, in previous studies (Diebold and Li (2006); Christensen et al.

(2011), and Diebold and Rudebusch (2013)), the parameters were defined to be constant throughout time, as the fluctuation of the values is small in each time point and to ensure the stability of the estimation results. Therefore, we adopted the averages of $\hat{\lambda}_t$, $\hat{\lambda}_{1,t}$, and $\hat{\lambda}_{2,t}$ as the parameters λ, λ_1, and λ_2, respectively. From the estimation, we obtained the estimated parameters11 λ = 0.004, λ_1 = 0.016, and λ2 = 0.012. Moreover, suppose that the value λ of the DNS model was identical to λ in the AFNS model[12]. From the estimated λ, λ_1, and λ_2, the factor loadings on β_1, β_2, β_3, and β_4 of the DNS and DNSS models were calculated. Then, X_1, X_2, X_3, and X_4 was simply a rewriting of each β.The dynamics of the factor loadings through the maturity τ are described in Figure 2. Figure 2A shows the dynamics of the factor loadings of the DNS model and Figure 2B shows the dynamics of those of the DNSS model. The first factor loading is a constant through the maturity. The second factor loading starts at 1 and goes to 0 gradually, so that it affects short-term yields. The third factor loading starts at 0 and rises in the medium-term, after which it goes to 0 again. The fourth factor loading also starts 0. Its dynamics are initially the same as the third factor. However, the fourth factor loading affects longer-term yields than the third factor loading. As we can see from the above results, the second factor loading was a short-term factor, the third was a medium-term factor, and the fourth was a medium- to long-term factor.

(A)

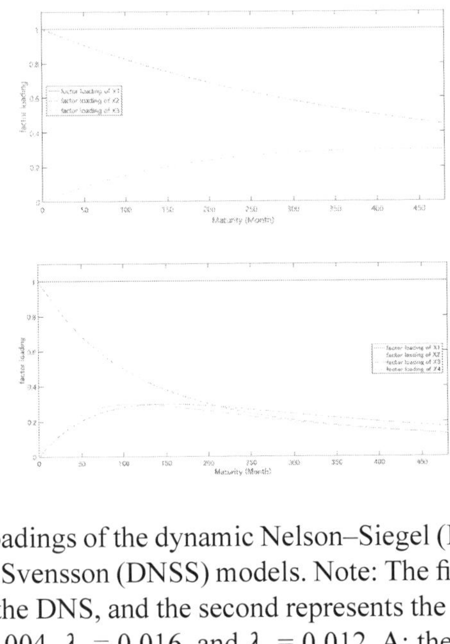

Figure 2. Factor loadings of the dynamic Nelson–Siegel (DNS) and the dynamic Nelson–Siegel– Svensson (DNSS) models. Note: The first plot represents the factor loadings of the DNS, and the second represents the factor loadings of the DNSS when $\lambda = 0.004$, $\lambda_2 = 0.016$, and $\lambda_3 = 0.012$. A: the dynamics of the factor loadings of the DNS model; B: the dynamics of those of the DNSS model.

Estimation of Latent State Variables

In the previous sub-section, we specified the values of λ, λ_1, and λ_2. In this sub-section, we estimated the unobservable state variables of the DNS, DNSS, and AFNS models. However, the AFNS model includes a constant term. Hence, we cannot use the state variables of the DNS model13 for the AFNS model, due to the constant term bias. The estimation method for the AFNS model was described later.

Here, we carried on the estimation of the second step. If we estimated the parameters without separating between the first step and the second step (i.e., using one-step estimation), we needed to use a state space estimation, such as a Kalman filter. This method imposes a large burden numerically and it is possible that the estimation result is only locally convergent. On the other hand, if we used the multiple regression method, once the value of λ was determined, we could estimate the state variables easily using OLS and, so, we separated the estimation: The first step[14] estimated the scale parameter λ, and the second step estimated the state variables using the value obtained in the first step.

First, we considered the estimation of the state variables in the DNS and DNSS models. Taking the values of λ, λ_1, and λ_2 as above and substituting them into Equations (66) and (67), the parameters $\hat{\beta}_{1,t}, \hat{\beta}_{2,t}, \hat{\beta}_{3,t}$, and $\hat{\beta}_{4,t}$ were estimated using ordinary least-squares at any one point. Moreover, since the estimated $\hat{\beta}_{i,t}$ corresponds to the unobservable variables at each time-point, the time-series of each estimated parameter was regarded as the time series of the state variables. Therefore, we redefined the parameters, $(\hat{\beta}_{1,t}^{DNS}, \hat{\beta}_{2,t}^{DNS}, \hat{\beta}_{3,t}^{DNS})$ and $(\hat{\beta}_{1,t}^{DNSS}, \hat{\beta}_{2,t}^{DNSS}, \hat{\beta}_{3,t}^{DNSS}, \hat{\beta}_{4,t}^{DNSS})$ as $(\hat{X}_{1,t}^{DNS}, \hat{X}_{2,t}^{DNS}, \hat{X}_{3,t}^{DNS})$ and $(\hat{X}_{1,t}^{DNSS}, \hat{X}_{2,t}^{DNSS}, \hat{X}_{3,t}^{DNSS}, \hat{X}_{4,t}^{DNSS})$.

Next, we explained how we estimated the state variables in AFNS. Although the number of the state variables was the same as the DNS model, the AFNS includes the constant term C, as shown in Equation (63). Therefore, we could not directly apply the estimated state variables in the DNS to the AFNS, as such a method causes the constant term bias. To correct the bias, we needed to estimate the state variables after subtracting the constant term from the yield.

Defining $\hat{z}_t(\tau) \equiv \hat{y}_t(\tau) - \hat{C}$, we have:

$$\hat{z}_t(\tau) = \hat{X}_{1,t} + \hat{X}_{2,t}\left(\frac{1-e^{-\lambda\tau}}{\lambda\tau}\right) + \hat{X}_{3,t}\left(\frac{1-e^{-\lambda_t\tau}}{\lambda\tau} - e^{-\lambda\tau}\right), \tag{68}$$

from Equation (47). We obtained the estimate C^ from Equation (46), assuming that the variance–covariance matrix Q of the AFNS model was the same as that of the DNS model. As we assumed that the parameter λ of the AFNS model should be the same as that of the DNS model, we could apply the OLS method to Equation (68).

Table 1 reports the mean, standard deviation, and the 1- and 12-month lag autocorrelations of the estimated state variables for the DNS, DNSS, and AFNS models. In the state variables for the DNS and DNSS models, there was no difference between the independent or correlated cases, since the state variables were identical.We saw two remarkable features in the AFNS model. First, regardless of whether the yield adjustment term was independent or correlated, the state variable had a strong auto-correlation with the 1-month lag. Second, the auto-correlation decayed with the 12-month lag. The sign of the mean was the same between the first and second state variables of the DNS and AFNS models.

Estimation of the Transition Matrix

Now, we proceeded to the estimation of the transition matrix A in Equations (48), (51), and (54). In this estimation, the data set was the estimated state variables of the previous sub-section. When estimating Equation (48), the covariance matrix Q was estimated as the sample covariance matrix of the error terms. Table 2 reports the estimation results for the transition matrix and Table 3 reports the variance–covariance matrix.

Table 1. Mean, standard deviation, and auto-correlation of the estimated state variables.

	Mean	Std. Dev.	$\rho(1)$	$\rho(12)$
DNS				
X_1^{DNS}	0.028	0.016	0.936	0.564
X_2^{DNS}	−0.029	0.016	0.939	0.582
X_3^{DNS}	0.025	0.033	0.961	0.746
DNSS				
X_1^{DNSS}	0.022	0.003	0.783	−0.043
X_2^{DNSS}	−0.021	0.003	0.815	0.008
X_3^{DNSS}	−0.090	0.027	0.895	0.219
X_4^{DNSS}	0.080	0.038	0.932	0.531
AFNS (Ind)				
X_1^{AFNS}	0.345	0.016	0.936	0.564
X_2^{AFNS}	−0.342	0.016	0.939	0.582
X_3^{AFNS}	−0.364	0.033	0.961	0.746
AFNS (Corr)				
X_1^{AFNS}	0.210	0.016	0.936	0.564
X_2^{AFNS}	−0.210	0.016	0.939	0.582
X_3^{AFNS}	−0.195	0.033	0.961	0.746

Note: The values of $\rho(L)$ represent an autocorrelation of lag L. The state variables were estimated using OLS at any one point.

Table 2. The transition matrix A of the DNS, DNSS, and arbitrage-free Nelson–Siegel (AFNS) models.

	DNS (Independent)				DNS (Correlated)			
	$A_{.1}$	$A_{.2}$	$A_{.3}$	$A_{.4}$	$A_{.1}$	$A_{.2}$	$A_{.3}$	$A_{.4}$
$A_{1,.}$	0.936	0	0	-	−0.389	−1.151	−0.071	-
	(0.041)				(0.570)	(0.529)	(0.084)	
$A_{2,.}$	0	0.939	0	-	1.203	1.977	0.070	-
		(0.040)			(0.571)	(0.530)	(0.084)	
$A_{3,.}$	0	0	0.962	-	2.036	1.952	0.984	-
			(0.034)		(0.976)	(0.906)	(0.144)	

	DNSS (Independent)				DNSS (Correlated)			
$A_{1,.}$	0.782	0	0	0	0.419	−0.291	−0.024	−0.025
	(0.063)				(0.194)	(0.202)	(0.034)	(0.027)
$A_{2,.}$	0	0.815	0	0	0.379	1.060	0.034	0.034
		(0.064)			(0.204)	(0.212)	(0.035)	(0.028)
$A_{3,.}$	0	0	0.895	0	−1.075	−0.497	0.174	−0.603
			(0.049)		(1.134)	(1.184)	(0.201)	(0.155)
$A_{4,.}$	0	0	0	0.932	2.316	0.412	−0.840	1.608
				(0.041)	(1.372)	(1.431)	(0.243)	(0.188)

	AFNS (Independent)				AFNS (Correlated)			
$A_{1,.}$	0.936	0	0	-	−0.389	−1.151	−0.071	-
	(0.041)				(0.570)	(0.529)	(0.084)	
$A_{2,.}$	0	0.939	0	-	1.203	1.977	0.070	-
		(0.040)			(0.571)	(0.530)	(0.084)	
$A_{3,.}$	0	0	0.962	-	2.036	1.952	0.984	-
			(0.034)		(0.976)	(0.906)	(0.144)	

Note: Standard error is represented in parenthesis.

Table 3. The variance–covariance matrix Q of the DNS and DNSS models

	DNS (Independent)				DNS (Correlated)			
	$Q_{.1}$	$Q_{.2}$	$Q_{.3}$	$Q_{.4}$	$Q_{.1}$	$Q_{.2}$	$Q_{.3}$	$Q_{.4}$
$Q_{1,.}$	3.429×10^{-5}	0	0	-	3.213×10^{-5}	-3.212×10^{-5}	-5.147×10^{-5}	-
$Q_{2,.}$	0	3.394×10^{-5}	0	-	-3.212×10^{-5}	3.216×10^{-5}	5.141×10^{-5}	-
$Q_{3,.}$	0	0	0.941×10^{-5}	-	-5.147×10^{-5}	5.141×10^{-5}	9.406×10^{-5}	-
	DNSS (Independent)				DNSS (Correlated)			

	DNSS (Independent)				DNSS (Correlated)			
$Q_{1,.}$	3.409×10^{-6}	0	0	0	3.183×10^{-6}	3.276×10^{-6}	0.170×10^{-6}	-1.363×10^{-5}
$Q_{2,.}$	0	3.777×10^{-6}	0	0	-3.276×10^{-6}	3.511×10^{-6}	-1.091×10^{-5}	1.625×10^{-5}
$Q_{3,.}$	0	0	1.332×10^{-5}	0	8.170×10^{-7}	-1.091×10^{-5}	1.093×10^{-4}	-1.260×10^{-4}
$Q_{4,.}$	0	0	0	1.963×10^{-4}	-1.363×10^{-5}	1.625×10^{-5}	1.260×10^{-4}	1.597×10^{-4}

Note: The variance–covariance matrix Q of AFNS is identical to that of DNS.

As Table 2 indicates, almost all of the diagonal elements of the transition matrix were statistically significant in all models. Note that, by assumption, the transition matrix of each DNS model was identical to that of the AFNS model, and that the covariance matrix Q of the AFNS model was identical to that of the DNS model.

As already explained in Section 4.2, we could calculate the yield adjustment term C of the AFNS model from the estimated variance–covariance matrix Q. However, due to large volatility, the yield adjustment term of the AFNS model with independent error term became too huge. Each element of the variance–covariance matrix was from two to ten times larger than what Christensen et al. (2011) 15 obtained using U.S. data. For example, our estimated Q results in a yield adjustment term of about 20,000 bps for a 480-month maturity. As this is extremely unrealistic, in what follows, we omitted the results of the AFNS model with independent error terms.

RESULTS

Forecasting Performances of the Three Models In this section, we estimated the root mean square forecast error (RMSFE) of the 1-month, 3-month, 6-month, and 12-month forecasts for each of the three models.

Out-of-Sample Forecasting Comparison of Forecasting Performances

We defined the h period forecast of the yield with maturity tau at time t as yt+h,t (τ). As in Christensen et al. (2011), the forecasting measurement equations of the three models are written, respectively, as:

$$y_{t+h,t}^{DNS}(\tau) = \mathbb{E}^{\mathbb{P}}[y_{t+h}(\tau)] = \mathbb{E}^{\mathbb{P}}\left[\hat{X}^1{}_{t+h}\right] + \mathbb{E}^{\mathbb{P}}\left[\hat{X}^2{}_{t+h}\right]\left(\frac{1-e^{-\lambda\tau}}{\lambda\tau}\right) + \mathbb{E}^{\mathbb{P}}\left[\hat{X}^3{}_{t+h}\right]\left(\frac{1-e^{-\lambda\tau}}{\lambda\tau} - e^{-\lambda\tau}\right). \quad 69)$$

$$y_{t+h,t}^{DNSS}(\tau) = \mathbb{E}^{\mathbb{P}}\left[\hat{X}^1{}_{t+h}\right] + +\mathbb{E}^{\mathbb{P}}\left[\hat{X}^2{}_{t+h}\right]\left(\frac{1-e^{-\lambda_1\tau}}{\lambda_1\tau}\right) + \mathbb{E}^{\mathbb{P}}\left[\hat{X}^3{}_{t+h}\right]\left(\frac{1-e^{-\lambda_1\tau}}{\lambda_1\tau} - e^{-\lambda_1\tau}\right) + \mathbb{E}^{\mathbb{P}}\left[\hat{X}^4{}_{t+h}\right]\left(\frac{1-e^{-\lambda_1\tau}}{\lambda_1\tau} - e^{-\lambda_1\tau}\right), \quad (70)$$

$$y_{t+h,t}^{AFNS}(\tau) = \mathbb{E}^{\mathbb{P}}[y_{t+h}(\tau)] = \mathbb{E}^{\mathbb{P}}\left[\hat{X}^1{}_{t+h}\right] + \mathbb{E}^{\mathbb{P}}\left[\hat{X}^2{}_{t+h}\right]\left(\frac{1-e^{-\lambda\tau}}{\lambda\tau}\right) + \mathbb{E}^{\mathbb{P}}\left[\hat{X}^3{}_{t+h}\right]\left(\frac{1-e^{-\lambda\tau}}{\lambda\tau} - e^{-\lambda\tau}\right) - \frac{\hat{C}(\tau)}{\tau}, \quad (71)$$

where $-\frac{\hat{C}(\tau)}{\tau}$ is the yield adjustment term. As the transition dynamics are first-order, we can write the expectation of the state variables at t + h as:

$$\mathbb{E}^{\mathbb{P}}[X_{t+h}] = \left(\sum_{i=0}^{h-1} A^i\right)(I - A)\mu + A^h X_t, \quad (72)$$

where X_t denotes the column vector of the state variables. In this paper, we employed the random walk model (without a drift term) as a benchmark model. This paper investigated the 1-, 3-, 6-, and 12-month ahead forecast (h = 1,3,6,12). In out-of-sample forecast, these models were estimated only one time over the full sample periods (September 2009 to June 201516) and then used for forecasting h-month ahead horizon (h = 1,3,6,12).

Tables 4–7 report the root mean square forecast error (RMSFE)[17] of the 1-, 3-, 6-, and 12-month forecasts, respectively. In the tables, the smallest RMSFE is boxed for each maturity and horizon. As these tables show, DNSS was clearly the best forecasting model among the three models for 6- and 12-month forecast horizons. Between two types of the DNSS models, the DNSS model based on the transition equation of the independent (uncorrelated) state variable was superior to the correlated model. It is clear to see that all of the off-diagonal elements of the transition matrix were statistically highly significant in the DNSS model. This superiority is plausible, because the DNSS model is theoretically the generalized model of the DNS model. We can say that the generality of the DNSS model was confirmed by our result. On the other hand, the random walk model was superior to the Nelson–Siegel class of models in the short-period forecast horizons. Table 8 shows the Diebold and Mariano test (Diebold and Mariano (1995)) for testing forecast accuracy. We compared the forecast accuracy of the DNSS model with the random walk model for the 6- and 12-month forecast horizons. The positive bold value shows that the DNSS model outperformed the random walk model at a 5% statistically significant level.

Table 4. Root mean squared forecast error of a 1-month forecast

	RW	DNS		DNSS		AFNS
Maturity		Ind	Corr	Ind	Corr	Corr
3-month	0.20	13.56	12.10	4.71	4.77	2.84
6-month	0.05	12.19	10.73	2.03	2.40	1.97
1-year	0.50	8.98	7.51	1.73	0.80	1.09
2-year	0.80	1.91	0.40	4.74	2.94	1.33
3-year	0.20	3.73	5.28	5.14	2.73	1.80
4-year	0.20	6.73	8.33	5.19	2.07	1.12
5-year	1.51	9.40	11.05	3.42	0.33	1.40
6-year	3.02	13.44	15.14	1.24	4.49	4.16
7-year	3.73	14.53	16.29	3.97	7.30	4.92
8-year	4.46	11.72	13.55	3.43	6.80	2.55
9-year	4.16	10.17	12.07	4.50	7.86	2.08
10-year	3.76	8.65	10.62	5.74	9.08	2.14
15-year	3.62	1.90	0.47	6.87	10.09	3.41
20-year	2.79	19.25	16.42	5.84	2.58	1.50
25-year	3.07	13.21	9.92	1.64	1.84	13.07
3-year	1.39	4.35	0.57	1.21	5.00	22.73
40-year	2.31	17.987	22.69	5.96	10.39	7.28

Note: The unit of the values is the basis points (bps). The boxed value is the smallest RMSFE for each maturity. RW denotes the random walk model (the benchmark model).

Table 5. Root mean squared forecast error of a 3-month forecast

	RW	DNS		DNSS		AFNS
Maturity		Ind	Corr	Ind	Corr	Corr
3-month	1.96	13.18	10.63	6.90	6.97	4.79
6-month	0.83	12.34	9.72	3.22	3.87	3.21
1-year	0.71	9.85	7.12	1.94	0.47	1.50
2-year	1.28	3.05	0.25	5.99	2.85	1.59
3-year	0.33	3.46	6.28	6.23	2.12	2.85
4-year	1.60	7.42	10.42	5.55	1.39	3.43
5-year	3.63	10.83	13.94	3.22	2.66	4.64
6-year	5.01	14.81	18.01	1.40	7.05	7.26
7-year	7.23	17.28	20.63	5.56	11.35	9.61
8-year	8.52	15.05	18.55	5.79	11.57	8.21
9-year	8.20	13.53	17.16	6.90	12.68	7.82
10-year	7.29	11.56	15.32	7.75	13.46	7.39
15-year	6.44	2.22	5.67	9.01	14.34	8.14
20-year	4.83	18.14	13.39	4.63	3.46	3.77
25-year	4.20	12.95	7.53	1.62	5.80	16.21
3-year	3.29	3.85	4.44	3.50	9.59	26.54
40-year	3.84	17.97	26.54	7.51	14.84	4.85

Note: The unit of the values is the basis points (bps). The boxed value is the smallest RMSFE for each maturity. RW denotes the random walk model (the benchmark model).

Table 6. Root mean squared forecast error of a 6-month forecast.

	RW	DNS		DNSS		AFNS
Maturity		Ind	Corr	Ind	Corr	Corr
3-month	3.83	13.18	9.42	9.35	9.37	6.93
6-month	2.50	12.12	8.28	5.41	6.22	5.32
1-year	1.87	9.50	5.77	1.98	2.86	4.07
2-year	1.55	2.78	2.80	5.63	2.34	4.08
3-year	1.40	3.62	8.16	6.54	1.76	4.86
4-year	3.23	8.30	13.01	5.46	2.39	6.14
5-year	5.84	12.41	17.26	2.67	5.33	8.05
6-year	7.88	17.02	22.03	2.91	10.53	11.44
7-year	10.73	20.28	25.46	7.81	15.58	14.53
8-year	13.05	19.12	24.53	9.29	16.96	14.22
9-year	13.10	17.96	23.58	10.99	18.54	14.31
10-year	12.32	16.12	21.95	12.17	19.54	14.13
15-year	13.27	8.15	14.62	16.28	22.69	17.08
20-year	11.99	14.29	10.31	7.98	12.28	13.09
25-year	11.05	10.33	8.79	9.52	15.13	24.89
3-year	8.91	4.60	12.52	10.57	17.54	33.92
40-year	8.87	20.42	33.47	13.26	22.14	8.17

Note: The unit of the values is the basis points (bps). The boxed value is the smallest RMSFE for each maturity. RW denotes the random walk model (the benchmark model).

Table 7. Root mean squared forecast error of a 12-month forecast.

	RW	DNS		DNSS		AFNS
Maturity		Ind	Corr	Ind	Corr	Corr
3-month	13.07	11.15	10.05	19.61	19.29	16.64
6-month	14.32	10.48	11.09	17.96	18.55	17.51
1-year	15.64	10.36	13.54	15.51	17.44	18.73
2-year	16.05	13.04	18.75	12.80	16.21	20.04
3-year	17.91	18.68	25.12	13.25	17.64	22.33
4-year	20.00	23.34	30.16	14.24	19.67	24.14
5-year	23.56	28.53	35.63	17.44	23.90	27.36
6-year	26.92	34.47	41.82	23.09	30.24	32.19
7-year	30.90	39.10	46.69	29.14	36.42	36.61
8-year	35.61	40.82	48.63	33.57	40.50	39.08
9-year	37.73	41.92	49.95	37.67	44.19	41.38
10-year	39.02	42.29	50.53	41.18	47.23	43.34
15-year	51.59	46.73	55.91	58.44	62.41	58.24
20-year	62.88	47.83	56.93	62.85	65.42	67.12
25-year	69.03	56.74	67.71	72.27	75.30	84.70
3-year	73.22	66.78	79.86	79.36	83.34	98.00
40-year	78.08	84.73	102.04	85.58	91.80	79.54

Note: The unit of the values is the basis points (bps). The boxed value is the smallest RMSFE for each maturity. RW

denotes the random walk model (the benchmark model).

Table 8. Diebold–Mariano test for 6- and 12-month forecasting of each maturity of the zero-yield government bond (random walk model and DNSS (Ind) model).

6-Month	3-Month	6-Month	1-Year	2-Year	3-Year	3-Year	5-Year	6-Year	7-Year	8-Year
DM	−3.61	−3.13	−0.26	−4.81	−7.43	−4.32	**2.59**	**3.87**	**3.61**	**4.04**
p-value	0.000	0.001	0.791	0.000	0.000	0.000	0.010	0.000	0.000	0.000
	9-year	10-year	15-year	20-year	25-year	30-year	40-year			
DM	**3.57**	0.69	−3.44	**2.26**	**3.98**	−1.73	−3.03			
p-value	0.000	0.488	0.000	0.023	0.000	0.082	0.002			
12-Month	3-month	6-month	1-year	2-year	3-year	3-year	5-year	6-year	7-year	8-year
DM	−3.91	−3.35	1.494	**2.42**	**2.47**	**2.92**	**3.68**	**4.59**	**4.49**	**4.74**
p-value	0.000	0.001	0.134	0.015	0.013	0.004	0.000	0.000	0.000	0.000
	9y	10y	15y	20y	25y	30y	40y			
DM	3.45	−2.30	−3.17	0.04	−2.19	−2.73	−2.94			
p-value	0.001	0.021	0.002	0.965	0.030	0.001	0.003			

Next, we compared the forecasting performance of the DNS model with the AFNS model[18]. Theoretically, an arbitrage opportunity should be exploited by the market force. If there remains an arbitrage opportunity in the real world, the market is considered inefficient.

From the results for the 1-, 3-, and 6-month forecast horizons, the correlated AFNS was superior to each DNS model in almost all of the maturities. However, as Table 7 shows, the 12-month forecast performance of the AFNS model was inferior to that of the DNS model. The results suggest that the imposition of an arbitrage-free condition does not improve long forecast performance.

Our results provide evidence against those in Christensen et al. (2011) and Diebold and Rudebusch (2013). They argued that the AFNS model improved the out-of-sample forecast for almost all maturities and horizons for the U.S. bond market. Our results indicated that the imposition of an arbitrage-free condition restricts the long-forecasting performance of the AFNS model for JGB yields.

The first plot in Figure 3 depicts the 3-month forecast yields of each model.

Each model captured the actual yield well, as shown by the first and second plots for the 3- and 6-month forecast yields of each model. The forecast performance became poorer as the forecasting horizon becomes longer[19]

The correlated AFNS captured the convexity due to the yield adjustment term, while the DNS model and the DNSS model did not capture that effect. We could see a large difference between the two curve-fitting models and the affine term structure model. Moreover, the plot of the DNS model was U-shaped and that of the DNSS model was S-shaped; such a difference in shape between models arose from the fourth term of the DNSS model. The figures suggest that the DNSS model captured the dynamics of the yield more flexibly than the DNS model.

In addition to the above analysis of the forecast accuracy, this paper investigated the cumulative of the squared forecast errors to compare the forecast accuracy of the models with the benchmark model. Figure 4 shows the cumulative of the squared forecast errors of the JGB yields of the 3-month, 1-, 5-, and 10-year maturity. The cumulative of the squared forecast errors imply that the upward (downward) trend in the plot suggest that the models are inferior (superior) to the benchmark model (the random walk model). In the forecasting the JGB yields of 3-month maturity and 1-year maturity, the cumulative of the squared errors of the DNS model shows the downward trend for long-forecast horizon. On the other hand, the cumulative of the squared errors of the DNSS model shows the downward trend for the long forecast horizon. These results provide the evidence that the DNS and the DNSS model is more useful for forecasting the long-term horizon than the random walk model from the perspective of the cumulative of the squared forecast errors.

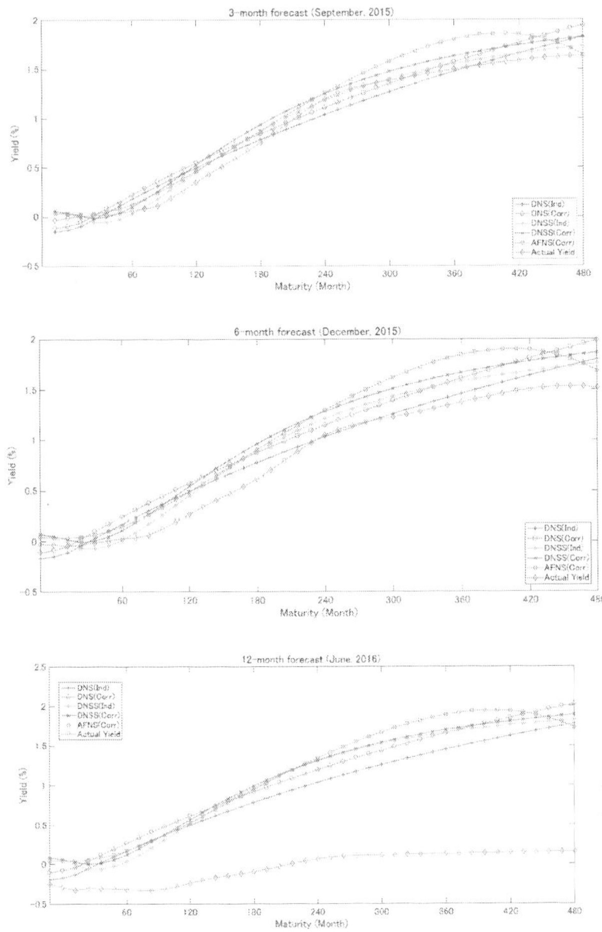

Figure 3. Forecast yields for 3-, 6-, and 12-month forecasts. Note: The notation "Ind" denotes the independent factor model and the notation "Corr" denotes the correlated factor model. The plots show the forecast yields of each model and the actual yield in September 2015; December 2015; and June 2016.

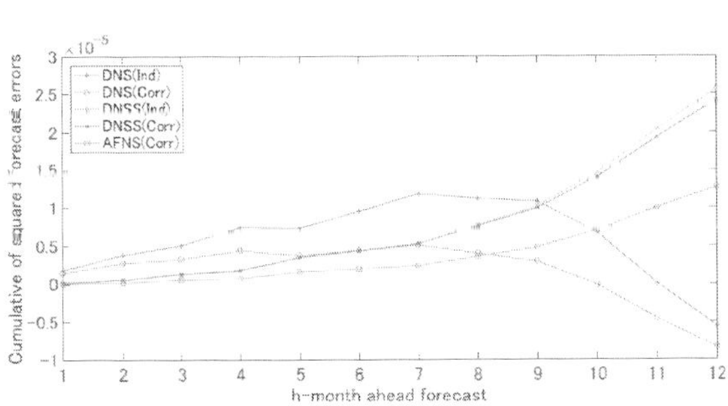

(A) 3-month maturity

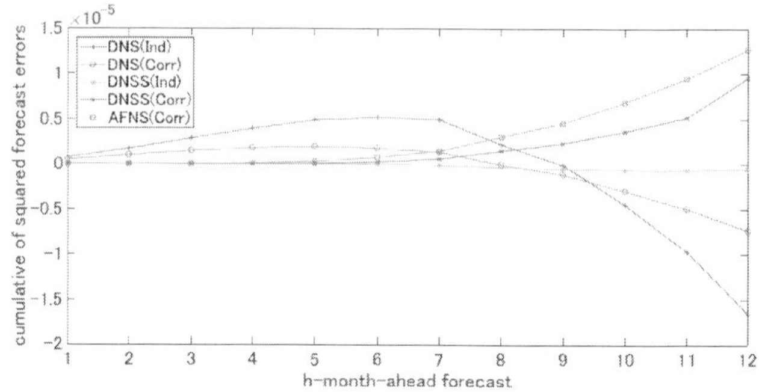

(B) 1-year maturity

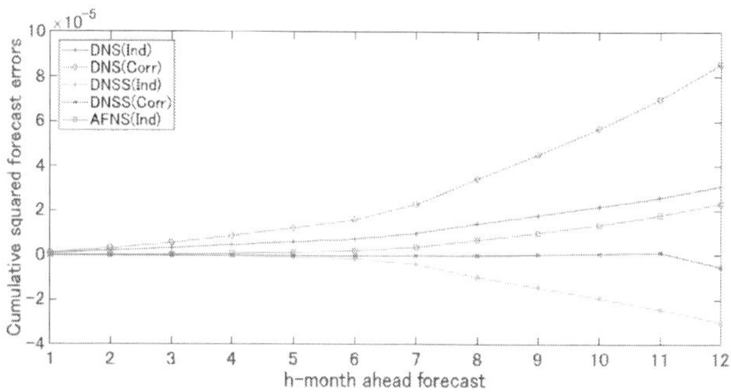

(C) 5-year maturity

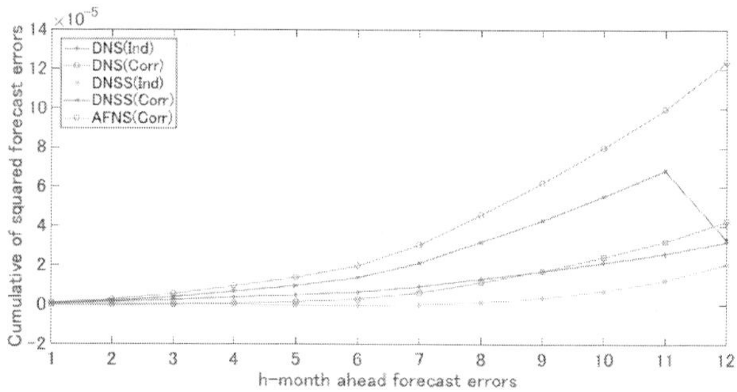

(D) 10-year maturity

Figure 4. Cumulative of the squared forecast errors. Note: Figure 5 shows the cumulative of the squared errors of the forecasting of the JGB yields of the 3-month, 1-, 5-, and 10-year maturity. The cumulative of the squared forecast errors imply that the upward (downward) trend in the plot suggest that the models are inferior (superior) to the benchmark model (the random walk model).

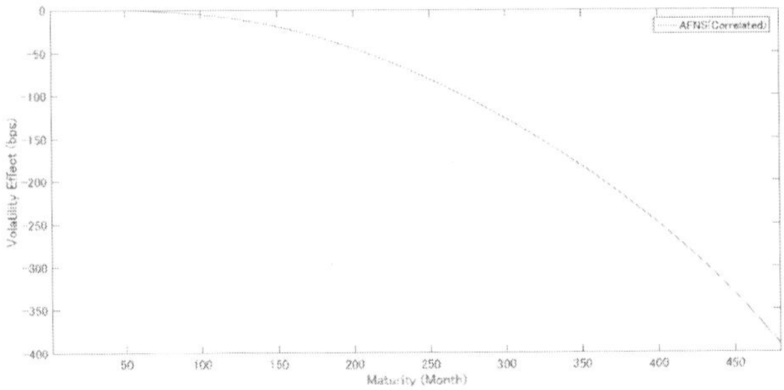

Figure 5. The volatility effect of the correlated AFNS. Note: The unit of the volatility effect is basis points (bps).

Volatility Effect

As the yield adjustment term depends on the volatility matrix, it is often called the volatility effect. In this sub-section, we refer to the yield adjustment term as the volatility effect and discuss it for JGB markets.

Diebold and Rudebusch (2013) argued that the volatility effect can be negligible in most periods, but may become larger in some periods. They also argued that the effect becomes larger with longer maturities.

The volatility effect generated a large difference between the DNS, DNSS, and AFNS models, since it affects the value of the state variables. As discussed in Section 4.1, it caused the constant term bias. The volatility effect arose from the non-linearity of the yield curve. It can be paraphrased mathematically as the Jensen's inequality term. In the economic literature, it has often been ignored, as a negligible quantity. However, with the exception of in finance research, this concept is typically not discussed at length.

Figure 4 shows the volatility effect, which was calculated[20] from the yield adjustment terms of the AFNS model. The volatility effect depended only on maturity, independent of the time. We can confirm that the magnitude of the volatility effect grew larger with longer maturities, in agreement with Diebold and Rudebusch (2013). The magnitude of the volatility effect was 9.8 bp for a 120-month maturity and 126 bp for a 300-month maturity. Hence, we might be able to ignore the effect for maturities from 3- to 120-months, which were practically well used. However, when we are interested in longer-term maturities, such as 300- and 480-month maturities, the volatility effect should be taken into account in the analysis.

CONCLUSIONS

This study investigated the forecasting performances for yields by the DNS, DNSS, and AFNS models. The DNS and DNSS models are curve-fitting models, which do not impose an arbitrage-free condition. On the other hand, the AFNS model is an affine term structure model based on the arbitrage-free theory. Following Christensen et al. (2011) and Diebold and Rudebusch (2013), we first introduced the theoretical and empirical framework of the three models. The AFNS was different from the other two models, in that includes a yield adjustment term in the yield curve equation.

Second, our study provided evidence that the DNSS model was the best model, in terms of forecast performance for yields in JGB markets. The forecast performance of the AFNS model was inferior to the DNS model for the 12-month forecast horizon. Our results were in contrast with those of Christensen et al. (2011) and Diebold and Rudebusch (2013), which showed that the AFNS model was superior to the DNS model for U.S. bond data. For JGB markets, the imposition of the arbitrage-free condition did not improve the forecast performance for relatively long forecasting horizons.

The volatility effect was negligibly small for the short maturities. On the other hand, the longer the maturity was, the bigger the volatility effect became. Therefore, when we consider yields at long maturities, we need to pay attention to the existence of the volatility effect.

Finally, we shortly discussed the remaining issues and the directions of future researches. This study adopted the period from August 2009 to June 2015. However, in February 2016, the Bank of Japan carried out the “minus” interest rate policy, which should have created a large, unpredictable shock in the JGB markets. As a large shock generates arbitrage opportunities, the analyses may be biased without taking this into account. Therefore, we need to replicate the analysis over another period.

In this paper, although we used only the three-factor AFNS model, it is also worth using the generalized model of the AFNS model. If so, we may compare the DNSS model with the arbitrage-free Nelson–Siegel model appropriately.

REFERENCES

1. Alfaro, Rodrigo A. 2011. Affine Nelson-Siegel model. Economics Letters 110: 1–3.
2. Al-Zoubi, Haitham A. 2009. Short-term spot rate models with nonparametric deterministic drift. Quarterly Review of Economics and Finance 49: 731–47.
3. Ang, Andrew, and Monika Piazzesi. 2003. A no-arbitrage vector autoregression of term structure dynamics with macroeconomi and latent variables. Journal of Monetary Economics 50: 745–87.
4. Bjork, Tomas, and Bent Jesper Christensen. 1999. Interest rate dynamics and consistent forward rate curves. Mathematical Finance 9: 323–48.
5. Chan, Ka-keung Ceajer, Andrew Karolyi, Francis A. Longstaff, and Anthony B. Sanders. 1992. An empirical comparison of alternative models of the short-term interest rate. Journal of Finance 47: 1209–27.
6. Chen, Yu-chin, and Kwok Ping Tsang. 2013. What does the yield curve tell us about exchange rate predictability? Review of Economics and Statistics 95: 185–205.
7. Christensen, Jens H. E., Francis X. Diebold, and Glenn D. Rudebusch. 2009. An arbitrage-free generalized Nelson-Siegel term structure model. Econometrics Journal 12: C33–C64.
8. Christensen, Jens H.E., Francis X. Diebold, and Glenn D. Rudebusch. 2011. The affine arbitrage-free class of Nelson-Siegel term structure models. Journal of Econometrics 164: 4–20.
9. Coroneo, Laura, Ken Nyholm, and Rositsa Vidova-Koleva. 2011. How arbitrage-free is the Nelson-Siegel model? Journal of Empirical Finance 18: 393–407.
10. Cox, John, Jonathan E. Ingersol, and Stephen A. Ross. 1985. A theory of the term structure of interest rates. Econometrica 53: 385–407.
11. De Rezende, Rafael B., and Mauro S. Ferreira. 2013. Modeling and forecasting the yield curve by an extended Nelson-Siegel class of models: A quantile autoregression approach. Journal of Forecasting 32: 111–23.
12. Diebold, Francis X., and Canlin Li. 2006. Forecasting the term structure of government bond yields. Journal of Econometrics 130: 337–64.
13. Diebold, Francis X., and Robert S. Mariano. 1995. Comparing

predictive accuracy. Journal of Business and Economic Statistics 13: 134–44.

14. Diebold, Francis X., and Glenn D. Rudebusch. 2013. Yield Curve Modeling and Forecasting. Princeton: Princeton University Press. Duffee, Gregory R. 2002. Term premia and interest rate forecasts in affine models. Journal of Finance 57: 405–43.
15. Duffie, Darrell, and Rui Kan. 1996. A yield-factor model of interest rates. Mathematical Finance 6: 379–406.
16. Fama, Eugene F. 2006. The behavior of interest rates. Review of Financial Studies 19: 359–79.
17. Fujii, Mariko, and Makoto Takaoka. 2007. Kinri no Kikannkozo to Macrokeizai: Nelson—Siegel Moderu wo Motiita Approach [Term Structure of Interest Rates and Macroeconomy: Nelson—Siegel Model Approach]. FSA Research Review. Tokyo: Financial Services Agency, pp. 219–48.
18. Gasha, Giancarlo, Ying He, Carlos Medeiros, Marco Rodriguez, Jean Salvati, and Jiangbo Yi. 2010. On the Estimation of Term Structure Models and an Application to the United State. IMF Working Paper WP/10/258. Washington: International Monetary Fund.
19. Gurkaynak, Refet S., Brian Sack, and Jonathan H. Wright. 2007. The U.S. treasury yield curve: 1961 to the present. Journal of Monetary Economics 54: 2291–304.
20. Gurkaynak, Refet S., and Jonathan H. Wright. 2012. Macroeconomics and the term structure. Journal of Economic Literature 50: 331–67.
21. Hamilton, James D., and Jing Cynthia Wu. 2012. Identification and estimation of Gaussian affine term structure models. Journal of Econometrics 168: 315–31.
22. Hördahl, Peter, Oreste Tristani, and David Vestin. 2005. A joint econometric model of macroeconomic and term-structure dynamics. Journal of Econometrics 131: 405–44.
23. Hull, John, and Alan White. 1990. Pricing interest rate derivative securities. Review of Financial Studies 3: 573–92.
24. Ishii, Hokuto. 2018. Modeling and predictability of exchange rate changes by the extended relative Nelson–Siegel class of models. International Journal of Financial Studies 6: 68.
25. Krippner, Leo. 2015. Zero Lower Bound Term Structure Modeling. New York: Palgrave Macmillan. Litterman, Robert, and Jose

Scheinkman. 1991. Common factors affecting bond returns. Journal of Fixed Income 1: 54–61.

26. 2008. Moench, Emanuel 2008. Forecasting the yield curve in a data-rich environment: a no-arbitrage factor augmented VAR approach. Journal of Econometrics 146: 26–43.
27. Niu, Linlin, and Gengming Zeng. 2012. The discrete-time framework of the arbitrage-free Nelson-Siegel class of term structure models. SSRN 2015858: 1–68.
28. Nelson, Charles, and Andrew Siegel. 1987. Parsimonious modeling of yield curves. Journal of Business 60: 473–89.
29. Piazzesi, Monika 2010. Affine term structure models. In Handbook of Financial Econometrics. Amsterdam, New York and Oxford: Elsevier, pp. 691–766.
30. Rudebusch, Glenn D., and Tao Wu. 2008. A macro-finance model of the term-structure, monetary policy, and the economy. Economic Journal 118: 906–26.
31. Svensson, Lars E. O. 1994. Estimating and Interpreting Forward Interest Rates: Sweden 1992–1994. NBER Working Paper #4871. Cambridge: National Bureau of Economic Research.
32. Vasicek, Oldrich. 1977. An equilibrium characterization of the term structure. Journal of Financial Economics 5: 177–88.
33. Wu, Tao. 2006. Macro Factors and the Affine Term Structure of Interest Rates. Journal of Money, Credit and Banking 38: 1847–75.

SECTION 3
PORTFOLIO THEORY

CHAPTER

9

Practical Aspects of Portfolio Selection and Optimisation on the Capital Market

Florin Turcas[a], Florin Dumiter[b], Petre Brezeanu[c], Pavel Farcas[b] and Sorina Coroiu[d]

[a]Bucharest Academy of Economic Studies, Bucharest, Romania;

[b]Accountancy and Finance, Vasile Goldis Western University of Arad, Arad, Romania;

[c]Finance, The Bucharest Academy of Economic Studies, Bucharest, Romania;

[d]University of Oradea, Oradea, Romania

Citation: Florin Turcas, Florin Dumiter, Petre Brezeanu, Pavel Farcas & Sorina Coroiu (2017) "Practical aspects of portfolio selection and optimisation on the capital market", Economic Research-Ekonomska Istraživanja, 30:1, 14-30, https://doi.org/10.1080/1331677X.2016.1265893

ABSTRACT

This article highlights some observations concerning the deficiencies in the application of statistics on the capital market, with special reference to Modern Portfolio Theory (MPT). The main point is the sensitivity of statistical parameters (especially the standard deviation of the daily rates of return) to subjective/random factors. For securities with similar patterns and quasi-identical charts, statistical results in contradiction to the evidence of the market can be obtained. This article makes a pledge in favour of the necessity for increased attention in constructing an optimal/efficient portfolio.

Keywords:- Stock exchange; statistics; technical analysis; Modern Portfolio Theory (MPT)

INTRODUCTION

For any investor and/or stock market speculator, the first thing taken into consideration is the return on investment and the associated risk. The No Arbitrage principle stipulates No Free Lunch (Arratia, 2014), meaning that money cannot be made of thin air (without an initial investment). Equally, you cannot earn money without risk, hence, the risk increases with the expected profits for securities located on the efficient frontier.

According to Modern Portfolio Theory (MPT), in order to create an optimal/efficient portfolio, a selection method is used based on statistical considerations of the rate of return, risk, and the correlation of securities. Securities can be stocks, bonds, indexes, futures/options, structured products, commodities, currencies; practically any asset traded to allow precise determination of rate of returns, on a predetermined time frame. The asset diversification of a portfolio aims at reducing non-systemic risks that are specific to a particular issuer or market.

Let $Y_0, Y_1, \ldots, Y_k, \ldots, Y_{T-1}, Y_T$, with $T \geq 2$, closing values of asset Y, traded on a market. Values can be daily, weekly, monthly, yearly, as long as they are available regularly and periodically.

For the time frame (interval) $[0,T]$, a characterisation of the evolution of the asset is made through:

- Periodic-time valuation of the average rate of return, assimilated to order I moment, attached to random variable that models the temporary evolution of the rate of return; and
- Momentary-periodic valuation of the dispersion of rate of return, assimilated to order II moment, of the random variable,

so by knowing/valuating main statistical parameters of a selection.

Also it is supposedly known – possible to estimate/approximate – the probability function of the random variable that is modelling the temporary evolution of the return of the asset.

For each moment **k**, the rate of return in moment k, 1 < k < T, can be defined in two ways $r'_k = \frac{Y_k - Y_{k-1}}{Y_{k-1}}$,

or $r''_k = \ln\left(\frac{Y_k}{Y_{k-1}}\right) = \ln(Y_k) - \ln(Y_{k-1})$ (1)

In the first form, the daily percentage rate of return is defined according to the daily quotations **Y** of current day **k**, compared with the previous day k-1. The second form of the equation (1), the log profitability is defined.

The second formula is a purely mathematical property of logarithms, having no sense applying logarithmic function for a physical quantity characterised by a unit of measurement. E.g., log($1)=0 ≠ log(¢100)=2, although 1 US dollar is obviously equal to 100 cents.

The probability function of the rate of returns can be considered normal – Gaussian, either lognormal, or approximated by a lognormal – log-Gaussian function. Therefore, we use the logarithmic expression, because statistically, it is necessary that a loss –x be counterbalanced by an equal and opposite gain, x, in order for the statistical mean to logically make sense. For small values of the $\mathbf{r}_k$ daily variation, the two rates of returns are close enough to be considered equal. For example, for daily variations of 5% between the values calculated by the two methods, there is a relative difference of only 2.42%.

LITERATURE REVIEW

One aspect of managing the financial instruments portfolio is to optimise portfolio administration based on various criteria; the most commonly used one is the profitability – risk criterion. In 1952, Markovitz proved that in a perfect market context, one can determine the set of efficient portfolios that lie on a hyperbolic curve, called the efficient frontier, where each portfolio that has a required return, the risk is minimal.

Sharpe (1964), based on the capital asset pricing model(CAPM) theory, has shown that when there is a risk free active in the portfolio, the efficient frontier is a straight line which can be determined. Considering the above, each portfolio manager tries to be on the efficient frontier. Analysing the literature in the field we were able to identify some studies that supplement the results of the two Nobel laureates.

Sortino and Satchell (2001) goes beyond MPT suggesting the downside risk as a better measure for portfolios risks. He mentions some divergences between authors regarding the theoretical and/or practical aspects, most of them because of the initial assumptions.

Sortino et al. (2010) developed their theory, introducing upside potential and the Desired Target Return. They also criticised the MPT, showing that, according to Fama, the CAPM market portfolio does not exist, no matter the value of alpha or beta considered.

Schulmerich, Leporcher, and Eu (2015) investigate the investor's behaviour in contrast between rational and irrational (behavioural). Crises are investigated, analysing if risk measurements are correct regarding the chosen investment strategies.

Marty (2013) systematise in a mathematical matter the basics of portfolio selection and analyse.

Elton et al. (2014) state that 'portfolio analysis is the recipe for one of the few "free lunches" in economics'. They discuss topics like market equilibrium, market efficiency, and investment valuation and analysis.

Baker and Filbeck (2013) synthesised the most recent breakthrough in portfolio theory and management, combining the theoretical background with the practical results.

Francis and Kim (2013) reviewed the relevant literature regarding MPT, as developed after Markowitz.

Sironi (2015) went beyond MPT and inspected the portfolio choice probabilistic scenario optimisation.

Our own studies, confirm the strong positive correlation between shares listed at the Bucharest Stock Exchange (BSE), making diversification less efficient against market crashes.

Levy (2012) demonstrates that CAPM must not be empirically rejected and can coexist with behavioural approaches.

Fărcaş and Cuzman (2002a, 2002b) studied the applicability of Markowitz and Sharpe's methods on the Romanian stock market as well as the methods of optimising the rate of return based on the risk-return criterion.

Fărcaş and Fărcaş (2008) bring forth practical and empirical literature on numerical methods for optimising equity portfolios with applications on the Romanian market. The paper had presented software which permits efficient frontier numerical construction in accordance with Markowitz theory and a case study applied to Romanian Bucharest Stock Exchange (BVB) market. The conclusion of this study highlights the optimisation methods via rentability – risk criteria, and their usefulness in the investment policy of the portfolio management, considering the confidence interval and error probabilities.

Fărcaş and Deac (2010) analyse the statistical relationship of changes in share prices and their influence on decisions regarding the optimisation of portfolio management and determining the efficient frontier portfolios.

Sortino and Price (1994) developed the concept of downside risk, in order to eliminate the disadvantages of risk measurement, through the dispersion variation around the average returns of financial instruments.

In our opinion, within empirical studies, on the application of optimisation theory equity portfolios, risk measurement is better represented by measuring the variation risk of dispersion of returns.

Arratia (2014) provides a qualitative overview regarding the techniques of using the statistical tools and methods available in free software R, for processing and analyzing real financial data.

Baker and Filbeck (2013) present a comprehensive analysis both for theory and practice regarding the structure, composition and optimization on an optimal portfolio.

Chen (2010) provides an important assessment regarding the technical analysis, it's principles and the basic techniques which underlies this special tactic.

Di Lorenzo (2013) presents the modern approach regarding the application of the technical analysis, especially regarding the specific protocols and techniques with a special focusing of the most efficient technical analysis queries

Elton et. al. (2014) assess one of the most comprehensive and complex analysis upon modern portfolio theory and investment with a special focus upon the perfect balance equilibrium in modern portfolio structures.

Vollmer (2015) provides and excting overview regarding the efficient portfolio optimisation using the CAPM, with in insightfull focus on the excess returns through active investment strategies.

RESEARCH METHODOLOGY

The main hypothesis of this study starts from the question: Why a coherent and complete theory, like MPT, does not fit with a practical result on the capital market. We have identified some of the reasons behind these weaknesses of the practical aspects of MPT.

Stock prices have been downloaded from specialised sites and taken into Microsoft Excel, as .csv or .xls files. Data processing was performed using Microsoft Excel Data Analysis module, and by using Crystal Ball, and the data was exported and processed in Mathcad or Mathlab.

The preliminary study of quotations, necessary to define the most relevant aspects of technical analysis, was performed on specialised sites, through special programs or on the stock exchanges' websites. Technical analyses were interpreted based on a vast bibliography available in this area (Thomsett, 2012a; Chen, 2010; Murphy, 2004; Di Lorenzo, 2013).

The selection and optimisation of a portfolio of **N** assets is performed the classical way using MPT:

- It is considered (Elton, 2014) a portfolio of N assets. Let $\mathbf{x}_i$ be the share of the asset **i** in the structure of the considered portfolio. The weights $\mathbf{x}_i$ satisfy the conditions:

$$\begin{cases} 0 \leq x_i < 1 \\ \sum_{i=1}^{N} x_i = 1 \end{cases}, \quad i = \overline{1, N}, \tag{2}$$

specifying that if the market sales short, weights can also be negative.

Also, the following sizes are known (have been evaluated):

$\overline{r_i}$ average rate of return of asset i, for an analysed period T;

σ_i standard deviation of i title;

σ_{ij} covariance between the rate of return on asset i and the rate of return of asset j;

$\sigma_{ii} = \sigma_i^2$

ρ_{ij} Pearson correlation coefficient.

The next relations take place:

$$-1 \le \rho_{ij} \le 1$$
$$\sigma_{ij} = \rho_{ij} \cdot \sigma_i \cdot \sigma_j \qquad (3)$$

For a portfolio of **N** assets, the average rate of return and risk are noted $\bar{r}$, respectively **σ**, and defined by the following relations:

Rate of return:

$$\bar{r} = \sum_{i=1}^{N} x_i \cdot \bar{r}_i; \qquad (4)$$

and Risk: $$\sigma = \sqrt{\sum_{i=1}^{N} \sum_{j=1}^{N} x_i \cdot x_j \cdot \sigma_{ij}} \qquad (5)$$

Selection and determination of efficient/optimal portfolio composition, in Markowitz's model, based on the rate of return – risk criterion, consists of determining weights $\mathbf{x}_1$, $\mathbf{x}_2, \ldots, \mathbf{x}_{N-1}$, $\mathbf{x}_N$ so that, for an imposed value of the rate of return of the portfolio the value of the associated risk should be minimal, or for given/accepted value of the risk, the value of the rate of return to be maximum.

Thus, determining the efficient/optimal portfolio composition, based on the rate of return – risk model, is made by solving the following optimisation problem:

$$\max_{Xi} \left\{ \frac{\bar{r}}{\sigma} \right\} = \max_{Xi} \left\{ \frac{\sum_{i=1}^{N} x_i \cdot \bar{r}_i}{\sqrt{\sum_{i=1}^{N} \sum_{j=1}^{N} x_i \cdot x_j \cdot \sigma_{ij}}} \right\}, \text{ with restriction } \sum_{i=1}^{N} x_i = 1 \qquad (6)$$

The above is a quadratic optimisation problem and can be solved through the method of Lagrange multipliers.

The graphical representation of the result set is the efficient frontier (Reilly & Brown, 2012; Francis & Kim, 2013), a curve, on which the most cost-effective/efficient – for a given risk – portfolio is found, respectively the least risky combination of assets that generate an expected rate of return. Although the mathematical approach is precise and rigorous, input data are estimates of statistical parameters. Hence, it starts from uncertain data, obtained with a certain level of confidence, irreproducible and affected by errors.

The optimisation problem shown in (6) is equivalent to Markowitz's (1952) optimisation method, based on the profitability – risk criterion.

RESULTS AND DISCUSSIONS

Data series invariance in logarithmic operation

Apparently, the plot analysis will be different depending on the adopted scale: normal or logarithmic. The technical analysis (Murphy, 1999) makes a clear distinction between these two types of graphs. In Figure 1 there is a difference between the two graphs, hence, each one will have to be analysed separately.

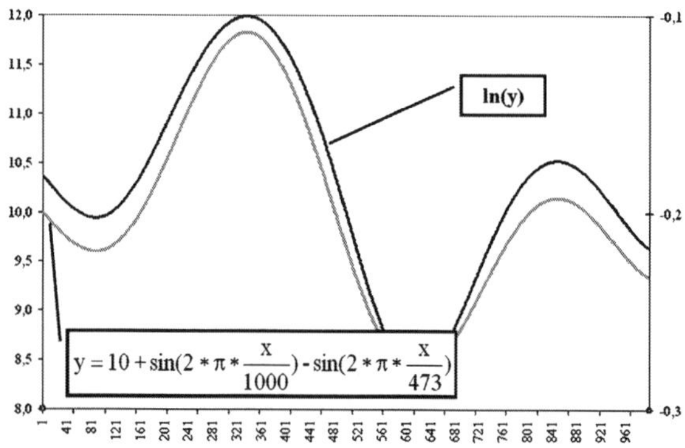

Figure 1. Difference between a normal and logarithmic chart. Source: Authors' estimation.

Under Association for Investment Management and Research (AIMR) standard rules, a risk measurement through the usage of standard deviation is recommended for total risk and the usage of the Sharpe's beta coefficient as a measure of market risk. However, according to Todea (2006) logarithmic returns have the property of being additive; property which does not check in the case of arithmetic returns (of percentage variations). Furthermore, there are differences in terms of symmetry on the usage of the two methods for calculating profitability a posteriori, through the method of percentage and logarithmic variations, the second method having symmetry property.

Therefore, it would be expected that through successive logarithmic or exponential operations, the graphical method will clarify particular aspects (details) of asset evolution. This conclusion is however contradicted by a logarithmic property of the function: for cyclical functions, the normal and logarithmic charts are overlapping, if represented at appropriate scales. The example in Figure 2 is conclusive: any technical analyst will draw the same conclusions, no matter which graphics are studied.

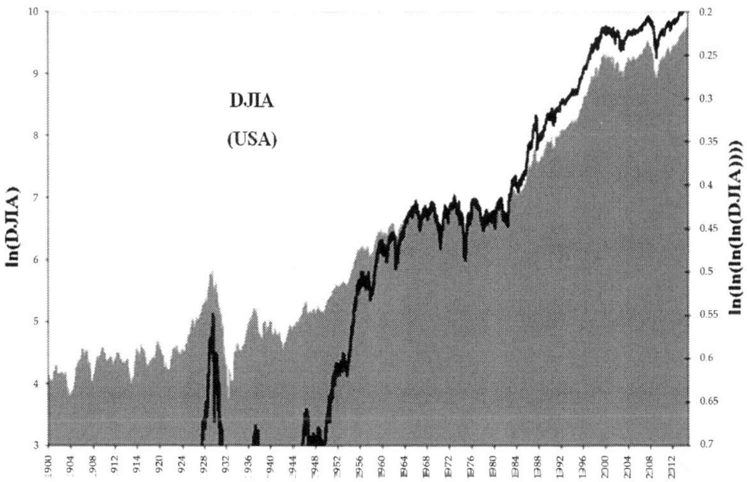

Figure 2. Similarity of a logarithmic graph compared to a temporal one. Source: Authors' estimation based on data collected from stooq.com.

In Figure 2 we intended to highlight only one graphical aspect of technical analysis. Figure 2 has an illustrative character – intuitive in order to highlight the correlation between Dow Jones Industrial Average (DJIA) rate variation and temporal variation of the same logarithmic index.

Following empirical studies, it was found that, generally speaking, the return's distributions are not, usually, normal or lognormal distributions.

The above-mentioned studies tackles, more or less, these distributions, which enables the application of optimisation methods according to the rentability – risk criteria (Markowitz's and Sharpe's model). Estimating future returns can be made with reasonable confidence only if we accept the assumption of normality.

Without accepting the normality hypothesis of the rate of return's variations, future estimates must rely on other techniques, which are not part of our study.

The checking of this invariance can be made by successively applying the logarithm operation to stock data series (according to Figure 3). The conclusion is valid also for other markets as shown in Figure 3 for DAX and SIF5 (Issuer from Romanian market, BVB).

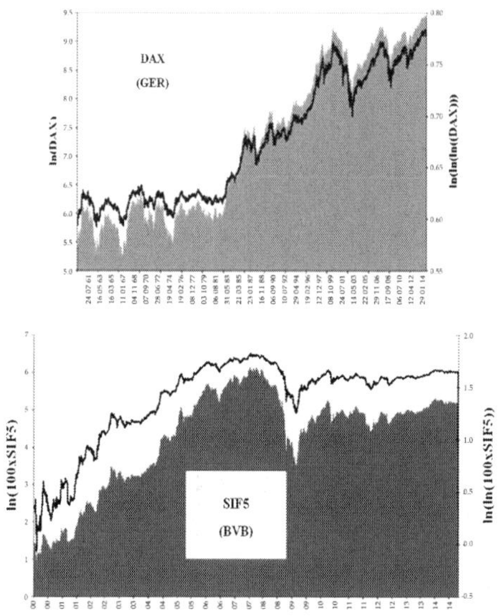

Figure 3. Stock charting invariance at logarithmic operation. Source: Authors' estimation based on data available from stooq.com and bvb.ro.

The results are as follows:

- Through successive logarithm (exponential) operations, the technical analysis of listed assets does not improve;
- Conclusions drawn on normal series are also available on logarithmic ones (and vice versa), the graphs being similar.
- Graphic invariance of the logarithmic operation may have important consequences for some of the capital markets theories, consequences which must be deepened by subsequent studies:
- Technical analysis deals differently with normal and logarithmic graphs. If, however, they have a similar structure, interchangeability would be one consequence, except for establishing the main trends,
- Fractal theories are based on double logarithmic graphs, possibly with the flexibility of time scale (Mandelbrot, 2008);
- Fractal aspects may be graphically pictured by successive logarithmic operations. Fractals are auto reproductive figures at scale, commonly found in the study of graphs, including stocks. Figure 4 shows that, without altering the time scale, the evolution of DJIA quotations, for short periods of time, are similar to long

periods. The possibility that successive logarithmic operations to reveal fractal and other parties with implications for predictability, remains a study objective as well as on stock market theories in general.

Figure 4. Fractal structures in Dow Jones chart. Source: Authors' processing.

Limitations of MPT application in practice

We will not dwell on application failures in real conditions, on the capital markets known to practitioners and mentioned on popular sites. We will only focus on those practical issues that make it difficult to use the theory models of the market even under ideal market conditions.

Distribution of daily rate of returns is not normal

The first and most obvious shortcoming of the practical application of MPT is that the daily rates of returns are not part of a normal statistical distribution. Considering the logarithmic rate of returns of the DJIA from 1900 to the date of writing (9 October 2014), in Figure 5 we can see that the distribution is different from a normal one.

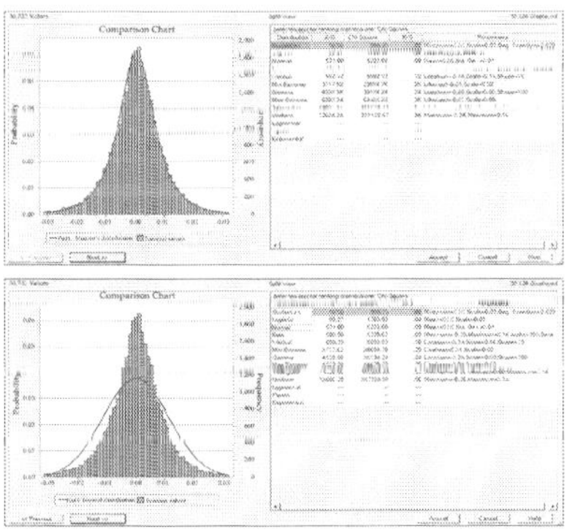

Figure 5. Statistical analysis of the daily DJIA. Source: Authors' estimation based on data from stooq.com.

Notes: Distribution is closer to Student (top) than normal (lower). Data processed by authors in CristallBall (Oracle).

The shape of the daily distribution differs from the monthly and weekly distributions, as it is highlighted in Figure 6. Therefore, the results obtained on daily series could differ from those on longer time series. This conclusion contradicts the main assumptions of the technical analysis, on the indifference of results against the selected time interval. It also draws attention to the credibility of such statements in the technical analysis, not statistically covered.

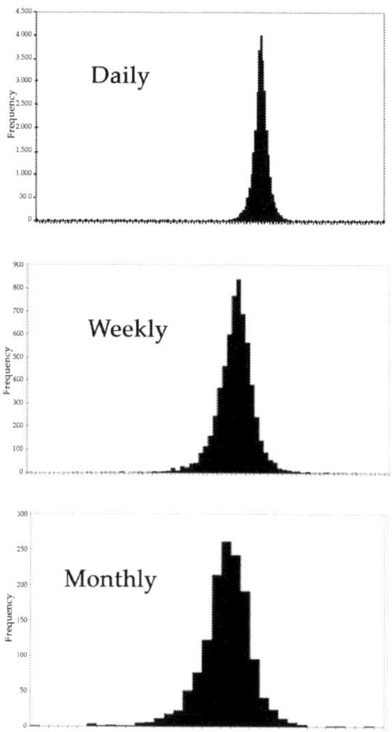

Figure 6. Statistical series depend on samplings (the period under review). Source: Authors' estimation based on stooq.com, from DJIA quotation series.

Courtesy access offered by the BSE, we were able to analyse all transactions on issuer SIF5, during the period 3 January 2006 to 20 January 2015. Only transactions on the Regular Market on BVB Stock Exchange (REGS) market were selected, so the statistical data base was no longer influenced by time frame. Statistical results show the majority data congested in the centre ($\Delta \ln \approx 0$), with little data, but large values, scattered to the end (long tails) according to Figure 7.

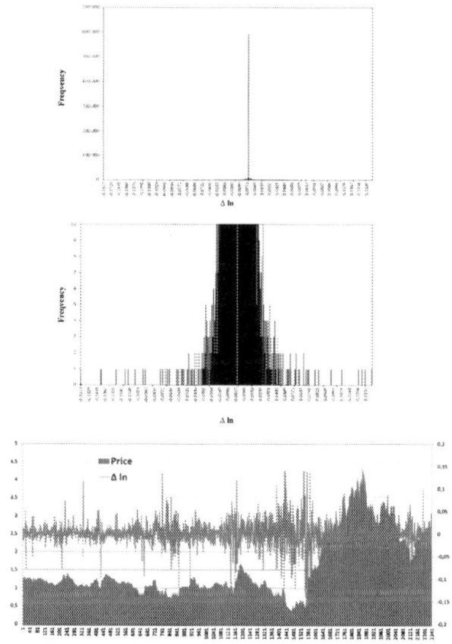

Figure 7. Distribution of all transactions on SIF5 issuer. Source: Authors' processing.

The implications of this result are important for capital market practitioners:

- It shows that market movements, even the wide ones, consist of several transactions with large variations followed by stabilising the course at the new price;
- The result demonstrates that in practice, price changes do not fall into any theoretical statistical distribution;
- The chosen time frame is essential for establishing the statistical distribution required by MPT, which is counter-intuitive for a practitioner.

Daily returns statistical series

It is considered a minor modification of the series of daily quotations, so:

$$\begin{cases} Y_1 \\ Y_2 \\ Y_3 \end{cases} \rightarrow \begin{cases} Y_3 \\ Y_1 \\ Y_2 \end{cases}, \qquad (7)$$

with the new series named Modified DJIA, MDJIA.

There were permutated three successive days' quotations, than the procedure was extended to the entire range formulas, in Microsoft Excel. As shown in Figure 8, the result (chart progress) does not change significantly. Any investor/speculator, except intraday, does not make any distinction between the two assets: DJIA and MDJIA.

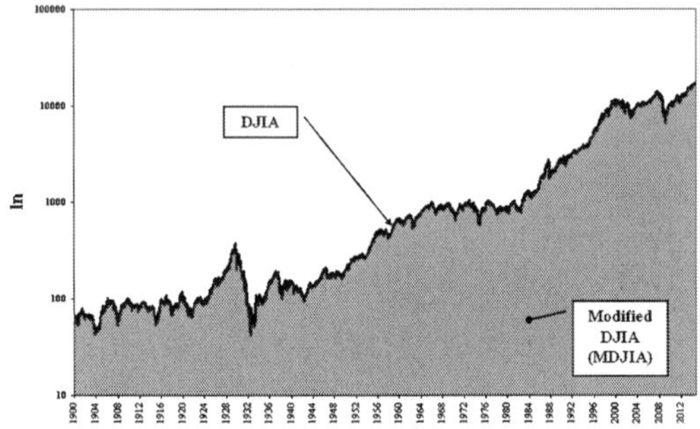

Figure 8. Similar charts for DJIA and MDJIA. Source: Authors' estimation.

Statistically speaking, permutations affect daily returns, so equation (1) is not invariant under the transformation (6). Analysing the data in Excel, the results are summarised in Table 1.

Table 1. Statistical analysis of DJIA and MDJIA.

	DJIA	MDJIA	Δ
Mean	**0.000184429**	**0.000185583**	**0.63%**
Standard Error	0.000064617	0.000101218	56.64%
Median	0.000484174	0.000325487	-32.77%
Standard Deviation	**0.011158064**	**0.017478251**	**56.64%**
Sample Variance	0.000124502	0.000305489	145.37%
Kurtosis	25.99698121	27.52381909	5.87%
Skewness	-0.818790328	-0.542106195	-33.79%
Range	0.399797287	0.669873139	67.55%
Minimum	-0.256324765	-0.366427657	42.95%
Maximum	0.143472522	0.303445482	111.50%
Sum	5.499303252	5.533699642	0.63%
Count	29818	29818	

Source: Authors' calculation based on data available from stooq.com.

The average value does not change substantially, a phenomenon easy to predict because the overall result is the same: the start and the end are the same. Instead, the standard deviation is radically different, so from the point of view of MPT, MDJIA appears much more risky than the DJIA. This is clearly contradicted in practice, however, being able to guarantee any technical analyst, on the graphic bases that both of them have equal absolute returns and risks.

Moreover, the two series give a completely surprising result: CORRELDJIA;MDJIA = -0.0158, meaning that the two data series are uncorrelated (independent). It follows that the same data series are correlated in terms of technical analysis (intuitive), but not correlated statistically and in terms of portfolio selection based on MPT.

This result leads us to conclude that MPT has a major deficiency: it was created as a way for portfolio diversification to overcome non-systemic risks. Or, this result is missed by mathematical formalism, as demonstrated by the following argument:

Don't put all your eggs in one basket is an idiom, including actions: if you drop the basket, all is lost. An important consequence would be the need to diversify, to disperse non-systemic risk; a prerequisite of MPT. Or, just this prerequisite is lost from the math of MPT.

From MPT it is known (Altăr, 2002) that an optimal Paretto portfolio corresponds to the following relationship between profitability and variance:
$$\sigma^2 = \frac{1}{D} \cdot \left[A\rho^2 - 2B\rho + C\right]$$

where:

$A = 1^T\Omega^{-1}1, B = 1^T\Omega^{-1}\mu, C = \mu^T\Omega^{-1}\mu, D = AC - B^2, \rho = x^T\mu$, with vectors **μ** of yields and **x** of weights. It is noted that the covariance matrix **Ω** must be reversible, i.e., it is not allowed for securities to be perfectly correlated, positive or negative $(\rho_{ij} \neq \pm 1)$.

Therefore, if the securities are perfectly correlated, MPT cannot make the selection, considering them identical and thus reducing the size of the matrix until it is reversed. But these securities are best suited for removing systemic risk: if an investor appreciates a security to be best suited for meeting his requirements, it will acquire spoken security. Hence, he exposes to non-systemic risks (depending only on the issuer), being exposed, for instance, to a particular problem of the issuer. But if he finds more issuers with identical risk-yield profile, he can build a diversified portfolio, in which the risk that one issuer has a problem is greatly diminished. This precise aspect is missed by the MPT, because it cannot work with the **Ω** singular covariance matrix.

In conclusion, a different approach to portfolio construction methodology is welcome, thanks to the theoretical and practical criticism of current theories. Standard statistical correlation imposed by the mathematical theory of MPT does not meet the needs of practitioners.

We are currently working on a new approach in what concerns the correlation measuring between two securities, based on measuring the rate of return in one security, determined by the variation of the other.

Implications in portfolio selection

By tracing the efficient frontier, it is confirmed that the above situation is confusing for investors.

Because daily statistical data are too small, annual data were considered, namely the average and standard deviation corresponding to 256 trading days. The results were processed in VisualMVO, and then plotted in Figure 9. The absurd character of the situation is clear: although the DJIA and MDJIA are equivalent in practical terms, MPT can build an effective border, wherefrom one can choose different portfolios.

Notes: Chart obtained by authors in VisualMVO. The result is not useful in practical terms, because the two titles are similar.

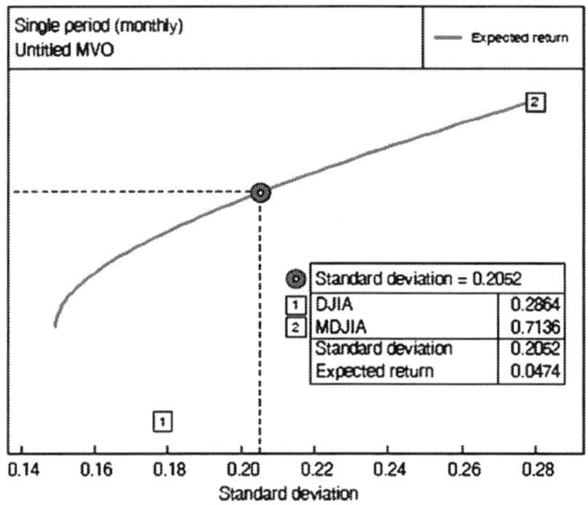

Figure 9. The DJIA+MDJIA portfolio. Source: Authors' estimation.

It may be objected that the example shown is unreal and specifically chosen to distort the results. For a technical analyst, using (Thomsett, 2012b) candlestick charting (see Figure 8, taken from the Wiener Börse site), the order of quotations matters; the daily quotation may not be permutated in accordance with Formula (5).

The criticism is not justified, if we analyse in detail the theoretical premises. MPT states that future estimates must be considered and not the historical values. Consider two analysts who judge a title similar evolution:

they forecast MDJIA and DJIA charts, which are similar. Their findings will be divergent: only the average is the same; the dispersion and the correlation are completely different, as shown in Figure 10, where Japanese formations depend on daily price movements. Therefore, they achieve results which behave differently in terms of MPT, which contradict common sense and the practical reality.

Figure 10. Japanese formations for technical analysis. Source: Authors' processing from data available on the Wiener Börse website.

CONCLUSION

The academic importance of the article is derived from our multiple attempts in signalling the limitations of the MPT in practice.

The practical importance of the article is derived from the observation that MPT has some limitations. We highlight some of these limitations.

Although MPT is a solid mathematical theory, its application is based on input data estimated from statistical selection of operations that can be sometimes precarious, subjective and can be easily manipulated.

In our opinion, the reasons of MPT's practical deficiencies are:

- Statistical consideration of returns. Based on our experience of the stock market, investors are concerned about the behaviour of the asset in relation to market trends.
- The statistical analysis of daily returns forms a static, rigid frame, for title or market characterisation (in intermarket portfolios). In

our opinion, the direction of analysis should be the mathematical dynamic expression of trends, in contrast with this static vision.

- According to the literature and the market experience standard deviation of the return is not a proper measure of risk.
- In a portfolio, volatility can be analysed regarding the investment period. Portfolio analysis must be performed frequently enough to cover the natural volatility of the security/market. But risk measure must also be related to trend and market crises.

The charts interpretation is straightforward, trends and sub-trends are obvious; nevertheless, their analytical description is difficult and complex. It is considered that the determination and analytical expression of the trend is a particular important problem that remains open, along with the measurement of asset's adaptability to the overall market trends.

Based on our experience regarding stock exchange markets, we have identified some innacuracies in applying MPT in practice. Some innacuracies can appear also due to the evaluation of the linkage between titles variation throughout the covariance matrix. We consider that this method is not an efficient tool for measuring the relationship between the titles variation.

The article brings a new result about the fact that graphs remain similiar in successive logarithmic operations and show that relatively small variations of statistical values have major effects on the results of the selected securities.

The study will be continued with graphic designs for selecting weights in portfolio, with studies on the correlation between securities and other econo-physical aspects that shape the capital market by similar physical phenomena and with mathematical modelling and simulation of trends and sub-trends value of securities and market indices.

REFERENCES

1. Altăr, Moisă (2002). *Portfolio theory*. Romania: The Bucharest Academy of Economic Studies.
2. Arratia, A. (2014). *Computational finance. An introductory course with R*. Barcelona: Atlantis Press. 10.2991/978-94-6239-070-6
3. Baker, H. K., & Filbeck, G. (2013). *Portfolio theory and management*. Oxford: Oxford University Press. 10.1093/acprof:oso/9780199829699.001.0001
4. Chen, J. (2010) *Essentials of technical analysis for financial markets*. New Jersey: John Wiley & Sons, Inc. [Google Scholar]
5. Di Lorenzo, R. (2013). *Basic technical analysis of financial markets. A modern approach*. Milano: Springer. 10.1007/978-88-470-5421-9
6. Elton, E. J., Gruber, M. J., Brown S. J., Goetzmann, W. N. (2014). *Modern portfolio theory and investment analysis. ninth edition*. New Jersey: John Wiley & Sons, Inc.
7. Fărcaş, P., & Cuzman, I. (2002a). *A selection model for efficient diversified equity portfolios. Bulletins for Applied & Computer Matematics (Bam-2035)*. Budapest: Pamm Centre, pp. 83–94.
8. Fărcaş, P., & Cuzman, I. (2002b). Comparative study of the sharpe and markowitz methods. Part I. Methods for portfolio optimization according to the risk – rentability criteria. *"Aurel Vlaicu" University of Arad Annals, Economics Series, 12*, 329–335.
9. Fărcaş, P., & Deac, D. (2010) Portfolio management of stocks quoted at the bucharest stock exchange and the statistical connections of the quotation variations. *Studia Universitatis "Vasile Goldi□" Arad, Economics Series*, Year 20/2010, Part II, *21*, 167–176, Arad, Romania.
10. Fărcaş, P., & Fărcaş, R.-C. (2008) Numerical methods for optimization of the financial assets portfolios. *Studia Universitatis "Vasile Goldiş" Arad, IT Series*, No. 1x/2008, "Vasile Goldiş" University Press. Vol. I, 150-159. [Google Scholar]
11. Francis, J. C., & Kim, D. (2013). *Modern portfolio theory. Foundations, analysis, and new developments*. Hoboken, New Jersey: John Wiley & Sons, Inc.
12. Levy, H. (2012). *The capital asset pricing model in the 21st century analytical, empirical, and behavioral perspectives*. Cambrigde, UK: Cambridge University Press.

13. Mandelbrot, B. B. (2008). *How fractals can explain what's wrong with wall street*. New York, NY: Scientific American.
14. Markowitz, H. M. (1952). Portfolio selection. *The Journal of Finance, 7*, 77–91. doi:10.2307/2975974. JSTOR 2975974.
15. Marty, W. (2013). *Portfolio analytics. An introduction to return and risk measurement*. Geneve: Springer International Publishing Switzerland. 10.1007/978-3-319-03509-3
16. Murphy, J. J. (1999). *Technical analysis of the financial markets. A comprehensive guide to trading methods and applications*. New York, NY: New York Institute of Finance.
17. Murphy, J. J. (2004). *Intermarket analysis. Profiting from global market relationships*. Hoboken, NJ: John Wiley & Sons, Inc.
18. Reilly, F., & Brown, K. (2012). *Investment analysis & portfolio management*. Mason, OH: FinancialEbooks.net, Business School Edition.
19. Schulmerich, M., Leporcher, Y.-M., & Eu, C.-H. (2015). *Applied asset and risk management. A guide to modern portfolio management and behavior-driven markets*. Berlin Heidelberg: Springer-Verlag. 10.1007/978-3-642-55444-5
20. Sharpe, W. F. (1964). Capital asset prices: A theory of market equilibrium under conditions of risk*. *The Journal of Finance, XIX* No. 3, 425–442.
21. Sironi, P. (2015). *Modern portfolio management: From Markowitz to probabilistic scenario optimisation. Goal-Based and long-term portfolio choice*. London: Incisive Media Investments Ltd.
22. Sortino, A., & Price, L. (1994). Performance measurement in a downside risk framework. *The Journal of Investing, 3*, 59–64. doi:10.3905/joi.3.3.59.
23. Sortino, F. A., & Satchell, S. E. (2001). *Managing downside risk in financial markets: Theory, practice and implementation*. Reed Educational and Professional Publishing Ltd.
24. Sortino, F., et al. (2010). *The Sortino framework for portfolio construction: Focusing on desired target return to optimize upside potential relative to downside risk*. Elsevier Inc.
25. Thomsett, M. C. (2012a). *Bloomberg visual guide to candlestick charting*. Hoboken, New Jersey: John Wiley & Sons Inc. 10.1002/9781119204923

26. Thomsett, M. (2012b). *Technical analysis of stock trends explained*. Ethan Hathaway: An Easy-to-Understand System for Trading Successfully.
27. Todea, A. (2006). *Investment. selection and financing the investment projects. Financial investments – portfolio management*. Cluj – Napoca: Casa Căr□ii de □tiin□ă Publisher.
28. Vollmer, M. (2015). *A beta-return efficient portfolio optimisation following the CAPM. An Analysis of International Markets and Sectors*. Wiesbaden: Springer Fachmedien. 10.1007/978-3-658-06634-5

CHAPTER

10

A General Framework for Portfolio Theory. Part III: Multi-Period Markets and Modular Approach

Stanislaus Maier-Paape[1], Andreas Platen[1], and Qiji Jim Zhu[2]

[1]Institut für Mathematik, RWTH Aachen University, D-52062 Aachen, Germany; maier@instmath.rwth-aachen.de

[2]Department of Mathematics, Western Michigan University, Kalamazoo, MI 49008, USA

ABSTRACT

This is Part III of a series of papers which focus on a general framework for portfolio theory. Here, we extend a general framework for portfolio theory in a one-period financial market as introduced in Part I [Maier-Paape and Zhu, Risks 2018, 6(2), 53] to multi-period markets. This extension is

Citation: Maier-Paape, S.; Platen, A.; Zhu, Q.J. "A General Framework for Portfolio Theory. Part III: Multi-Period Markets and Modular Approach". Risks 2019, 7, 60. https://doi.org/10.3390/risks7020060

reasonable for applications. More importantly, we take a new approach, the "modular portfolio theory", which is built from the interaction among four related modules: (a) multi period market model; (b) trading strategies; (c) risk and utility functions (performance criteria); and (d) the optimization problem (efficient frontier and efficient portfolio). An important concept that allows dealing with the more general framework discussed here is a trading strategy generating function. This concept limits the discussion to a special class of manageable trading strategies, which is still wide enough to cover many frequently used trading strategies, for instance "constant weight" (fixed fraction). As application, we discuss the utility function of compounded return and the risk measure of relative log drawdowns.

Keywords:- Portfolio theory; modular portfolio theory; efficient frontier; trading strategy; multi-period market model; arbitrage; bond replicating; risk-free; relative log drawdown

INTRODUCTION

This is Part III of a series of papers which focus on a general framework for portfolio theory. We laid out a general framework for portfolio theory in a one-period financial market for trading-off between reward and risk in Part I (Maier-Paape and Zhu 2018a) and addressed specifically drawdown risk measures in Part II (Maier-Paape and Zhu 2018b). Furthermore, a fourth part is planned where we provide a case study on how to implement the general framework in real financial markets. Here, in Part III, we extend the general framework for one-period financial markets to multi-period financial markets and go beyond the setting of a finite sample space.

In addition to extending the framework in Part I (Maier-Paape and Zhu 2018a) to more general settings, we now take a modular approach in organizing this more general framework for portfolio/trading theory. We recognize the problem of trading-off between higher reward and lower risk using portfolio/trading strategies within four modular blocks (modular portfolio theory): (a) multi-period market model; (b) trading strategies; (c) risk and utility functions; and (d) the optimization problem.

The multi-period market is assumed to consist of one risk-free and $M \in \mathbb{N}$ risky assets. The trading strategy is parameter dependent and specifies how the investor, once started the investment, wants to trade the portfolio over time. Based on this, a (convex) risk function r and a (concave) reward/utility function u can be defined, which manifest in an optimization problem of the form

$$\min_{x \in A} \mathrm{r}(x) \quad \text{subject to } \mathrm{u}(x) \geq \mu,\ S_0^{\top} x = \beta, \tag{1}$$

where $A \subset R^{M+1}$, $\mu \in R$ and $\beta > 0$ are fixed and $S_0 \in R^{M+1}$ gives the initial price values of the M+1 assets. This just sketches the situation discussed here. Especially the role of the trading strategy and the meaning of the vector $x \in R^{M+1}$ are discussed in more detail below.

Using a one-period market, there is nothing to do in Block (b) because the portfolio just consists of a simple portfolio vector x, which gives the weights. This one-period case is extensively studied in the literature. One of the first discussions on optimization problems of the form of Equation (1) was done by Markowitz (1952, 1959), the so-called modern portfolio theory, and afterwards with the capital asset pricing model (CAPM) by Lintner (1965); Mossin (1966); Sharpe (1964). In both settings, the risk function is defined by the standard deviation of the returns of the portfolio and the utility function is the mean return of the portfolio. The only part which can be chosen in this work could be the specific one-period market model because Block (b) is trivial, Block (c) is already fixed and Block (d) is of the form of Equation (1), in general with $A = R^{M+1}$. Decades later, Rockafellar et al. (2006) and also Part I (Maier-Paape and Zhu 2018a) discussed a more general setting where Block (c) gets more degrees of freedom regarding the choice of the risk function and the utility function. In Rockafellar et al. (2006), the risk function is allowed to be more general with some specific assumptions, so-called deviation measures, but the utility function still is the (arithmetic) mean return. In Maier-Paape and Zhu (2018a), in addition, the utility function is of more general form with some reasonable assumptions and the one-period market model is assumed to be defined on a finite probability space. In both cases the optimization problem is of the form of Equation (**1**) as well. The idea of using a multi-period market model together with trading strategies as building blocks for a modular portfolio theory was firstly introduced by Platen (2018). Accordingly, we here enhance on this idea and develop an in itself complete and compact approach to this new aspect of portfolio theory. Of course, multi-period market models have been used before, though not in the context of portfolio theory. For an introduction to this topic we refer to (Föllmer and Schied 2016, Section 5.1) and also (Carr and Zhu 2018, Chapter 3).

The generalization to multi-period markets for portfolio purposes is important in applications. In practice, investors and regulators always need to make decisions at different phases of financial markets under different policy environments. Moreover, many important market operations such as hedging and pricing of options and other contingent claims have to be dealt with in a multi-period financial market setting. Finally, the multi-period financial market model is crucial in adequately modeling certain important reward and risk measures such as compounded return and drawdown related risk measures. The absolute drawdown was already discussed, e.g., by Chekhlov et al. (2003, 2005); Goldberg and Mahmoud (2017) and Zabarankin et al. (2014). The risk functions used therein are based on the ideas of the value at risk but applied to the absolute drawdown. The relative drawdown is much more involved and rarely discussed in the literature. Grossman and Zhou (1993) studied an optimization problem using the maximum relative drawdown and a geometric Brownian motion with drift as market model where just one risky and also one risk-free asset are assumed. Cvitanić and Karatzas (1995) extended the results for more than one risky asset and Cherny and Obłój (2013) discussed the setting using an abstract semimartingale financial market and more general utility and risk functions. Properties on the mean of the logarithm of the relative drawdown are discussed in Part II (Maier-Paape and Zhu 2018b), where a one-period market model is used with a finite probability space and independent and identically distributed returns. The trick to construct the drawdown measure is to generate short equity curves by random drawings from the return distribution and to subsequently calculate the drawdowns of the equity curves. The relative drawdown for a more general market model was also discussed by Platen (2018).

A more technical challenge of our extension is that the space of random variables on the sample space that represents the payoff is no longer a finite dimensional space and, therefore, no longer enjoys local compactness properties. We circumvent this difficulty by introducing a trading strategy generating function. Doing so, we limit ourselves to a special class of manageable trading strategies. We illustrate by examples that the class of trading strategies we study here is wide enough to include many frequently used trading strategies such as "buy and hold" or "constant weight" (fixed fractions). Another strategy could be to fix the amount of money invested over time, which was discussed by Platen (2018). Although the here used strategies seem to be simple, it still shows the potential behind this new building block in portfolio theory.

The paper is arranged as follows. In Section 2, we layout the multi-period market models (Building Block (a)) and trading strategies (Building Block (b)) and derive several basic properties such as the fundamental characterization of a multi-period market with no nontrivial risk-free trading strategy (see Theorem 1). In Section 3, we discuss our main results according to the modular approach. After giving examples for the risk and utility functions (Building Block (c)) based on Blocks (a) and (b) in Section 3.1, the optimization problem (Building Block (d)) is introduced in Section 3.2. The corresponding notion of efficient frontier is extensively studied, e.g., in terms of graphs (see Section 3.3) and the main theorems for the existence (and uniqueness) of solutions are derived in Section 3.4. An application of the theory for the compounded return and the expected log relative drawdown is discussed in detail in Section 4. To measure the return in risk and utility function as relative log returns has two reasons. Firstly, it yields the necessary convexity and concavity, respectively. Secondly, it guarantees that drawdowns and runups are measured equally. For instance, a drawdown of 50% needs a runup of 100% for compensation. Taking log relative returns, the absolute value of both movements is equal. The paper ends with some conclusions in Section 5.

Before we continue, we would like to emphasize once more that this Part III is an advancement of Parts I and II Maier-Paape and Zhu (2018a, 2018b). Although this Part III is self-contained, complications stemming from the multi-period market model and the general probability space make this Part III more involved. Therefore, the reader might it find helpful to compare results here with the "easy" case in Parts I and II (see Maier-Paape and Zhu (2018a, 2018b)). For instance, the "core" theory here on optimization problems such as in Equation (1) and efficient portfolios is very much in the spirit of Part I (Maier-Paape and Zhu 2018a) (cf. Section 3.2, Section 3.3 and Section 3.4). Similarly, Section 3.1 and Section 4, where we lift the drawdown risk measures to multi-period markets, use many ideas of Part II (Maier-Paape and Zhu 2018b). On the other hand, the contents of Section 2 on theory of multi-period markets and trading strategies has no counterpart in Parts I and II Maier-Paape and Zhu (2018a, 2018b) but many connections to classical financial mathematics (e.g., Föllmer and Schied (2016)). As a matter of fact, Section 2 is a bit technical. However, since it only provides the framework for the core theory later on, it might help at first reading to concentrate oneself in Section 2 solely on the definitions.

MULTI-PERIOD MARKET AND TRADING STRATEGIES

In this section, we describe a multi-period (financial) market model. In such a model, investment decisions are made over several periods with potentially different investment environments characterized by different economic, financial and policy situations. The role of portfolios is replaced by trading strategies which can be viewed as a sequence of portfolios varying in time according to an a priori given, but possibly random, strategy. The information on the investment environment is revealed with the progress of time and the action of the trading strategy is contingent on the existing information. The availability of the information is modeled by a filtration. This section lays a foundation for the subsequent analysis.

Definitions

The following notion of a multi-period market is closely related to (Föllmer and Schied 2016, Section 5.1) and (Platen 2018, Sections 2.1.2, 2.2.1, and 2.2.2). We assume that M+1 financial instruments (one risk-free asset with index 0 and M∈N risky assets with indexes 1,…,M) are given. Their initial prices are denoted by $S_0 := (S_0^0, S_0^1, \ldots, S_0^M) \in \mathbb{R}_{>0}^{M+1}$. A model for N∈N future time steps is of the following form:

Let (Ω,Σ,P) be a probability space. By L2:=L2(Ω,Σ,P):=L^2(Ω,Σ,P;R) we denote the set of all random variables X:Ω→R with finite norm $\|X\|_{L2}:=(E[X^2])^{1/2}$, where $\langle X,Y\rangle_{L2}:=E[XY]$ for X,Y∈L^2 is the inner product. For a set of M+1 assets, we define $L^2(\Omega,\Sigma,P;R^{M+1})$ where each of the M+1 components of the elements are in L^2. This could model a one-period market. For a multi-period market model, let $(\Omega,\Sigma,\{Fn\}_{0\leq n\leq N},P)$ be a filtered probability space with filtration $\{Fn\}_{0\leq n\leq N}$ which satisfies

$$\{\emptyset,\Omega\} =: \mathcal{F}_0 \subset \mathcal{F}_1 \subset \cdots \subset \mathcal{F}_{N-1} \subset \mathcal{F}_N := \Sigma.$$

Define

$$\left(N;\mathbb{R}^M\right) := \mathcal{L}^2\left(\Omega,\Sigma,\{\mathcal{F}_n\}_{0\leq n\leq N},\mathrm{P};\mathbb{R}^M\right) := \bigtimes_{n=0}^{N} \mathcal{L}^2\left(\Omega,\mathcal{F}_n,\mathrm{P};\mathbb{R}^M\right)$$

and $L^2(N):=L^2(N;R)$ with corresponding set of positive processes

$$\mathcal{L}^2\left(N;\mathbb{R}^M_{>0}\right) := \mathcal{L}^2\left(\Omega,\Sigma,\{\mathcal{F}_n\}_{0\leq n\leq N},\mathrm{P};\mathbb{R}^M_{>0}\right) := \bigtimes_{n=0}^{N} \mathcal{L}^2\left(\Omega,\mathcal{F}_n,\mathrm{P};\mathbb{R}^M_{>0}\right) \subset \mathcal{L}^2\left(N;\mathbb{R}^M\right).$$

Analogously, L^0 denotes the set of all random variables and L^1 the random variables with finite (absolute) expectation. In most cases, we use L^2, which, however, often is not required. In these cases, one could also use, e.g., L0.

Firstly, we define the notion of risk-free which means that there is no uncertainty and the price development is (not necessarily strictly) monotone increasing.

Definition 1 (Risk-free asset).

The stochastic process $Z:=(Z_0,Z_1,\ldots,Z_N)\in L^2N$ *is called* risk-free *if* Zn *is constant almost surely (a.s.) for* n=0,1,…,N *and* Zn≥Zn−1>0 *a.s. for* n=1,…,N.

Definition 2

(Multi-period market model, cf. (Föllmer and Schied 2016, Section 5.1)). *For* M∈N *let* $S := (S_n)_{0\leq n\leq N} \in \mathcal{L}^2(N;\mathbb{R}^{M+1}_{>0})$ *with*

$$S_n := \left(S_n^0, S_n^1, \ldots, S_n^M\right)^{\top} \in \mathbb{R}_{>0} \times \mathcal{L}^2\left(\Omega, \mathcal{F}_n, \mathrm{P}; \mathbb{R}^M_{>0}\right) \subset \mathcal{L}^2\left(\Omega, \mathcal{F}_n, \mathrm{P}; \mathbb{R}^{M+1}_{>0}\right), \tag{2}$$

where $(S_0^0, S_1^0, \ldots, S_N^0) \in \mathcal{L}^2(N)$*, i.e., the asset with index zero, is risk-free. The stochastic process S is called a* multi-period market model *of size* M+1 *with N time steps.*

A portfolio in a one-period market model just contains of a single vector which gives the weights for each asset. In a multi-period market model the situation is much more complex. After each time step we can change the weights. We even can change the weights, say after time step *n*, based on information of all past time steps up to step *n*. Hence, in our situation, we denote a series of time varying portfolios by a trading strategy as follows.

Definition 3 (Trading strategy).

For M,N∈N *let* $S := (S_n)_{0\leq n\leq N} \in \mathcal{L}^2(N;\mathbb{R}^{M+1}_{>0})$ *be a market model. A time dependent vector*

$$X := (x_n)_{1\leq n\leq N} \in \mathcal{L}^0\left(N-1;\mathbb{R}^{M+1}\right) := \bigtimes_{n=0}^{N-1} \mathcal{L}^0\left(\Omega, \mathcal{F}_n, \mathrm{P}; \mathbb{R}^{M+1}\right),$$

for the same filtered probability space $(\Omega,\Sigma,\{F_n\}_{0\leq n\leq N},P)$ *from the market model, with*

$$x_n := \left(x_n^0, x_n^1, \ldots, x_n^M\right)^{\top} \in \mathcal{L}^0\left(\Omega, \mathcal{F}_{n-1}, \mathrm{P}; \mathbb{R}^{M+1}\right), \tag{3}$$

for n=1,…,N, *is called a* trading strategy.

We still need to give the trading strategy a meaning and a real connection to the market model *S*. The values of a trading strategy $X=(x_n)_{1\leq n\leq N}$ have the following interpretation.

- x_n may depend on $S_0,\ldots,S_{n-1}$ but not on later prices;
- S_n^i absolute price of the *i*th asset at time *n*;
- x_n^i: number of shares invested into the *i*th asset from time step n−1 to *n*;

- $S^i_{n-1}x^i_n$: amount of money invested into the *i*th asset;
- $S^i_n x^i_n$: absolute value of this investment after the time step from n−1 to *n*; and
- $S_n^\top x_n = \sum_{i=0}^{M} S^i_n x^i_n$: absolute value of all investments after the time step from n−1 to *n*.

Note that Equations (2) and (3) imply that S_n is F_n measurable while x_n is F_{n-1} measurable. The reason is that x_n are the number of shares for each asset hold from time step n−1 to *n*. This must be known at time step n−1 where the shares have to be bought. Hence, it must be F_{n-1} measurable. The prices S_n, of course, are known not before time *n*, i.e., it must be F_n measurable. Using this, we can define the wealth process realized by a trading strategy applied to the market model.

Definition 4 (Wealth of trading strategy).

Let $S \in \mathcal{L}^2(N;\mathbb{R}^{M+1}_{>0})$ *be a market model and* $W_0 \in R>0$ *the investor's fixed initial wealth. For a trading strategy* $X = (x_n)_{1\le n\le N} \in \mathcal{L}^0(N-1;\mathbb{R}^{M+1})$*, the* wealth process $W(X) \in L^0(N)$ *is defined by*

$$\mathcal{W}_0(X) := \mathcal{W}_0$$

$$\mathcal{W}_n(X) := \mathcal{W}_n(x_1,\dots,x_n) := \mathcal{W}_{n-1}(X) + (S_n - S_{n-1})^\top x_n = \mathcal{W}_0 + \sum_{k=1}^{n} (S_k - S_{k-1})^\top x_k \quad (4)$$

for n=1,…,N. *Hence,* $W: L^0(N-1;R^{M+1}) \to L^0(N)$ *is an affine linear functional.*

Note, that, whenever $X \in L^2(N-1;R^{M+1})$, then, by the Cauchy–Schwarz inequality, we obtain that $W(X) \in L1(N)$.

Even though the market model consists of positive stochastic processes, we may open short positions using a trading strategy. Hence, total ruin may occur. Since we always try to avoid a ruin, we define the set of admissible trading strategies.

Definition 5 (Admissible trading strategy).

A trading strategy X which satisfies $W_n(X)>0$ *a.s. for* n=1,…,N *is called* admissible. *The set of all admissible trading strategies is denoted by* $A=A(S) \subset L^0(N-1;R^{M+1})$.

Note that A is a convex set.

Properties of the Multi-Period Market Model

The most important scenario we try to avoid is total ruin. This strongly depends on the trading strategy. However, the opposite should also be impossible, i.e., it should not be possible to gain money without risk, namely arbitrage.

This property strongly depends on the market model itself. The literature mostly discusses the notion of arbitrage, where an arbitrage opportunity beats the risk-free asset with positive probability while it is never worse than the risk-free asset. In Part I (see (Maier-Paape and Zhu 2018a, Section 3.1)), the notion of a risk-free portfolio is introduced for the one-period market and a finite probability space. A risk-free portfolio is almost the same as an arbitrage opportunity but does not have to beat the risk-free asset with positive probability. We discuss this kind of extension for the more general case with multi-period market models (cf. also (Platen 2018, Section 2.2.3 and Section 2.2.4) for more details).

When discussing arbitrage, the notion of self-financing is often used (see, e.g., (Föllmer and Schied 2016, Definition 5.4)), which means that all money which has been invested initially stays invested and no fresh money is invested afterwards.

Definition 6

(Self-financing, see (Föllmer and Schied 2016, Definition 5.4)). *Let* $S \in \mathcal{L}^2(N;\mathbb{R}^{M+1}_{>0})$ *be a market model for* N≥2. *A trading strategy* $X := (x_n)_{1 \le n \le N}$ *with*

$$S_n^\top x_n = S_n^\top x_{n+1} \tag{5}$$

for all n=1,…,N−1 *is called* self-financing.

The space of self-financing trading strategies is linear and simplifies the wealth process as follows.

Proposition 1.

Let $S \in \mathcal{L}^2(N;\mathbb{R}^{M+1}_{>0})$, $N \ge 2$, *be a market model. A trading strategy* X *is self-financing if and only if*

$$\mathcal{W}_n(X) = \mathcal{W}_0 + \left(S_n^\top x_n - S_0^\top x_1\right) \tag{6}$$

for all n=1,…,N. *If* $S_0^\top x_1 = \mathcal{W}_0$, *then Equation (6) becomes* $\mathcal{W}_n(X) = S_n^\top x_n$.

Proof.

If X is self-financing, then Equation (**4**) becomes a telescoping sum and directly gives Equation (**6**). On the other hand, if Equation (**6**) holds true, then, by Equation (**6**), we get

$$\mathcal{W}_{n+1}(X) = \mathcal{W}_0 + \left(S_{n+1}^{\top} x_{n+1} - S_0^{\top} x_1\right)$$

and, by Equation (**4**) together with Equation (**6**), we obtain

$$\mathcal{W}_{n+1}(X) = \mathcal{W}_n(X) + (S_{n+1} - S_n)^{\top} x_{n+1} = \mathcal{W}_0 + \left(S_n^{\top} x_n - S_0^{\top} x_1\right) + (S_{n+1} - S_n)^{\top} x_{n+1}$$

for n=1,…,N−1. Equating both expressions gives $S_n x_n = S_n x_{n+1}$ for all n=1,…,N−1, i.e., *X* is self-financing.

Remark 1 (Bond).

Let $Z := (z_n)_{1 \le n \le N}$ *be the trading strategy which represents the bond, i.e.,*

$$z_n := \left(\frac{\mathcal{W}_0}{S_0^0}, 0, \ldots, 0\right)^{\top} \quad \text{for } n = 1, \ldots, N. \tag{7}$$

Of course, Z is self-financing with $S_0^{\top} z_1 = \mathcal{W}_0$. *Therefore, Proposition 1 gives*

$$\mathcal{W}_n(Z) = S_n^{\top} z_n = \mathcal{W}_0 \frac{S_n^0}{S_0^0} \quad \text{for } n = 1, \ldots, N, \tag{8}$$

A trading strategy *X* is called *trivial*, if $\widehat{X} \equiv 0$ a.s., where $\widehat{X} := (\widehat{x}_n)_{1 \le n \le N}$ with $\widehat{x}_n := (x_n^1, \ldots, x_n^M)^{\top}$ denotes the risky part of *X*. Analogously, we define the risky part of by $(\widehat{S}_n)_{0 \le n \le N}$, where $\widehat{S}_n := (S_n^1, \ldots, S_n^M)^{\top}$.

The following notion of arbitrage opportunity, bond replicating and risk-free trading strategy is related to (Föllmer and Schied 2016, Definition 5.10), see (Platen 2018, Remark 2.2.17).

Definition 7 (Arbitrage opportunity, bond replicating, and risk-free).

Let $S \in \mathcal{L}^2(N; \mathbb{R}_{>0}^{M+1})$ *for* M,N∈N *be a market model and* $X := (x_n)_{1 \le n \le N}$ *a trading strategy.*

(a). We say X is risk-free *if*

$$S_{n-1}^{\top} x_n \le \mathcal{W}_{n-1}(X) \quad \text{a.s. for all } n = 1, \ldots, N \text{ and } \mathcal{W}_N(X) \ge \mathcal{W}_0 \frac{S_N^0}{S_0^0} \quad \text{a.s.} \tag{9}$$

We say market model S has no nontrivial risk-free trading strategy *if there does not exist a risk-free trading strategy X with* $\widehat{X} \not\equiv 0$ *(i.e., besides the trivial ones with* $\widehat{X} = 0$ *a.s. there are no risk-free trading strategies).*

(b) We say X is an arbitrage opportunity *if*

$$S_{n-1}^{\top} x_n \le \mathcal{W}_{n-1}(X) \quad \text{a.s. for all } n = 1, \ldots, N, \qquad \mathcal{W}_N(X) \ge \mathcal{W}_0 \frac{S_N^0}{S_0^0} \quad \text{a.s.,}$$

and

$$\mathrm{P}\left(\mathcal{W}_N(X) > \mathcal{W}_0 \frac{S_N^0}{S_0^0}\right) > 0.$$

We say market model S is arbitrage-free, *if there does not exist any arbitrage opportunity.*

(c) . We say X is bond replicating *if*

$S_{n-1}^{\top} x_n \leq W_{n-1}(X)$ *a.s. for all* $n = 1, \ldots, N$ *and* $W_N(X) = W_0 \frac{S_N^0}{S_0^0}$ *a.s.*

We say market model S has no nontrivial bond replicating trading strategy, *if there does not exist a bond replicating trading strategy X with* $\widehat{X} \neq 0$ *(i.e., besides the trivial ones with* $\widehat{X} = 0$ *a.s. there are no bond replicating trading strategies).*

Remark 2 (Interpretation of Definition 7).

The first property of a risk-free trading strategy in Equation (9) says that, at time step n−1, *no more than the available capital is invested. The second property in Equation (9) means that the final wealth of the trading strategy is always at least as much as the final wealth of the bond strategy according to Remark 1.*

An arbitrage opportunity has the same properties, but on top of that the strategy wins strictly more than the bond strategy with positive probability.

A bond replicating trading strategy is also not allowed to invest more than the available capital. In this case, the final wealth has to be exactly the same as for the bond strategy.

The next result gives necessary and sufficient conditions for a market model having no nontrivial risk-free trading strategy. Those conditions are important when looking at properties for risk and utility measures on such market models. Another essential property regarding uniqueness is that two different trading strategies should result in two different wealth processes, which is ensured by the addition in the next result.

Theorem 1 (Multi-period market model with no nontrivial risk-free trading strategy).

The following assertions are equivalent:

(a) *S has no nontrivial risk-free trading strategy.*

(b) *S is arbitrage-free and has no nontrivial bond replicating trading strategy.*

(c) *For all* $n=1,\ldots,N$ *and all* $\eta \in L^0(\Omega, \mathcal{F}_{n-1}, P; \mathbb{R}^{M+1})$ *with* $\hat{\eta} \neq 0$ *it is*

$$P\left(\left(S_n - \frac{S_n^0}{S_{n-1}^0} S_{n-1}\right)^{\top} \eta < 0\right) > 0. \quad (10)$$

(d) *S is arbitrage-free and the following holds for all trading*

strategies X and Y:

$$\mathcal{W}(X) = \mathcal{W}(Y) \text{ a.s. implies } \widehat{X} = \widehat{Y} \text{ a.s.} \tag{11}$$

If in addition $S_n^0 > S_{n-1}^0$ *for all* n=1,…,N *and one of the (a), (b), (c) or (d) holds, then the mapping* W *is injective, i.e.,* W(X)=W(Y) *a.s. implies* X=Y *a.s.*

Proof.

The equivalence of (a) and (b) directly follows from Definition 7.

Proof of implication from (a) to (c): Assume this implication is wrong, i.e., assume there exist $n_0 \in \{1,\dots,N\}$ and $\eta^* \in L^0(\Omega, F_{n0-1}, P; R^{M+1})$ with $\widehat{\eta}^* \neq 0$ such that

$$1 = \mathrm{P}\left(\left(S_{n_0} - \frac{S_{n_0}^0}{S_{n_0-1}^0} S_{n_0-1}\right)^{\top} \eta^* \geq 0\right) = \mathrm{P}\left(\left(\widehat{S}_{n_0} - \frac{S_{n_0}^0}{S_{n_0-1}^0} \widehat{S}_{n_0-1}\right)^{\top} \widehat{\eta}^* \geq 0\right), \tag{12}$$

or, equivalently, $S_{n_0}^{\top} \eta^* \geq (S_{n_0}^0 / S_{n_0-1}^0) S_{n_0-1}^{\top} \eta^*$ a.s. Let trading strategy $Z := (z_n)_{1 \leq n \leq N}$ represent the bond (see Remark 1). Define $Y := (y_n)_{1 \leq n \leq N}$ by $y_{n0} := \eta*$ and $y_n := z_n$ for $n \neq n_0$. Observe that for $1 \leq n \leq n_0 - 1$ by Equation (8) now $W_n(Y) = W_n(Z)$=W0S0n/S00 holds true. Since the property in Equation (12) of η∗ is independent on its risk-free part, we can choose, without loss of generality (w.l.o.g.), the bond part $\eta^0 \in \mathcal{L}^0(\Omega, \mathcal{F}_{n_0-1}, \mathrm{P})$ of η∗ such that $S_{n_0-1}^{\top} y_{n_0} = S_{n_0-1}^{\top} \eta^* = S_{n_0-1}^0 \eta^0 + \widehat{S}_{n_0-1}^{\top} \widehat{\eta}^* = \mathcal{W}_{n_0-1}(Y) = \mathcal{W}_{n_0-1}(Z) = S_{n_0-1}^{\top} z_{n_0}$. For $n_0 \leq n \leq N$, it follows from Equations (4), (7) and (8) that

$$\begin{aligned}
\mathcal{W}_n(Y) &= \mathcal{W}_n(Z) - (S_{n_0} - S_{n_0-1})^{\top} z_{n_0} + (S_{n_0} - S_{n_0-1})^{\top} \eta^* \\
&= \mathcal{W}_0 \frac{S_n^0}{S_0^0} - \mathcal{W}_0 \frac{S_{n_0}^0 - S_{n_0-1}^0}{S_0^0} + S_{n_0}^{\top} \eta^* - S_{n_0-1}^{\top} \eta^* \\
&\geq \mathcal{W}_0 \frac{S_n^0}{S_0^0} - \mathcal{W}_0 \frac{S_{n_0}^0 - S_{n_0-1}^0}{S_0^0} + \frac{S_{n_0}^0}{S_{n_0-1}^0} S_{n_0-1}^{\top} \eta^* - S_{n_0-1}^{\top} \eta^* = \mathcal{W}_0 \frac{S_n^0}{S_0^0} = \mathcal{W}_n(Z)
\end{aligned} \tag{13}$$

holds a.s. and using $S_{n-1}^{\top} y_n = S_{n-1}^{\top} z_n = \mathcal{W}_{n-1}(Z)$ one easily obtains with Equation (13) that $S_{n-1}^{\top} y_n \leq \mathcal{W}_{n-1}(Y)$ a.s. for all n=1,…,N. In particular, using Equation (13) for n=N, trading strategy *Y* must be risk-free and nontrivial. This contradicts (a). Hence, there cannot be such an η∗, i.e., (c) must hold.

Proof of implication from (c) to (a): Let (c) hold and assume there exists a nontrivial risk-free trading strategy X:=(xn)1≤n≤N, i.e., *X* satisfies Equation (9) and $\widehat{X} \neq 0$. Let $n_0 \in \{1,\dots,N\}$ be minimal with the property $\widehat{x}_{n_0} \neq 0$. Before time n_0, trading strategy *X* can at most invest into the bond. Hence,

$$S_{n_0-1}^{\top} x_{n_0} \leq \mathcal{W}_{n_0-1}(X) \leq \mathcal{W}_0 S_{n_0-1}^0 / S_0^0 \quad \text{a.s.} \tag{14}$$

because of the first property in Equation (9). From (c) we get that x_{n_0} satisfies Equation (10) for $\eta := x_{n_0}$ and $n = n_0$. Hence, using Equations (4), (10) and (14), we obtain that the following holds true with positive probability:

$$\mathcal{W}_{n_0}(X) = \mathcal{W}_{n_0-1}(X) + (S_{n_0} - S_{n_0-1})^\top x_{n_0} < \mathcal{W}_{n_0-1}(X) + \left(\frac{S^0_{n_0}}{S^0_{n_0-1}} - 1\right) S^\top_{n_0-1} x_{n_0} \leq \mathcal{W}_0 \frac{S^0_{n_0}}{S^0_0}.$$

Because of the second property in Equation (**9**), it must be $n_0<N$ and there must exist a maximal $n_1 \in \{n_0+1,\ldots,N\}$ such that

$$\mathrm{P}\left(\mathcal{W}_{n_1-1}(X) < \mathcal{W}_0 \frac{S^0_{n_1-1}}{S^0_0}\right) > 0 \quad \text{and} \quad \mathrm{P}\left(\mathcal{W}_{n_1}(X) \geq \mathcal{W}_0 \frac{S^0_{n_1}}{S^0_0} \,\middle|\, \mathcal{W}_{n_1-1}(X) < \mathcal{W}_0 \frac{S^0_{n_1-1}}{S^0_0}\right) = 1.$$

Define

$$\eta^\# := \begin{cases} x_{n_1}, & \text{if } \mathcal{W}_{n_1-1}(X) < \mathcal{W}_0 \frac{S^0_{n_1-1}}{S^0_0}, \\ (\mathcal{W}_0/S^0_0, 0, \ldots, 0)^\top, & \text{otherwise.} \end{cases}$$

Observe that $\widehat{x}_{n_1} \not\equiv 0$ and hence $\widehat{\eta}^\# \not\equiv 0$. Using Equations (4) and (9) it can then be shown that

$$\left(S_{n_1} - \frac{S^0_{n_1}}{S^0_{n_1-1}} S_{n_1-1}\right)^\top \eta^\# = (S_{n_1} - S_{n_1-1})^\top \eta^\# - \frac{S^0_{n_1} - S^0_{n_1-1}}{S^0_{n_1-1}} S^\top_{n_1-1} \eta^\# \geq 0 \quad \text{a.s.}$$

which contradicts (c). Hence, *S* has no nontrivial risk-free trading strategy.

Proof of implication from (a)–(c) to (d): We just need to show Equation (11) for an arbitrage-free market *S*. Let $X := (x_n)_{1\leq n\leq N}$ and $Y := (y_n)_{1\leq n\leq N}$ fulfill W(X)=W(Y) a.s. From Equation (4) it follows that $(S_n - S_{n-1})^\top (x_n - y_n) = 0$ a.s. for n=1,…,N. Now, let n∈{1,…,N} be arbitrary. W.l.o.g. it is $S^\top_{n-1}(x_n - y_n) \leq 0$. Define $\eta^\dagger := x_n - y_n + (c, 0, \ldots, 0)^\top$ with $c := -S^\top_{n-1}(x_n - y_n)/S^0_{n-1} \geq 0$. We have STn−1η†=0 and therefore

$$0 \leq (S_n - S_{n-1})^\top \eta^\dagger = \left(S_n - \frac{S^0_n}{S^0_{n-1}} S_{n-1}\right)^\top \eta^\dagger \quad \text{a.s.}$$

Because of (c), we then must have $\widehat{\eta}^\dagger = \widehat{x}_n - \widehat{y}_n \equiv 0$ a.s. Since *n* was arbitrary $\widehat{X} \equiv \widehat{Y}$ a.s. must hold, which proves Equation (11).

It remains to show the implication from (d) to (c): Since *S* is arbitrage-free, we firstly can show that for all $\eta \in \mathcal{L}^0(\Omega, \mathcal{F}_{n-1}, \mathrm{P}; \mathbb{R}^{M+1})$ with $\widehat{\eta} \not\equiv 0$ it is

P/\(Sn−S0nS0n−1Sn−1)Tη<0\/>0or(Sn−S0nS0n−1Sn−1)Tη=0a.s.(15)

$$\mathrm{P}\left(\left(S_n - \frac{S^0_n}{S^0_{n-1}} S_{n-1}\right)^\top \eta < 0\right) > 0 \quad \text{or} \quad \left(S_n - \frac{S^0_n}{S^0_{n-1}} S_{n-1}\right)^\top \eta = 0 \quad \text{a.s.} \tag{15}$$

Assume not, then there exists an $n_0 \in \{1,\ldots,N\}$ and η∗ with $\widehat{\eta}^* \not\equiv 0$ such that

$$1 = \mathrm{P}\left(\left(S_{n_0} - \frac{S^0_{n_0}}{S^0_{n_0-1}} S_{n_0-1}\right)^{\top} \eta^* \geq 0\right) \quad \text{and} \quad \mathrm{P}\left(\left(S_{n_0} - \frac{S^0_{n_0}}{S^0_{n_0-1}} S_{n_0-1}\right)^{\top} \eta^* > 0\right) > 0.$$

We can proceed as in the proof for the implication from (a) to (c) if we replace Equation (12) by Equation (16). Then, the so-constructed *Y* is still risk-free and nontrivial. In particular, Equation (13) for n=N still holds true and due to Equation (16) it even holds true with a strict inequality, at least with positive probability. This implies that the corresponding *Y* is an arbitrage opportunity. Since this is a contradiction, there cannot be such an η^*.

To show (c), i.e., to show that Equation (10) must hold true, we need to exclude the second property in Equation (15) by using Equation (11). We proof this indirectly: Assume there exist $n_0 \in \{1,\ldots,N\}$ and $\eta^\# \in \mathcal{L}^0(\Omega, \mathcal{F}_{n_0-1}, \mathrm{P}; \mathbb{R}^{M+1})$ with $\widehat{\eta}^\# \neq 0$ such that

$$\left(S_{n_0} - \frac{S^0_{n_0}}{S^0_{n_0-1}} S_{n_0-1}\right)^{\top} \eta^\# = 0 \quad \text{a.s.,}$$

i.e., $S_{n_0}^{\top}\eta^\# = (S^0_{n_0}/S^0_{n_0-1})S_{n_0-1}^{\top}\eta^\#$ a.s. Using this $\eta^\#$, we can build a trading strategy *Y* exactly as in the proof for the implication from (a) to (c) where again, w.l.o.g., $S_{n_0-1}^{\top}\eta^\# = \mathcal{W}_{n_0-1}(Y) = \mathcal{W}_{n_0-1}(Z) = \mathcal{W}_0 S^0_{n_0-1}/S^0_0$. Then, Equation (13) holds true for $\eta^*=\eta^\#$ with equality for all $n_0 \leq n \leq N$ but in particular for n=N, i.e., *Y* is nontrivial and bond replicating. We conclude that even Wn(Y)= $\mathcal{W}_0 S^0_n/S^0_0 = \mathcal{W}_n(Z)$ a.s. for n=1,…,N, i.e., W(Y)=W(Z) a.s. Then, Equation (11) implies $\widehat{Y} = \widehat{Z}$ a.s., which is a contradiction, because $\widehat{Y} \neq 0 \equiv \widehat{Z}$. Hence, (c) must hold true.

The additional result in the case $S^0_n > S^0_{n-1}$ for all n=1,…,N remains to be proved: Let W(X)=W(Y) a.s.,$\widehat{X} = \widehat{Y}$ a.s. and assume X≢Y. Then, using Equation (**4**), we get $0 = W_n(X) - \mathcal{W}_n(Y) = (S^0_n - S^0_{n-1})(x^0_n - y^0_n)$ for all n=1,…,N, a contradiction. Hence, whenever W(X)=W(Y) a.s. and$\widehat{X} = \widehat{Y}$ a.s. it must be X=Y a.s., which completes the proof.

Remark 3

(Connection to (Maier-Paape and Zhu 2018a, Section 3.1)). In the one-period case N=1, we can define $R := S^0_1/S^0_0$ *and* $x := x_1$. *If we have* $S_0^{\top} x = \mathcal{W}_0$, *then we obtain from Equation (4) that*

$$\mathcal{W}_N(X) - \mathcal{W}_0 \frac{S^0_N}{S^0_0} = \mathcal{W}_0 + (S_1 - S_0)^{\top} x - R\mathcal{W}_0 = (S_1 - RS_0)^{\top} x.$$

Hence, for all x such that ST0x=W0 *Definition 7 is equivalent to the definitions in (Maier-Paape and Zhu 2018a, Definition 4).*

Moreover, Theorem 1 implies the result (Maier-Paape and Zhu 2018a, Theorem 2) in the one-period case with finite probability space for the case $R \geq 1$ *and therefore can be seen as a generalization of (Maier-Paape and Zhu 2018a, Theorem 2). Note, that Assertion (ii) in (Maier-Paape and Zhu 2018a, Theorem 2), which corresponds to (c) in Theorem 1, includes the assumption that S is arbitrage-free. This assumption is not required in (c) of Theorem 1. See also (Platen 2018, Corollary 2.2.24) for more details.*

Trading Strategy Generating Function

In most cases, an investor already has a fixed strategy to trade the M risky assets and the bond when the initial weights vector is known. For instance, one could want to freeze the fractions of capital invested in the portfolio assets. The investor's strategy then is to reallocate the portfolio after each time step such that these fixed fractions are reestablished. Hence, we are not interested in finding the "optimal" trading strategy over all possibilities, but in the "optimal" initial weights for our fixed and well-known strategy. To have a mathematical formalism for this, we make the following definition.

Definition 8 (Trading strategy generating function).

Let $M, N \in \mathbb{N}$, $A \subset \mathbb{R}^{M+1}$ *and a market model* $S \in \mathcal{L}^2(N; \mathbb{R}^{M+1}_{>0})$ *be given. We call a function* $v\colon A \to \mathcal{L}^0(N-1; \mathbb{R}^{M+1})$, *which maps a vector* $y \in A$ *to a trading strategy, a* trading strategy generating function $X = v(y)$, *where*

$$v_n(y) \in \mathcal{L}^0\left(\Omega, \mathcal{F}_{n-1}, P; \mathbb{R}^{M+1}\right) \text{ for } n = 1, \ldots, N.$$

We say the set $A \subset \mathbb{R}^{M+1}$ *is* admissible, *if* $v(y) \in \mathcal{A}(S)$ *is an admissible trading strategy for all* $y \in A$, *i.e.,* $v(A) \subset \mathcal{A}(S)$ *(see Definition 5).*

When dealing with a one-period market model, there are always some constraints. One of the most reasonable conditions is to require that all wealth is invested into the M+1 assets and there is no cash (or the bond may simulate the cash position). In a multi-period market, the same holds true, i.e., the initial investment $x_1 = v_1(y)$ should also be fixed by, e.g., $\beta \in \mathbb{R}$ (e.g., $\beta = W^0$), such that $S_0^\top v_1(y) = \beta$. If the trading strategy generating function v in addition always gives self-financing portfolios (see Definition 6), and $\beta = W_0$, we know that after each time step the complete wealth is invested.

Under some reasonable assumptions, the following result gives the boundedness of admissible sets under the constraint $S_0^\top v_1(y) = \beta$. Note that (Maier-Paape and Zhu 2018a, Lemma 2) shows a related result for a one-period market using a general class of expected utility functions. Here, we only focus on a general trading strategy and its admissible sets. Such a result is also shown in (Platen 2018, Lemma 2.2.29).

Lemma 1.

Assume the market model $S \in \mathcal{L}^2(N;\mathbb{R}^{M+1}_{>0})$ *has no nontrivial risk-free trading strategy. Let* $v:R^{M+1} \to L_0(N-1;R^{M+1})$ *be a trading strategy generating function and assume there is a matrix* $\mathbf{B} \in \mathbb{R}^{(M+1)\times(M+1)}$ *with full rank such that* $v_1(y)=By$ *for all* $y \in R^{M+1}$. *Define* $A_\beta := \{y \in \mathbb{R}^{M+1} : S_0^\top(\mathbf{B}y) = \beta\}$ *for some fixed* $\beta>0$. *Then, each admissible subset* $A \subset A\beta$ *is bounded.*

Proof.

We use an indirect proof. Assume the assertion does not hold and A is unbounded. Then, there must be a sequence $(y^m)_{m\in\mathbb{N}} \subset A$ with $S_0^\top(\mathbf{B}y^m) = \beta$ and $\|y^m\| \to \infty$ as $m \to \infty$. Then, for $(x^m)_{m\in\mathbb{N}} := (\mathbf{B}y^m)_{m\in\mathbb{N}}$, we also have $\|x^m\| \to \infty$ as $m \to \infty$, because $\|y^m\| = \|\mathbf{B}^{-1}\mathbf{B}y^m\| \le \|\mathbf{B}^{-1}\|\|x^m\|$. The assumption of v, the definition of admissible in Definition 5, and Equation (4) give

$$\mathcal{W}_1(v(y^m)) = \mathcal{W}_0 + (S_1 - S_0)^\top v_1(y^m) = \mathcal{W}_0 + (S_1 - S_0)^\top x^m > 0 \tag{17}$$

a.s. for all $m \in N$.

Property $S_0^\top x^m = \beta$ implies that $\|\widehat{x}^m\| \to \infty$ as $m \to \infty$. Then, there exists a subsequence (w.l.o.g. the original sequence) such that $x^m/\|\widehat{x}^m\| \to x^* = (x_0^*, (\widehat{x}^*)^\top)^\top \in \mathbb{R}^{M+1}$ as $m \to \infty$ where $\|\widehat{x}^*\| = 1$ and $x_0^* = -S_0^\top \widehat{x}^*/S_0^0$. Consequently, we have $S_0^\top x^* = 0$. Dividing Equation (17) by $\|\widehat{x}^m\|$ and taking the limit as $m \to \infty$ yields $S_1^\top x^* \ge 0$ a.s. Therefore, it must be

$$\mathrm{P}\left(\left(S_1 - \frac{S_1^0}{S_0^0} S_0\right)^\top x^* < 0\right) = \mathrm{P}\left(S_1^\top x^* < 0\right) = 0,$$

which is a contradiction, because by assumption the market has no nontrivial risk-free trading strategy (cf. Theorem 1 (c)).

Now, we give two examples for trading strategy generating functions.

Example 1 (Buy and hold; constant number of shares).

The buy and hold (bnh) strategy simply buys the assets at the beginning and does not change the number of shares for each asset in the subsequent time steps. Hence, the corresponding trading strategy generating function is defined by

$$v_{\text{bnh}}: \mathbb{R}^{M+1} \to \mathcal{L}^0\left(N-1;\mathbb{R}^{M+1}\right), \quad y \mapsto X = (x_n)_{1\le n\le N} \quad \textit{with } x_n := y \textit{ for } n = 1,\ldots,N.$$

Obviously, the trading strategy $X=v_{\text{bnh}}(y)$ *is self-financing for each* $y \in \mathbb{R}^{M+1}$ *(cf. Definition 6). Therefore, Equation* (**4**) *in Definition 4 and Proposition 1 give*

$$\mathcal{W}_n(v_{\text{bnh}}(y)) = \mathcal{W}_0 + \sum_{k=1}^{n} (S_k - S_{k-1})^\top y = \mathcal{W}_0 + (S_n - S_0)^\top y \tag{18}$$

for $n=1,\ldots,N$. *The largest admissible set for* v_{bnh} *according to Definition 8 is given by*

$$A_{\text{bnh}} := \left\{ y \in \mathbb{R}^{M+1} \; : \; S_0^\top y - \mathcal{W}_0 < S_n^\top y \quad a.s. \quad for\ n = 1,\ldots,N \right\}.$$

For this example, Lemma 1 can directly be applied using $B=\mathrm{Id}\in\mathbb{R}^{(M+1)\times(M+1)}$ *if the market model S has no nontrivial risk-free trading strategy. Then, for* $\beta>0$ *fixed, the set* $A_{\text{bnh}} \cap \{y\in\mathbb{R}^{M+1}: S_0^\top y=\beta\}$ *is bounded.*

Example 2 (Constant weight/fixed fraction).

Constant weights means, that the fractions invested into the assets stay constant in time. For this, it is needed that the portfolio is reallocated after each time step.

First, we define the rates of return T of the multi-period market model S by

$$T_n := \left(T_n^0, T_n^1, \ldots, T_n^M\right)^\top, \; T_n^i := \frac{S_n^i}{S_{n-1}^i} - 1 > -1 \quad a.s., i = 0,1,\ldots,M, \tag{19}$$

for time steps $n=1,\ldots,N$ *(cf. Definition 2).*

For the corresponding trading strategy generating function, which we denote by v_{twr}, *we need to make sure that after each time step, the same fractions of wealth, given by some fixed* $f\in\mathbb{R}^{M+1}$, *are invested into the assets. Using trading strategy* $X=v_{\text{twr}}(f)$, *this should result into a wealth process*

$$\mathcal{W}_n(v_{\text{twr}}(f)) - \mathcal{W}_{n-1}(v_{\text{twr}}(f)) \cdot \left(1 + T_n^\top f\right) = \mathcal{W}_0 \prod_{k=1}^{n} \left(1 + T_k^\top f\right) \tag{20}$$

for $n=1,\ldots,N$, *which is related to the* terminal wealth relative *(TWR) (see, e.g., Vince (2009)). To achieve this, we first define* $f \mapsto X=v_{\text{twr}}(f)$ *by*

$$v_{\text{twr}}(f) := ((v_{\text{twr}})_1(f), \ldots, (v_{\text{twr}})_N(f)), \quad (v_{\text{twr}})_n(f) := \left((v_{\text{twr}})_n^0, (v_{\text{twr}})_n^1, \ldots, (v_{\text{twr}})_n^M\right)$$

for $n=1,\ldots,N$ *where*

$$(v_{\text{twr}})_1^i(f) := \frac{f_i}{S_0^i}\mathcal{W}_0, \qquad (v_{\text{twr}})_n^i(f) := \frac{f_i}{S_{n-1}^i}\mathcal{W}_{n-1}(v_{\text{twr}}(f)) = \frac{f_i}{S_{n-1}^i}\mathcal{W}_0 \prod_{k=1}^{n-1}\left(1 + T_k^\top f\right) \tag{21}$$

for $n=2,\ldots,N$ *and* $i=0,1,\ldots,M$. *Here, for instance,* $(v_{\text{twr}})_1^i(f)$ *denotes the amount of shares of the ith asset that have to be bought initially to invest the fraction* f_i *of the initial wealth* W_0 *into this asset for the first time step.*

Now, we need to show that this indeed yields Equation (20). *Inserting Equation* (21) *into the definition of the wealth (see Definition 4) and using Equation* (19), *we obtain*

$$\begin{aligned} \mathcal{W}_n(v_{\mathrm{twr}}(f)) &= \mathcal{W}_{n-1}(v_{\mathrm{twr}}(f)) + (S_n - S_{n-1})^\top (v_{\mathrm{twr}}(f))_n \\ &= \mathcal{W}_{n-1}(v_{\mathrm{twr}}(f)) \cdot \left(1 + \sum_{i=1}^{M} \left(S_n^i - S_{n-1}^i\right) \frac{f_i}{S_{n-1}^i}\right) \\ &= \mathcal{W}_{n-1}(v_{\mathrm{twr}}(f)) \cdot \left(1 + T_n^\top f\right) \end{aligned}$$

for $n=1,\ldots,N$ *and Equation* (20) *follows by induction. Of course, this only makes sense for admissible trading strategies. Therefore, we define*

$$A_{\mathrm{twr}} := \left\{ f \in \mathbb{R}^{M+1} \ : 1 + T_n^\top f > 0 \ a.s. \ for \ n = 1, \ldots, N \right\}. \tag{22}$$

Note that, in general, v_{twr} *is nonlinear for* $N \geq 2$.

In addition, in this case, we can apply Lemma 1 directly if the market model S has no nontrivial risk-free trading strategy using a diagonal matrix B *with diagonal entries* $b_{ii} = \mathcal{W}_0 / S_0^i > 0$ *for* $i=0,1,\ldots,M$.

Remark 4.

Let the situation of Example 2 for $N \geq 2$ *be given and* $f \in \mathbb{R}^{M+1}$ *be fixed. Assume there exists some* $n \in \{1,\ldots,N-1\}$ *such that* $P(1 + T_n^\top f \neq 1) > 0$. *Then, one can show that* $v_{\mathrm{twr}}(f)$ *is self-financing if and only if* $\sum_{i=0}^{M} f_i = 1$. *In addition, note that* $\sum_{i=0}^{M} f_i = 1$ *if and only if* $\mathcal{W}_0 = S_0^\top (v_{\mathrm{twr}})_1(f)$. *A proof can be found in (Platen 2018, Proposition 2.2.32).*

EFFICIENT PORTFOLIOS

Having the multi-period financial market set up in the previous section, we are ready to focus on the main theme of the paper. In this section, we extend the general framework for portfolio theory from Part I (Maier-Paape and Zhu 2018a) to the setting of multi-period financial markets. We derive a characterization of the efficient frontier for trading-off risk and reward using admissible trading strategies. Furthermore, we also discuss the relationship between points on this efficient frontier and their corresponding trading strategies. We do so using the modular approach alluded to in the introduction. The general portfolio/trading strategy trade-off problem is considered in the light of the interaction among four related modules. While we already discussed Blocks (a) (multi-period market) and (b) (trading strategies) in the last section, we now want to concentrate on Blocks (c) (risk and utility function) (performance criteria; see Section 3.1) and (d) (the optimization problem) (including discussion of efficient frontier and efficient portfolios; see Section 3.2, Section 3.3 and Section 3.4).

Performance Criteria

In Part I (Maier-Paape and Zhu 2018a), we chose to introduce risk and utility functions to measure performance criteria in an axiomatic way. This is not necessary here. Indeed, for our modular portfolio theory, it suffices to assume the risk functions to be (closed) proper convex and the utility function to be (closed) proper concave. Clearly, this is more general than often used assumptions like for instance positive homogeneous risk functions.

One reason to choose a multi-period market model over a one-period market model could be the possibility to involve complex trading strategies. Another and possibly more important reason could be path-dependent risk measures, which cannot be directly used on a one-period market. One well-known path-dependent risk measure is the drawdown, which can be defined in different ways and different variants.

Definition 9 (Absolute/relative drawdown process).

Assume we have a model for a wealth process $W := (W_n)_{0 \leq n \leq N} \in \mathcal{L}^2(N)$, *e.g.*, $W = \mathcal{W}(X)$ *for some trading strategy X and a multi-period market model S (see Definitions 3 and 4). The* absolute drawdown process $\mathcal{D}_{\mathrm{abs}} = ((\mathcal{D}_{\mathrm{abs}})_1, \ldots, (\mathcal{D}_{\mathrm{abs}})_{\mathrm{N}})$ *is defined by*

$$(\mathcal{D}_{\mathrm{abs}})_{\mathrm{n}}(W) := \max_{0 \leq \ell \leq n} \{W_\ell\} - W_n \geq 0, \tag{23}$$

for n=0,1,...,N. *The* relative drawdown process $\mathcal{D}_{\mathrm{rel}} = ((\mathcal{D}_{\mathrm{rel}})_1, \ldots, (\mathcal{D}_{\mathrm{rel}})_{\mathrm{N}})$ *is defined for positive wealth processes (e.g., when using admissible trading strategies) by*

$$(\mathcal{D}_{\mathrm{rel}})_{\mathrm{n}}(W) := \frac{(\mathcal{D}_{\mathrm{abs}})_{\mathrm{n}}(W)}{\max\limits_{0 \leq \ell \leq n} \{W_\ell\}} = 1 - \frac{W_n}{\max\limits_{0 \leq \ell \leq n} \{W_\ell\}} = 1 - \min_{0 \leq \ell \leq n} \left\{ \frac{W_n}{W_\ell} \right\} \in [0, 1), \tag{24}$$

for n=0,1,...,N. *Both* Dabs *and* Drel *are stochastic processes and no risk measures up to now.*

Chekhlov et al. (2003, 2005) studied the absolute drawdown for a simple trading strategy and a finite probability space. The risk measure they defined is called conditional drawdown at risk (CDaR) and can be seen as a conditional value at risk of the absolute drawdown process. Later, Zabarankin et al. (2014) proposed using the absolute drawdown but this time on a rolling frame of size τ∈N, i.e., they use

$$(\mathcal{D}_{\text{abs},\tau})_n(W) := \max_{n_\tau \leq \ell \leq n} \{W_\ell\} - W_n$$

for n=1,...,N, where n_τ:=max{1,n−τ}. Again, they used the concept of the conditional value at risk.

Goldberg and Mahmoud (2017) defined the so-called conditional expected Drawdown (CED), which is similarly defined as CDaR. The CED is the conditional value at risk of the maximum absolute drawdown over all scenarios, where the market model is defined in a continuous time setting.

Maier-Paape and Zhu (2018b) studied the expected value of the logarithm of the relative drawdown at time step *N* (called current drawdown) in a finite probability space. Therein, the multi-period market is constructed using a one-period market model by K∈N iid drawings. We want to use this variant, but in our more general setting with a multi-period market model and using a general trading strategy generating function. It is defined as follows:

Definition 10 (Multi-path expected log drawdown).

Let $S \in \mathcal{L}^2(N;\mathbb{R}^{M+1}_{>0}$ *be the market model and* v *be a trading strategy generating function with domain* $A \subset \mathbb{R}^{M+1}$ *and with wealth process* W(v(x)), x∈A. *Then, the* multi-path expected log drawdown *is defined by*

$$\rho_{\ln}: A \to [0,\infty], \quad x \mapsto \begin{cases} \mathrm{E}\left[-\ln\left(1-(\mathcal{D}_{\text{rel}})_N(\mathcal{W}(v(x)))\right)\right], \\ \infty, \end{cases}$$ if v(x) is admissible

(cf. Definition 5), otherwise.

Remark 5.

Assuming the range of S is bounded (which is reasonable for real markets) and the trading strategy generating function v is continuous as a function from A to $L^2(N)$, *then* x↦W(v(x)) *is continuous and, therefore, so is* $\rho_{\ln}$.

A reasonable utility function (corresponding to the drawdown in Definition 10) may have the form

$$u: A \to \mathbb{R}, \qquad x \mapsto \mathrm{E}\left[\mathcal{W}_N(v(x)) - \mathcal{W}_0\right]. \tag{25}$$

Using the buy and hold strategy, we obtain

$$u_{\text{bnh}}(x) := \mathrm{E}\left[\mathcal{W}_N(v_{\text{bnh}}(x)) - \mathcal{W}_0\right] = \mathrm{E}\left[S_N - S_0\right]^\top x, \tag{26}$$

which is linear. Another variant uses the terminal wealth relative (TWR), which, in our setting, is defined by

$$\mathrm{TWR}(f) := \prod_{n=1}^{N}\left(1 + T_n^\top f\right) \in \mathcal{L}^0(\mathbb{R}_{>0}), \qquad f \in A_{\text{twr}} \text{ (cf. Equation (22))},$$

with the rates of return T from Equation (19). Note that, because of Equation (20) in Example 2, we have

$$\mathrm{TWR}(f) = \frac{\mathcal{W}_N(v_{\mathrm{twr}}(f))}{\mathcal{W}_0}, \tag{27}$$

i.e., it is the quotient of end and start capital and the v_{twr} trading strategy is strongly related to the variant used in Vince (2009). As a utility function for the TWR, we define

$$u_{\log\mathrm{TWR}}: \mathbb{R}^{M+1} \to [-\infty, \infty], \qquad f \mapsto \begin{cases} \mathrm{E}\left[\ln\left(\mathrm{TWR}(f)\right)\right], & f \in A_{\mathrm{twr}}, \\ -\infty, & f \notin A_{\mathrm{twr}}, \end{cases} \tag{28}$$

with Atwr from Equation (22). Inserting the above characterizations of TWR gives

$$u_{\log\mathrm{TWR}}(f) - \mathrm{E}\left[\ln\left(\mathcal{W}_N(v_{\mathrm{twr}}(f))\right) - \ln\left(\mathcal{W}_0\right)\right] = \mathrm{E}\left[\ln\left(\prod_{n=1}^{N}(1+T_n^\top f)\right)\right] = \sum_{n=1}^{N} \mathrm{E}\left[\ln(1+T_n^\top f)\right] \tag{29}$$

for f∈A_{twr}. A corresponding risk function would be the drawdown in Definition 10 with v:=v_{twr}. With

$$\mathrm{TWR}_a^b(f) := \prod_{n=a}^{b}\left(1+T_n^\top f\right) \in \mathcal{L}^0(\mathbb{R}_{>0}), \qquad \text{for } 1 \le a \le b \le N, \tag{30}$$

and using Equation (20), we get

$$\begin{aligned} \rho_{\ln}(f) &= \mathrm{E}\left[-\ln\left(\min_{0\le \ell\le N}\left\{\frac{\mathcal{W}_N(v_{\mathrm{twr}}(f))}{\mathcal{W}_\ell(v_{\mathrm{twr}}(f))}\right\}\right)\right] \\ &= \mathrm{E}\left[-\ln\left(\min\left\{1, \min_{1\le \ell\le N}\left\{\mathrm{TWR}_\ell^N(f)\right\}\right\}\right)\right] \\ &= \mathrm{E}\left[\max\left\{0, \max_{1\le \ell\le N}\left\{-\ln\left(\mathrm{TWR}_\ell^N(f)\right)\right\}\right\}\right]. \end{aligned} \tag{31}$$

for f∈A_{twr} (see also (Maier-Paape and Zhu 2018b, Definition 6 and Theorem 8)). Under reasonable assumptions on the market, we show below (see Section 4) that ρln is proper convex and can therefore be used as a risk function. Similarly, we find that ulogTWR is proper concave and use it as a utility function.

Optimization

At the core of our framework for the portfolio/trading strategy theory is an optimal trade-off between the two competing performance criteria risk and reward. This subsection discusses two related optimization problems: either minimizing the risk with a lower bound for the reward or maximizing the reward with an upper bound for the risk under the setting below.

Setting 1.

Assume we have the following:

(*i*). Multi-period market model $S \in \mathcal{L}^2(N;\mathbb{R}^{M+1}_{>0})$, M,N∈N *(see Definition 2).*

(*ii*). Trading strategy, *which is defined by a given trading strategy generating function* v:A→$L^0(N-1;R^{M+1})$ *as in Definition 8 with non-empty and convex domain* A⊂R^{M+1}.

(*iii*). Utility function u:A→R∪{−∞}, *which is assumed to be proper concave.*

(*iv*) Risk function r:A→R∪{∞}, *which is assumed to be proper convex.*

We always assume that dom(u)∩dom(r)≠∅ *holds, where* dom(u)= {x∈A:u(x)>−∞} *and* dom(r)={x∈A:r(x)<∞} *are both convex sets.*

Here, technically both u and r are defined on A. In practice, they are functions of the trading strategy payoff, i.e., they depend on the trading strategy generating function v. Thus, the properties of u and r in fact may require, e.g., continuity of v.

Problem 1.

Assume we have given Setting 1. We are looking at the two following problems:

(a) *Let* β>0 *and* μ∈R *be fixed. The* minimum risk optimization problem *is defined by*

$$\min_{x\in A} \mathfrak{r}(x) \quad \textit{subject to } \mathfrak{u}(x) \geq \mu,\ S_0^{\top} v_1(x) = \beta. \qquad \textbf{(MinR)}$$

(b). *Let* β>0 *and* r∈R *be fixed. The* maximum utility optimization problem *is defined by*

$$\max_{x\in A} \mathfrak{u}(x) \quad \textit{subject to } \mathfrak{r}(x) \leq r,\ S_0^{\top} v_1(x) = \beta. \qquad \textbf{(MaxU)}$$

Note that $v_1(x)$ represents the initial portfolio allocation at time t=0 and thus by $S_0^{\top} v_1(x) = \beta$ the initial investment size is fixed.

Efficient Frontier

In this section, we define the efficient frontier related to Problem 1 and develop several helpful characterizations of this frontier. This generalizes several results known for the one-period model (see, e.g., Maier-Paape and Zhu (2018a)) to multi-period markets with trading strategy (see also (Platen

2018, Section 2.4.2)).

Definition 11 (Risk utility space).

Let Setting 1 be given. The sublevel and superlevel sets of r *and* u *for thresholds* r,μ∈R *are denoted by*

$$\mathcal{B}_{\mathfrak{r},A}(r) := \{x \in A : \mathfrak{r}(x) \le r\} \subset \mathrm{dom}(\mathfrak{r}) \quad \textit{and} \quad \mathcal{B}_{\mathfrak{u},A}(\mu) := \{x \in A : \mathfrak{u}(x) \ge \mu\} \subset \mathrm{dom}(\mathfrak{u}),$$

respectively. For its intersection, we write

$$\mathcal{B}_{\mathfrak{r},\mathfrak{u},A}(r,\mu) := \{x \in A : \mathfrak{r}(x) \le r \textit{ and } \mathfrak{u}(x) \ge \mu\} = \mathcal{B}_{\mathfrak{r},A}(r) \cap \mathcal{B}_{\mathfrak{u},A}(\mu) \subset A.$$

Then,

$$\mathcal{G}(\mathfrak{r},\mathfrak{u};A) := \{(r,\mu) : \mathcal{B}_{\mathfrak{r},\mathfrak{u},A}(r,\mu) \ne \emptyset\} \subset \mathbb{R}^2 \qquad (32)$$

is the set of valid risk and utility levels *in the risk utility space.*

Remark 6.

We need u *and* r *to be upper and lower semi-continuous, respectively, where both functions, in practice, should be defined on top of a trading strategy generating function v. Note that then it is reasonable that v is continuous. Otherwise, it might be impossible for* u *and* r *to have these semi-continuity properties.*

Remark 7.

Instead of (Maier-Paape and Zhu 2018a, Assumption 4), which states that either $B_{r,A}(r)$ *or* $B_{u,A}(\mu)$ *is compact for all* r,μ∈R, *respectively, we here often require in the following that* $B_{r,u,A}(r,\mu)$ *is compact for all* r,μ∈R, *which is less, see Proposition 2 (b) in the following.*

The following is an analog result to (Maier-Paape and Zhu 2018a, Proposition 7) and (Platen 2018, Proposition 2.4.6).

Proposition 2 (Properties in risk utility space).

Let Setting 1 be given. Then, the following holds true:

(a) r *is closed proper convex if and only if* Br,A(r) *is closed for all* r∈R.
u *is closed proper concave if and only if* Bu,A(μ) *is closed for all* μ∈R.

(b) *Assume* $B_{u,A}(\mu)$ *and* $B_{r,A}(r)$ *are closed for all* μ,r∈R. *If either* $B_{u,A}(\mu)$ *is compact for all* μ∈R *or* $B_{r,A}(r)$ *is compact for all* r∈R, *then* $B_{r,u,A}(r,\mu)$ *is convex and compact for all* r,μ∈R.

(c) G(r,u;A) *is convex and* (r,μ)∈G(r,u;A) *implies that, for any* k>0, *we have* (r+k,μ)∈G(r,u;A) *and* (r,μ−k)∈G(r,u;A).

(d) *If* $B_{r,u,A}(r,\mu)$ *is compact for all* $r,\mu \in R$, *then* $G(r,u;A)$ *is closed.*

Proof.

Proof of (a): Note that r is by definition closed proper convex, if it is proper convex and moreover its epigraph $epi(r)=\{(x,s) \in A \times R : r(x) \leq s\}$ is closed. Thus, the claim here follows from a classical result from convex analysis, see (Rockafellar 1972, Theorem 7.1). The same holds true for $-u$ which gives the statement for u.

Proof of (b): The compactness of $B_{r,u,A}(r,\mu)$ follows directly. The convexity of $B_{r,u,A}(r,\mu)$ follows from convexity of $B_{r,A}(r)$ and $B_{u,A}(\mu)$ (see (Rockafellar 1972, Theorem 4.6)).

Proof of (c): Clearly, $(r,\mu) \in G(r,u;A)$ implies directly from the definition that, for any $k>0$, we have $(r+k,\mu) \in G(r,u;A)$ and $(r,\mu-k) \in G(r,u;A)$. Furthermore, convexity of $G(r,u;A)$ follows directly from the convexity of r and the concavity of u.

Proof of (d): Let $((r_n,\mu_n))_{n \in N} \subset G(r,u;A)$ be an arbitrary convergent sequence with $(r_n,\mu_n) \to (r,\mu) \in R^2$ as $n \to \infty$. Then, there exists a sequence $(x^n)_{n \in N} \subset A$ with $x_n \in B_{r,u,A}(r_n,\mu_n) \subset dom(u) \cap dom(r)$, i.e., $r(x_n) \leq r_n$ and $u(x_n) \geq \mu_n$. For all $\varepsilon>0$, there exists $n_0 \in N$ such that $r_n < r+\varepsilon$ and $\mu_n > \mu-\varepsilon$ and, therefore, $x_n \in B_{r,u,A}(r+\varepsilon,\mu-\varepsilon)$ for all $n \geq n0$. By assumption, $Br,u,A(r+1,\mu-1)$ is compact.

Then, there must be a convergent subsequence, w.l.o.g. the original sequence, with $x^n \to x$ as $n \to \infty$ with $x \in B_{r,u,A}(r+1,\mu-1)$. Moreover, compactness of $B_{r,u,A}(r+\varepsilon,\mu-\varepsilon)$ yields that r restricted to $\{x \in A : u(x) \geq \mu-\varepsilon\}$ is lower semi-continuous (cf. (Rockafellar 1972, Theorem 7.1)). Similarly, u restricted to $\{x \in A : r(x) \leq r+\varepsilon\}$ is upper semi-continuous. Thus, $r(x) \leq r+\varepsilon$ and $u(x) \geq \mu-\varepsilon$ follow for all $\varepsilon>0$. For this, it must be $x \in B_{r,u,A}(r,\mu)$ and, hence, $(r,\mu) \in G(r,u,;A)$.

As in (Maier-Paape and Zhu 2018a, Definition 5), we define the efficient frontier.

Definition 12 (Efficient portfolio and efficient frontier).

In the situation of Setting 1, we say an element $x \in A$ *is called* efficient *provided that there does not exist any* $x' \in A$ *such that either*

$$[r(x') \leq r(x) and u(x') > u(x)] or [r(x') < r(x) and u(x') \geq u(x)].$$

$$\left[r(x') \leq r(x) \quad and \quad u(x') > u(x)\right] \quad or \quad \left[r(x') < r(x) \quad and \quad u(x') \geq u(x)\right].$$

We call set

$$\mathcal{G}_{eff}(\mathfrak{r},\mathfrak{u};A) := \left\{(\mathfrak{r}(x),\mathfrak{u}(x)) \in \mathbb{R}^2 : x \in A \text{ is efficient}\right\} \subset \mathcal{G}(\mathfrak{r},\mathfrak{u};A)$$

the efficient frontier.

An important property is that the efficient frontier lies on the boundary of the set of valid risk and utility levels, which is shown next; see (Maier-Paape and Zhu 2018a, Theorem **3**) for the one-period case with finite probability space and also (Platen 2018, Theorem 2.4.8) for more details.

Theorem 2 (Properties of efficient frontier).

Assume we are in the situation of Setting 1.

(a) *The efficient frontier* Geff(r,u;A) *is located in the boundary of* G(r,u;A) *and has no vertical and no horizontal line segments.*

(b) *If* $_{B_{r,u,A}}$(r,μ) *is compact for all* r,μ∈R, *then* G_{eff}(r,u;A) *is non-empty and equals to the non-vertical and non-horizontal part of the boundary of* G(r,u;A), *i.e.,*

$$\mathcal{G}_{eff}(\mathfrak{r},\mathfrak{u};A) = \{(r,\mu) \in \partial\mathcal{G}(\mathfrak{r},\mathfrak{u};A) : (r-k,\mu),(r,\mu+k) \notin \partial\mathcal{G}(\mathfrak{r},\mathfrak{u};A)\ \forall k>0\}, \quad (33)$$

where ∂G(r,u;A) *denotes the boundary of* G(r,u;A) *in* R^2.

(c) *If* B⊂A *is convex, then* G_{eff}(r,u;A)∩G(r,u;B)⊂G_{eff}(r,u;B).

Proof.

For proof of (a) and (c), see the proof in (Maier-Paape and Zhu 2018a, Theorem 3). It remains to prove (b) (see (Platen 2018, Theorem 2.4.8) for more details):

1. "⊂" follows from (a).
2. Show "⊃": Let (r_0,μ_0)∈{(r,μ)∈∂G(r,u;A):(r−k,μ),(r,μ+k)∉∂G (r,u;A)∀k>0} be arbitrary. Then, since G(r,u;A) is closed by Proposition 2 (d), it has to be (r0,μ0)∈G(r,u;A). Hence, there must exist an x0∈A such that $r(x_0)\leq r_0$ and $u(x_0)\geq \mu_0$. In addition, it must be (r,μ)∉G(r,u;A) for all $r\leq r_0$ and $\mu\geq\mu_0$ with $(r,\mu)\neq(r_0,\mu_0)$, because G(r,u;A) is convex and unbounded from below and unbounded to the right by Proposition 2 (c). Consequently, even $r(x_0)=r_0$ and $u(x_0)=\mu_0$ must hold and x_0 thus is efficient, i.e., x^0∈G_{eff}(r,u;A).
3. Show G_{eff}(r,u;A)≠∅: Because of Setting 1, we have dom(u)∩dom(r)≠∅, i.e., there exists x_1∈dom(u)∩dom(r) and it is $(r_1,\mu_1):=(r(x_1),u(x_1))$∈G(r,u;A). Since r is convex

and u is concave, r is bounded below and u is bounded above on each compact set. The set $B_{r,u,A}(r_1,\mu_1)$ is compact by assumption. Hence, by definition of $B_{r,u,A}(r_1,\mu_1)$, the function r on $B_{r,u,A}(r_1,\mu_1)$ is contained in say [r∗,r1] and the function u on $B_{r,u,A}(r_1,\mu_1)$ is contained in say $[\mu_1,\mu*]$. Therefore, the image of (r,u) restricted on $B_{r,u,A}(r_1,\mu_1)$ is a subset of G(r,u;A) and $\emptyset \neq G(r,u;A) \cap Q \subset [r^*,r_1]\times[\mu 1,\mu^*]$ for $Q=\{(r',\mu'):r' \leq r_1, \mu' \geq \mu_1\}$, see Figure 1. Clearly, there must be a point $(r_2,\mu_2)\in\partial G(r,u;A)\cap Q$ such that (r_2-k,μ_2) and $(r_2,\mu+k)$ do not belong to G(r,u;A) for all k>0. Since G(r,u;A) is closed, by Equation (33), the point (r_2,μ_2) belongs to $G_{eff}(r,u;A)$, i.e., $G_{eff}(r,u;A)\neq\emptyset$.

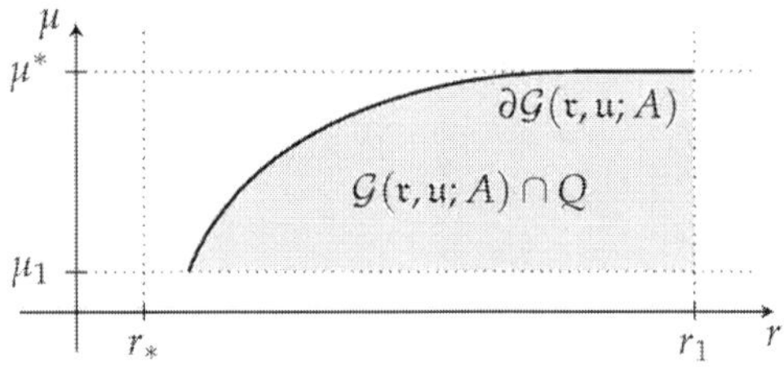

Figure 1. Illustration to show Geff(r,u;A)≠∅.

This completes the proof.

As indicated by the last theorem, the efficient frontier $G_{eff}(r,u;A)$ is not necessarily the whole boundary of G(r,u;A). As a consequence, $G_{eff}(r,u;A)$ might be bounded. The corresponding bounds are defined as follows.

Definition 13.

In the situation of Setting 1, assume that $G_{eff}(r,u;A)$ *is non-empty. Define the bounds for risk and utility of efficient elements by*

$$r_{\min} := \inf_{(r,\mu)\in\mathcal{G}_{eff}(\mathfrak{r},\mathfrak{u};A)} \{r\}, \qquad \mu_{\min} := \inf_{(r,\mu)\in\mathcal{G}_{eff}(\mathfrak{r},\mathfrak{u};A)} \{\mu\}, \tag{34a}$$

$$r_{\max} := \sup_{(r,\mu)\in\mathcal{G}_{eff}(\mathfrak{r},\mathfrak{u};A)} \{r\}, \qquad \mu_{\max} := \sup_{(r,\mu)\in\mathcal{G}_{eff}(\mathfrak{r},\mathfrak{u};A)} \{\mu\}, \tag{34B}$$

respectively.

The following alternative representation is similar as the one in (Maier-

Paape and Zhu 2018a, Proposition 9) for the one-period market (see also (Platen 2018, Lemma 2.4.10)).

Lemma 2 (Infima/Suprema of $G_{eff}(r,u;A)$).

Let Setting 1 be given and assume $B_{r,u,A}(r,\mu)$ *is compact for all* $r,\mu \in R$, *so that by Theorem 2 (b) in particular* $G_{eff}(r,u;A)$ *is non-empty. Then,*

$$r_{\min} = \inf_{x \in \mathrm{dom}(u) \cap \mathrm{dom}(r)} \{r(x)\} < \infty,$$

$$\mu_{\max} = \sup_{x \in \mathrm{dom}(u) \cap \mathrm{dom}(r)} \{u(x)\} > -\infty,$$

and, depending on $r_{\min}$ *and* $\mu_{\max}$, *we have*

$$r_{\max} = \begin{cases} \min_{x \in \mathcal{B}_{u,A}(\mu_{\max})} \{r(x)\}, & \text{if } \mu_{\max} < \infty \text{ and } \mathcal{B}_{u,A}(\mu_{\max}) \cap \mathrm{dom}(r) \neq \emptyset, \\ \infty, & \text{otherwise.} \end{cases}$$

$$\mu_{\min} = \begin{cases} \max_{x \in \mathcal{B}_{r,A}(r_{\min})} \{u(x)\}, & \text{if } r_{\min} > -\infty \text{ and } \mathcal{B}_{r,A}(r_{\min}) \cap \mathrm{dom}(u) \neq \emptyset, \\ -\infty, & \text{otherwise,} \end{cases}$$

If $B_{r,A}(r)$ *is compact for all* $r \in R$, *then* $r_{\min} > -\infty$ *and* $B_{r,A}(r_{\min}) \neq \emptyset$. *If* $B_{u,A}(\mu)$ *is compact for all* $\mu \in R$, *then* $\mu_{\max} < \infty$ *and* $B_{u,A}(\mu_{\max}) \neq \emptyset$.

Proof.

We define $B := \mathrm{dom}(u) \cap \mathrm{dom}(r)$. By assumption, $B \neq \emptyset$. Since the vertical part of $\partial G(r,u;A)$, if it exists, does not change the infimum in r, we get

$$r_{\min} = \inf_{(r,\mu) \in \mathcal{G}_{eff}(r,u;A)} \{r\} = \inf_{(r,\mu) \in \partial \mathcal{G}(r,u;A)} \{r\} = \inf_{(r,\mu) \in \mathcal{G}(r,u;A)} \{r\} = \inf_{x \in B} \{r(x)\} < \infty. \tag{35}$$

Analogously, the horizontal part of $\partial G(r,u;A)$, if it exists, does not change the supremum in μ and therefore

$$\mu_{\max} = \sup_{(r,\mu) \in \mathcal{G}_{eff}(r,u;A)} \{\mu\} = \sup_{(r,\mu) \in \partial \mathcal{G}(r,u;A)} \{\mu\} = \sup_{(r,\mu) \in \mathcal{G}(r,u;A)} \{\mu\} = \sup_{x \in B} \{u(x)\} > -\infty. \tag{36}$$

We next show the properties of $r_{\max}$ and $\mu_{\min}$. Since the properties of $\mu_{\min}$ can be shown similarly, we only show it for $r_{\max}$.

Firstly, assume that $\mu_{\max} < \infty$ and $B_{u,A}(\mu_{\max}) \cap \mathrm{dom}(r) \neq \emptyset$. Then, there exists $x_0 \in B_{u,A}(\mu_{\max})$ such that $r(x_0) \in R$. Of course, $(r(x_0),u(x_0)) = (r(x_0),\mu_{\max})$ is on the horizontal part of $\partial G(r,u;A)$, see Figure 2. By assumption, $G_{eff}(r,u;A) \neq \emptyset$ and hence the set $\{(r,\mu_{\max}) : r \in R\}$ cannot be a subset of $G(r,u;A)$. Therefore, and since $G(r,u;A)$ is closed by Proposition 2 (d), we obtain $r^* := \min\{r : (r,\mu_{\max}) \in G(r,u;A)\} > -\infty$. Using Equation (33), we get $(r^*,\mu_{\max}) \in G_{eff}(r,u;A)$ yielding an efficient portfolio $x_1 \in A$ with $(r(x_1),u(x_1)) = (r^*,\mu_{\max})$. From Equation (34b), we conclude $r_{\max} = r^* = \min\{r(x) : x \in B_{u,A}(\mu_{\max})\}$ and the assertion is proved.

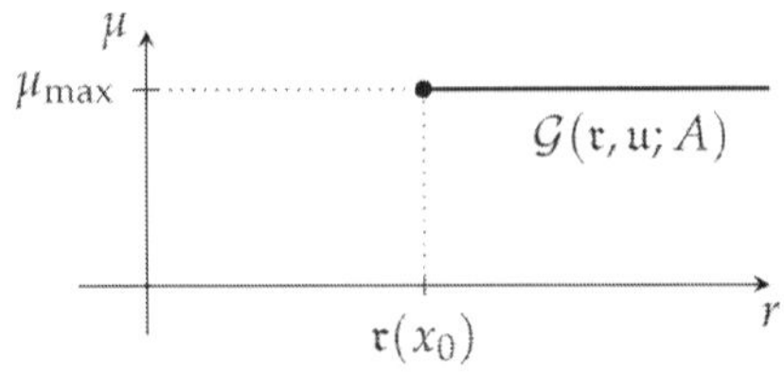

Figure 2. Illustration for the proof of Lemma 2.

Now, assume that $\mu_{max}=\infty$ or $B_{u,A}(\mu_{max})\cap dom(r)=\emptyset$. In both cases, the supremum μ_{max} of the μ values of $G_{eff}(r,u;A)$ is not attained in the risk utility space. Since G(r,u;A) is closed and convex by Proposition 2 (c) and (d), there cannot be a horizontal part of $\partial G(r,u;A)$. In addition, $G_{eff}(r,u;A)\neq\emptyset$ by Theorem 2 (b) and because of Equation (**33**) there is a sequence $[(r_n,\mu_n)]$ $n\in N\subset G_{eff}(r,u;A)$ such that $\mu_n\rightarrow\mu_{max}$ as $n\rightarrow\infty$, which is, w.l.o.g., strictly increasing in μ_n. Then, this sequence must be strictly increasing in r_n as well, otherwise, (r_n,μ_n) would not belong to an efficient element in *A*. If $\mu_{max}=\infty$, it then must be $r_{max}=\infty$ because G(r,u;A) is convex. If $B^{u,A}(\mu_{max})\cap dom(r)=\emptyset$ (and $\mu max<\infty$) it must be $r_{max}=\infty$ as well, because otherwise, $(r,\mu_{max})\in\partial G(r,u;A)$ for all r>rmax but $(r,\mu_{max})\notin G(r,u;A)$, which contradicts that G(r,u;A) is closed.

It remains to show the result in the special situation when $B_{r,A}(r)$ is compact for all r∈R. Then, r is lower semi-continuous (see, e.g., (Rockafellar 1972, Theorem 4.6 and 7.1)). Let x′∈dom(r) be arbitrary. Since $B_{r,A}(r(x'))$ is compact, the minimum of r is attained in $B_{r,A}(r(x'))$ (see, e.g., (Barbu and Precupanu 2012, Theorem 2.8)).

The case when Bu,A(μ) is compact again can be shown similarly.

Related to the bounds, we define next all relevant risk and utility levels of the efficient frontier.

Definition 14.

For Setting 1, we define the projection of $G_{eff}(r,u;A)$ *to the r- and μ-axis by*

$$I := \{r \in \mathbb{R} : \exists \mu \in \mathbb{R} \text{ s.t. } (r,\mu) \in \mathcal{G}_{eff}(\mathfrak{r},\mathfrak{u};A)\},$$
$$J := \{\mu \in \mathbb{R} : \exists r \in \mathbb{R} \text{ s.t. } (r,\mu) \in \mathcal{G}_{eff}(\mathfrak{r},\mathfrak{u};A)\},$$ *respectively.*

From Lemma 2, we already obtain the possibilities for the intervals *I* and *J*, see (Maier-Paape and Zhu 2018a, Corollary 2) for a related result in the one-period case.

Corollary 1.

In the situation of Lemma 2, we have r_{min}=inf(I), r_{max}=sup(I), μ_{min}=inf(J) *and* μ_{max}=sup(J). *Furthermore, exactly one of the following situations holds true depending on the situation:*

- $I=[r_{min},r_{max}]$ *and* $J=[\mu_{min},\mu_{max}]$;
- $I=[r_{min},\infty)$ *and* $J=[\mu_{min},\mu_{max})$, *where* $\mu_{max}=\infty$ *is possible;*
- $I=(r_{min},r_{max}]$ *and* $J=(-\infty,\mu_{max}]$, *where* $r_{min}=-\infty$ *is possible; or*
- $I=(r_{min},\infty)$ *and* $J=(-\infty,\mu_{max})$, *where* $\mu_{max}=\infty$ *and/or* $r_{min}=-\infty$ *is possible.*

In particular, I and J are non-empty intervals. Figure 3 shows some examples.

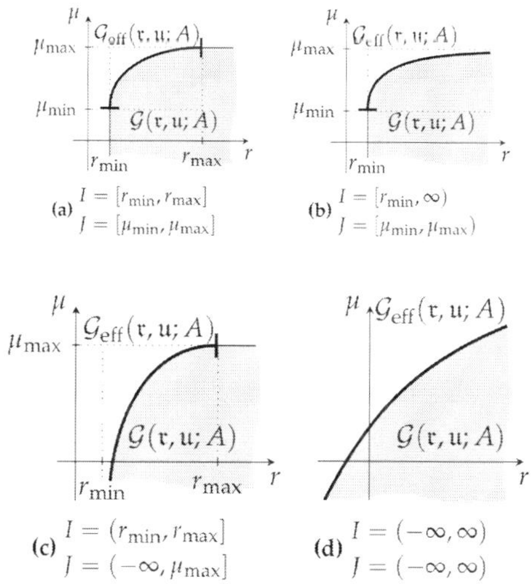

Figure 3. Illustration of efficient frontiers for different cases of the intervals *I* and *J*.

Proof.

This is a direct consequence from Lemma 2.

In Definition 14, we define all valid *r* and μ values (separated from each other and not the combinations of them) of the efficient frontier G_{eff}(r,u;A), which must be on the boundary of G(r,u;A) according to Theorem 2. Within the valid *r* and μ area, this boundary is defined by the two functions

$$v: I \to \mathbb{R}, \qquad r \mapsto \sup_{(r,\mu)\in \mathcal{G}(r,u;A)} \{\mu\} = \sup_{x\in \mathcal{B}_{r,A}(r)} \{u(x)\}, \tag{37a}$$

$$\gamma: J \to \mathbb{R}, \qquad \mu \mapsto \inf_{(r,\mu)\in\mathcal{G}(\mathfrak{r},\mathfrak{u};A)} \{r\} = \inf_{x\in\mathcal{B}_{\mathfrak{u},A}(\mu)} \{\mathfrak{r}(x)\}. \tag{37b}$$

where I and J are from Definition 14. Next, we show some important properties for both functions, see (Maier-Paape and Zhu 2018a, Proposition 8) for the one-period case and see also (Platen 2018, Proposition 2.4.15).

Proposition 3 (Functions related to efficient frontier).

In the situation of Setting 1, assume that $B_{r,u,A}(r,\mu)$ *is compact for all* $r,\mu\in R$. *Then, the functions* $v:I\to R$ *and* $\gamma:J\to R$ *from Equation (37) are well-defined and continuous. Furthermore, we have*

$$v(r) = \max_{(r,\mu)\in\mathcal{G}(\mathfrak{r},\mathfrak{u};A)} \{\mu\} = \max_{x\in\mathcal{B}_{\mathfrak{r},A}(r)} \{\mathfrak{u}(x)\}, \qquad \textit{for all } r \in I, \tag{38a}$$

$$\gamma(\mu) = \min_{(r,\mu)\in\mathcal{G}(\mathfrak{r},\mathfrak{u};A)} \{r\} = \min_{x\in\mathcal{B}_{\mathfrak{u},A}(\mu)} \{\mathfrak{r}(x)\}, \qquad \textit{for all } \mu \in J, \tag{38b}$$

while v is increasing and concave and γ is increasing and convex.

Proof.

We show only the properties of γ. The proof for v can be done similarly.

Let $\mu\in J$ be arbitrary. From μmax in Lemma 2 and Definition 14, we know that there must exist $x*\in\text{dom}(u)\cap\text{dom}(r)$ (note that $\text{dom}(u)\cap\text{dom}(r)\neq\emptyset$) such that $\mu\leq u(x^*)\leq\mu\text{max}$ and $r^*:=r(x^*)\in R$. We then have $x*\in B_{r,u,A}(r*,\mu)\neq\emptyset$ and

$$\gamma(\mu) = \inf_{(r,\mu)\in\mathcal{G}(\mathfrak{r},\mathfrak{u};A)} \{r\} = \inf_{x\in\mathcal{B}_{\mathfrak{u},A}(\mu)} \{\mathfrak{r}(x)\} = \inf_{x\in\mathcal{B}_{\mathfrak{r},\mathfrak{u},A}(r^*,\mu)} \{\mathfrak{r}(x)\}. \tag{39}$$

The function r restricted to the compact set $B_{r,u,A}(r^*,\mu)\subset\text{dom}(r)$ must have closed (even compact) sublevel sets and hence is lower semi-continuous on $B_{r,u,A}(r*,\mu)$ (see (Rockafellar 1972, Theorem 7.1)). Consequently the infimum in Equation (39) becomes a minimum. Hence, Equation (38b) follows and γ is well-defined. The function γ is increasing which directly follows from the definition in Equation (37b) because $B_{u,A}(\mu_1)\supset B_{u,A}(\mu_2)$ for all $\mu_1<\mu_2$.

Obviously, $(\gamma(\mu),\mu)\in\partial G(r,u;A)$ for all $\mu\in J$. Hence, convexity of γ follows from convexity of G(r,u;A). Then, we already know that γ is continuous in the interior of the domain J (see (Rockafellar 1972, Theorem 10.1)). Closedness of G(r,u;A) (see Proposition 2 (d)), together with the possibilities for I and J (see Corollary 1), implies closedness of the epigraph of γ. Therefore, γ must be lower semi-continuous (see (Rockafellar 1972, Theorem 7.1)). Since γ is convex, it must even be continuous on J.

An important consequence of the last result is the representation of $G_{eff}(r,u;A)$ as graph of ν and (after interchanging coordinates) of γ, see (Maier-Paape and Zhu 2018a, Theorem 4) for the one-period case and also (Platen 2018, Corollary 2.4.16).

Corollary 2 (Parametrization of efficient frontier as graph).

Let Setting 1 be given and assume that $B_{r,u,A}(r,\mu)$ *is compact for all* $_{r,\mu}\in R$. *Then, the efficient frontier has the representation*

$$\mathcal{G}_{eff}(r,u;A) = \{(r,\nu(r)) : r \in I\} = \{(\gamma(\mu),\mu) : \mu \in J\}. \quad (40)$$

Moreover, ν and γ are strictly increasing and $\nu=\gamma^{-1}$, *i.e.,* $\nu(\gamma(\mu))=\mu$ *for all* $\mu\in J$ *and* $\gamma(\nu(r))=r$ *for all* $r\in I$.

Proof.

Theorem 2 (b) (see Equation (33)) and the definitions of *I* and *J* (see Definition 14) imply $G_{eff}(r,u;A)=\partial G(r,u;A)\cap(I\times J)$. Because of Proposition 2 (c) together with Equation (33), there is exactly one element $(r*(\mu),\mu)\in G_{eff}(r,u;A)$ for each fixed $\mu\in J$ and there is exactly one element $(r,\mu^*(r))\in G_{eff}(r,u;A)$ for each fixed $r\in I$. Obviously, it must be $r^*(\mu)=\gamma(\mu)$ and $\mu^*(r)=\nu(r)$. Uniqueness of the elements implies Equation (40).

Because of Equation (40) it directly follows that $\nu=\gamma-1$. Hence, ν and γ are bijective and, because of Proposition 3, increasing. Consequently, they must even be strictly increasing.

This gives many reasonable results which we can use to show solvability of the two optimization problems in Equations (MinR) and (MaxU).

Efficient Portfolios

This final subsection links points on the efficient frontier to their corresponding portfolio/trading strategy. The first result gives the existence of solutions (see (Platen 2018, Theorem 2.4.19) for similar results). Note that from now on we formally "hide" the side condition $S_0^\top v_1(x) = \beta$ of Equations (MinR) and (MaxU) in the set *A*.

Theorem 3 (Existence for Problem 1).

Let Setting 1 and $\beta>0$ *be given and assume* $A\subset\{x\in R^{M+1}: S_0^\top v_1(x) = \beta\}$ *is non-empty and convex. Suppose* $Br_{,u,A}(r,\mu)$ *is compact for all* $r,\mu\in R$ *and let* $I,J\subset R$ *be the intervals from Definition 14.*

(a) *For each* $\mu \in J$, *there exists an efficient element* $x_\mu \in A$ *with* $u(x_\mu)=\mu$. *The element* x_μ *also solves Equation* (MinR).

(b) *For each* $r \in I$, *there exists an efficient element* $y_r \in A$ *with* $r(y_r)=r$. *The element* y_r *also solves Equation* (MaxU).

(c) *Each solution of Equation* (MinR) *for* $\mu \in J$ *and each solution of Equation* (MaxU) *for* $r \in I$ *is efficient. Moreover, each efficient element* $x^* \in A$ *solves Equation* (MinR) *for* $\mu=u(x^*)$ *and Equation* (MaxU) *for* $r=r(x^*)$.

Proof.

Statements (a) and (b) follow from Corollary 2. For instance, by Equation (40), for every $r \in I$, there exists some $y_r \in A$ with $r(y_r)=r$ and $u(y_r)=v(r)$. Clearly, $S_0^\top v_1(y_r) = \beta$ by assumption on A. Using Equation (38a), we conclude

$$u(y_r) = \nu(r) = \max_{x \in A} \{u(x) : r(x) \leq r\}$$

yielding that y_r solves Equation (MaxU) and, moreover, each efficient element $x^* \in A$ with risk value r solves Equation (MaxU) as well. Conversely, any (other) solution y'_r of Equation (MaxU) for $r \in I$ satisfies $u(y'_r)=v(r)=u(y_r)$ and $r(y'_r) \leq r$. Since y_r is efficient, $r(y'_r)<r$ is not possible, i.e., we must have $r(y'_r)=r$. Therefore, y'_r is efficient as well. The claim for $\mu \in J$ follows similarly.

For uniqueness, more assumptions are required. If either u is strictly concave or r is strictly convex, the uniqueness is guaranteed, see (Maier-Paape and Zhu 2018a, Theorem 5) for the one-period case with finite probability space and also (Platen 2018, Theorem 2.4.20) for a similar result.

Theorem 4 (Uniqueness and efficient portfolio path)**.**

Let the situation in Theorem 3 be given. Furthermore, assume that either u *is strictly concave in* dom(u) *or* r *is strictly convex in* dom(r). *Then, the following holds.*

(a) *For each* $\mu \in J$, *there is exactly one efficient element* $x_\mu \in A$ *with* $u(x_\mu)=\mu$, *which in addition is the unique solution of Equation* (MinR).

Furthermore, the mapping $\tilde{\gamma}: J \to A,\ \mu \mapsto x_\mu$ *is continuous.*

For each $\mu \notin J$ *and* $\mu \geq \mu_{max} = \sup J$, *there does not exist any solution of Equation* (MinR).

If $\mu_{min} > -\infty$, *then for* $\mu \notin J$ *and* $\mu \leq \mu_{min}$ *(i.e.,* $\mu < \mu_{min}$, *see Corollary 1) the solution of Equation* (MinR) *is not necessarily unique and can be an element in A which is not efficient.*

(b) *For each* r∈I, *there is exactly one efficient element* y_r∈A *with* $r(y_r)=r$, *which in addition is the unique solution of Equation* (MaxU).

Furthermore, the mapping $\tilde{\nu}: I \to A,\ r \mapsto y_r$ *is continuous.*

For each r∉I *and* $r \le r_{min}$=inf I, *there does not exist any solution of Equation* (MaxU).

If $r_{max}<\infty$, *then for* r∉I *and* $r \ge r_{max}$ *(i.e.,* $r>r_{max}$*) the solution of Equation* (MaxU) *is not necessarily unique and can be an element in A, which is not efficient.*

Proof.

The existence of efficient elements is already guaranteed by Theorem 3. Let $(r^*,\mu^*)\in G_{eff}(r,u;A)$ be arbitrary. The uniqueness of an efficient element $x*\in A$ with $(r(x^*),u(x^*))=(r^*,\mu^*)$ follows from strict convexity of r or strict concavity of u, respectively, which we show next: Assume the solution x∗ is not unique. Then, there is an efficient element $x'\in A$ with $x'\neq x^*$ and $(r(x'),u(x'))=(r*,\mu^*)=(r(x^*),u(x^*))$. For $x_0:=(x*+x')/2$ we have

$$r(x_0) \le \frac{1}{2}r(x^*) + \frac{1}{2}r(x') = r^* \quad \text{and} \quad u(x_0) \ge \frac{1}{2}u(x^*) + \frac{1}{2}u(x') = \mu^*. \tag{41}$$

Since either r is strictly convex or u is strictly concave, one of the two inequalities in Equation (41) must be strict, which contradicts $(r^*,\mu^*)\in G_{eff}(r,u;A)$. Hence, the efficient portfolio for $(r^*,\mu^*)\in G_{eff}(r,u;A)$ is unique.

Furthermore, $\tilde{\gamma}$ and $\tilde{\nu}$ are well-defined. Next, we show continuity. We only show this for $\tilde{\gamma}$ (continuity for $\tilde{\nu}$ can be shown similarly). Suppose $\tilde{\gamma}$ is discontinuous at some point $\mu_0\in J$. Then, there exist c>0 and a sequence $(\mu_n)_n\in N\subset J$ with $\mu_n\to\mu_0$ as $n\to\infty$ and $\|\tilde{\gamma}(\mu_n) - \tilde{\gamma}(\mu_0)\| \ge c$ for all n∈N. Since γ is continuous, see Proposition 3, we obtain that $(r(\tilde{\gamma}(\mu_n)), u(\tilde{\gamma}(\mu_n))) = (\gamma(\mu_n), \mu_n) \to (\gamma(\mu_0), \mu_0)$ as $n\to\infty$. Hence, for all ε>0 there exists $n_0(\varepsilon)\in N$ such that $\gamma\tilde{}(\mu_n)\in B_{r,u,A}(\gamma(\mu_0)+\varepsilon,\mu_0-\varepsilon)$ for all $n\ge n_0(\varepsilon)$. Since, e.g., $B_{r,u,A}(\gamma(\mu_0)+1,\mu_{0-1})$ is compact, there exists a convergent subsequence of $(\tilde{\gamma}(\mu_n))_{n\in N}$ with limit $x*\in B_{r,u,A}(\gamma(\mu_0)+1,\mu_{0-1})$. Using again lower semi-continuity of r restricted to $B_{r,u,A}(\gamma(\mu_0)+1,\mu_0-1)$ and upper semi-continuity of u restricted to $B_{r,u,A}(\gamma(\mu_0)+1,\mu_0-1)$ as in the proof of Proposition 2 (d) gives $x*\in B_{r,u,A}(\gamma(\mu_0),\mu_0)$. Then, x^* must be efficient, because $(\gamma(\mu_0),\mu_0)\in G_{eff}(r,u;A)$, and we have $(r(x^*),u(x^*))=(\gamma(\mu_0),\mu_0)$, i.e.,

$x^* = \widetilde{\gamma}(\mu_0)$. This is a contradiction, because it must be $\|x^* - \widetilde{\gamma}(\mu_0)\| \geq c > 0$. Consequently, γ˜ is continuous.

The situations where $\mu \notin J$ or $r \notin I$ follow easily: For instance, in case $\mu \geq \mu_{max}$ and $\mu \notin J$, there is no portfolio $x \in A$ such that $u(x) \geq \mu$, see Equation (36). In the case $-\infty < \mu < \mu min$, there is an efficient element (which also solves Equation (MinR)), namely $\widetilde{\gamma}(\mu_{\min})$, but there might also be a solution of Equation (MinR), e.g., $x' \in A$, such that $\mathfrak{r}(x') = \mathfrak{r}(\widetilde{\gamma}(\mu_{\min}))$ and $\mu \leq \mathfrak{u}(x') < \mathfrak{u}(\widetilde{\gamma}(\mu_{\min})) = \mu_{\min}$. However, this element x′ is not efficient.

Remark 8

(Connection to (Maier-Paape and Zhu 2018a, Theorem 5)). *In (Maier-Paape and Zhu 2018a, Theorem 5), a related result is shown for the one-period case* N=1 *for a finite probability space. The utility function therein is of the form* $\mathfrak{u}(x) = \mathrm{E}\left[u(S_1^\top x)\right]$, *for some concave function* $u: \mathbb{R} \rightarrow \mathbb{R} \cup \{-\infty\}$, *and the risk function* r *must be non-negative, convex and independent of* x0. *Additional assumptions are that* $B_{r,A}(r)$ *is compact for all* $r \in \mathbb{R}$ *or* $B_{u,A}(\mu)$ *is compact for all* $\mu \in \mathbb{R}$. *This implies that* $B_{r,u,A}(r,\mu)$ *is compact for all* $r, \mu \in \mathbb{R}$ *(cf. Proposition 2 (b)). Since (Maier-Paape and Zhu 2018a, Theorem 5) assumes moreover unit initial cost (i.e.,* ST0x=1*), this already gives all assumptions for Theorem 3 in the case that* v1(x):=x *for all* $x \in A$.

However, the result in (Maier-Paape and Zhu 2018a, Theorem 5) is also a uniqueness result and therefore requires additional assumptions on u *and/ or* r. *For this, either u must be strictly concave or* r *must be strictly convex in the risky part (note that (c3) in (Maier-Paape and Zhu 2018a, Theorem 5) implies that* r2 *is strictly convex in the risky part). (Maier-Paape and Zhu 2018a, Theorem 5) then gives uniqueness.*

Since Theorems 3 and 4 with v1(x)=x *are restricted to the set* $A_\beta := \{x \in \mathbb{R}^{M+1} : S_0^\top x = \beta\}$, *this additional assumption on* r *(being strictly convex in the risky part) implies, that the function* r *restricted to the set* $A\beta \cap A$ *is strictly convex (and not only strictly convex on the risky part). Hence, the assumptions of (Maier-Paape and Zhu 2018a, Theorem 5) are stronger than the assumptions in Theorem 4 and give a similar result. Therefore, Theorem 4 is a full generalization of (Maier-Paape and Zhu 2018a, Theorem 5).*

Note that the assumption in (Maier-Paape and Zhu 2018a, Theorem 5) that u is strictly concave, is not enough to obtain strict concavity of u *in the setting of (Maier-Paape and Zhu 2018a, Theorem 5). Hence, Assumption (c1) in (Maier-Paape and Zhu 2018a, Theorem 5) may not be enough to obtain uniqueness (other than falsely stated there). However, e.g., if S has no nontrivial risk-free portfolio, then* $\mathrm{u}(x) = \mathrm{E}\left[u(S_1^\top x)\right]$ *is strictly concave (see (Maier-Paape and Zhu 2018a, Proposition 6)), and uniqueness follows.*

APPLICATION

Let us focus on Example 2 with the trading strategy generating function v_{twr} which ensures that the portfolio weights are constant after each time step. Our admissible set is given by

$$A_{\text{twr}} = \left\{ f \in \mathbb{R}^{M+1} \; : \; 1 + T_n^\top f > 0 \text{ a.s. for } n = 1, \ldots, N \right\} \tag{42}$$

see Equation (22).

Looking at Problem 1 for some special risk and utility functions, we also need to ensure the second constraint. Using Equation (21), this constraint reads

$$S_0^\top (v_{\text{twr}})_1(f) = W_0 \sum_{i=0}^{M} f_i = \beta.$$

The risk and utility functions we are looking at in the following are independent on W0. Hence, w.l.o.g., we may set W_0:=1. The set of all vectors fulfilling the second constraint in Problem 1 is then given by

$$A_\beta := \left\{ f \in \mathbb{R}^{M+1} \; : \; S_0^\top (v_{\text{twr}})_1(f) = \beta \right\} = \left\{ f \in \mathbb{R}^{M+1} \; : \; \sum_{i=0}^{M} f_i = \beta \right\}. \tag{43}$$

Lemma 3 (utility function; logarithm of TWR).

Let the multi-period market model S be given and assume that $T_n \in \mathcal{L}^1(\Omega, \mathcal{F}_n, \mathrm{P}; \mathbb{R}^{M+1})$ *for* n=1,...,N, *where* T_n *is from Equation* (19) *in Example 2. Define* $\mathrm{ulog}_{\mathrm{TWR}}$ *as in Equation* (28), *i.e.*,

$$\mathrm{u}_{\log\mathrm{TWR}}(f) = \mathrm{E}\left[\ln\left(\mathrm{TWR}(f)\right)\right] = \sum_{n=1}^{N} \mathrm{E}\left[\ln\left(1 + T_n^\top f\right)\right] \tag{44}$$

for $f \in A_{\text{twr}}$ *and* $\mathrm{u}_{\log\mathrm{TWR}}(f) = -\infty$ *for all* $f \notin A_{twr}$.

Then, $\mathrm{ulog}_{\mathrm{TWR}}$ *is proper concave and* $\mathrm{ulog}_{\mathrm{TWR}} < \infty$. *Furthermore, if S has no nontrivial risk-free trading strategy, then* ulogTWR *restricted to* $\mathrm{dom}(\mathrm{ulog}_{\mathrm{TWR}}) \cap A_\beta$, *with* A_β *from Equation* (43), *is strictly concave and* $B_{\mathrm{ulogTWR}, A\beta}(\mu)$ *is bounded for all* $\mu \in \mathrm{R}$ *and all* $\beta > 0$.

Proof.

Since $\ln(1+s) \leq s$ for all $s > -1$ and T_n has a finite expectation by assumption, we have for all $f \in A_{twr}$ that

$$u_{\log TWR}(f) = \sum_{n=1}^{N} \mathrm{E}\left[\ln\left(1 + T_n^\top f\right)\right] \leq \sum_{n=1}^{N} \mathrm{E}\left[T_n\right]^\top f < \infty \tag{45}$$

Of course, we also have $u_{\log TWR}(f) = -\infty < \infty$ for all $f \notin A_{twr}$.

The mapping $f \mapsto \ln(1 + T_n(\omega)^\top f)$ is concave for each $\omega \in \Omega$. Because of linearity and monotonicity of the expectation, the mapping $f \mapsto \ln(1 + T_n(\omega)^\top f)$ is concave. The same holds true for $u_{\log TWR}$. Obviously, $0 \in \mathrm{dom}(u_{\log TWR})$ because $u_{\log TWR(0)} = 0$ and therefore we obtain that the function $u_{\log TWR}$ is proper concave.

Now, assume that S has no nontrivial risk-free trading strategy. Because of Theorem 1 (d), W is injective in the risky part. Using the definition of v_{twr} in Equation (21), we obtain that $W(v_{twr}(f)) = W(v_{twr}(f'))$ a.s. for $f, f' \in A_\beta$ implies $f^{\frown} = f^{\frown\prime}$. Since $f, f' \in A_\beta$, it even must be $f = f'$ if $W(v_{twr}(f)) = W(v_{twr}(f'))$ a.s.

Consequently, for arbitrary $f, f' \in \mathrm{dom}(u_{\log TWR}) \cap A_\beta$ with $f \neq f'$ there exists $n \in \{1, \ldots, N\}$ such that $T_n^\top f \not\equiv T_n^\top f'$, i.e., $T_n^\top f \neq T_n^\top f'$ with positive probability (see Equation (20)). Therefore, for all $\lambda \in (0,1)$, we obtain from strict concavity of ln that

$$\mathrm{P}\left(\ln\left(1 + T_n^\top(\lambda f + (1-\lambda) f')\right) > \lambda \ln\left(1 + T_n^\top f\right) + (1-\lambda)\ln\left(1 + T_n^\top f'\right)\right) > 0.$$

It follows that

$$\mathrm{E}\left[\ln\left(1 + T_n^\top(\lambda f + (1-\lambda) f')\right)\right] > \mathrm{E}\left[\lambda \ln\left(1 + T_n^\top f\right) + (1-\lambda)\ln\left(1 + T_n^\top f'\right)\right].$$

This implies strict concavity for at least one summand of $u_{\log TWR}$ which directly gives strict concavity of ulogTWR restricted to $(u_{\log TWR}) \cap A_\beta$.

The boundedness of $\mathcal{B}_{u_{\log TWR}, A_\beta}(\mu) \subset \mathrm{dom}(u_{\log TWR}) \cap A_\beta$ directly follows from Lemma 1, because $\mathrm{dom}(u_{\log TWR}) \cap A_\beta \subset A_{twr} \cap A_\beta$ is admissible for the trading strategy generating function v_{twr} (see Definition 8 and Equation (20)) and the corresponding matrix B in Lemma 1 for this example is a diagonal matrix with positive entries $b_{ii} = W_0 / S_0^i > 0,$, for $i = 0, 1, \ldots, M$, on the diagonal (see Equation (21)).

Lemma 4 (risk function; logarithm of TWR).

As in Lemma 3, let the multi-period market model S be given and assume that $T_n \in \mathcal{L}^1(\Omega, \mathcal{F}_n, \mathrm{P}; \mathbb{R}^{M+1})$ *for* n=1,…,N, *where* T_n *is from Equation* (19) *in Example 2. Define the log drawdown function* ρln *(see Equation* (31)*), by*

ρln(f)=E[max{0,max1≤ℓ≤N{−ln(TWRNℓ(f))}}]=E[max{0,max1≤ℓ≤N{−∑n=ℓNln(1+TTnf)}}],

$$\rho_{\ln}(f) = \mathrm{E}\left[\max\left\{0, \max_{1\le\ell\le N}\left\{-\ln\left(\mathrm{TWR}_\ell^N(f)\right)\right\}\right\}\right] = \mathrm{E}\left[\max\left\{0, \max_{1\le\ell\le N}\left\{-\sum_{n=\ell}^{N}\ln\left(1+T_n^\top f\right)\right\}\right\}\right],$$

for $f \in A_{twr}$ *and* $\rho_{\ln}(f)=\infty$ *for all* $f \notin A_{twr}$. *Then,* $\rho_{\ln}$ *is proper convex,* $\rho_{\ln} \geq 0$ *and* $\mathrm{dom}(\rho_{\ln})=\mathrm{dom}(u_{\log TWR})$. *If S has no nontrivial risk-free trading strategy, then* $B_{\rho\ln, A\beta}(r)$ *is bounded for all* $r \in \mathbb{R}$ *and all* $\beta>0$.

Proof.

The property $\rho l_n \geq 0$ is obvious. Since $f \mapsto -\ln(1+T_n^\top f)$ is convex and the maximum of convex functions again is convex, it follows that $\rho_{\ln}$ is convex as well. In addition, $\rho_{\ln}(0)=0$ and therefore $0 \in \mathrm{dom}(\rho_{\ln})$. Hence, $\rho_{\ln}$ is proper convex.

Inserting the known characterizations of $u_{\log TWR}$ and $\rho_{\ln}$ from above and using the properties of the logarithm yield for $f \in A_{twr}$ that

$$\begin{aligned} u_{\log\mathrm{TWR}}(f) + \rho_{\ln}(f) &= \mathrm{E}\left[\sum_{n=1}^{N}\ln\left(1+T_n^\top f\right) + \max\left\{0, \max_{1\le\ell\le N}\left\{-\sum_{n=\ell}^{N}\ln\left(1+T_n^\top f\right)\right\}\right\}\right] \\ &= \mathrm{E}\left[\max\left\{0, \max_{1\le\ell\le N}\left\{\sum_{n=1}^{\ell}\ln\left(1+T_n^\top f\right)\right\}\right\}\right] \\ &\le \mathrm{E}\left[\sum_{n=1}^{N}\left|T_n^\top f\right|\right] < \infty, \end{aligned}$$

because $T_n \in \mathcal{L}^1(\Omega, \mathcal{F}_n, \mathrm{P}; \mathbb{R}^{M+1})$. Of course, we directly see from this that we also have $u_{\log TWR}(f)+\rho_{\ln}(f) \geq 0$. Hence, whenever $u_{\log TWR}(f) \in \mathbb{R}$, it must be $\rho_{\ln}(f) \in \mathbb{R}$ and vice versa. It directly follows that $\mathrm{dom}(\rho_{\ln})=\mathrm{dom}(u_{\log TWR})$. As in the proof of Lemma 3, the boundedness of $B\rho_{\ln}, A_\beta(r) \subset A\beta$ directly follows from Lemma 1.

It is worth noting that $\rho_{\ln}$ may not be strictly convex.

Remark 9

(Connection to Maier-Paape and Zhu (2018b)). *Maier-Paape and Zhu (2018b) proved properties such as convexity for risk functions involving the relative drawdown but for a one-period market model. The function* rcur *discussed therein corresponds to* ρln *from Lemma 4 in the case we have a finite and discrete market model where the rates of returns are iid.*

Assume we want to solve an optimization such as Equation (MinR) or Equation (MaxU) using the utility and risk functions from Lemmas 3 and 4, respectively, and the corresponding trading strategy generating function v_{twr}. All requirements for Setting 1 are then fulfilled (note that $0 \in dom(u_{logTWR}) = dom(\rho_{ln}) \neq \emptyset$). To be able to apply Theorem 3 or Theorem 4, we need that $B_{\rho ln,ulogTWR,A\beta(r,\mu)}$ is compact for all $r,\mu \in R$. From Lemma 3, we obtain boundedness in case *S* has no nontrivial risk-free trading strategy. However, in general, it is not clear whether or not the superlevel sets of u_{logTWR} are closed. Moreover, we do not know whether $dom(\rho_{ln}) = A_{twr}$ holds true. Before we discuss the solutions of the corresponding optimization problems in Equations (MinR) and (MaxU), we firstly need to take care of these assumptions. We start with a more specific situation where we can ensure the compactness of $B_{\rho ln,ulogTWR,A\beta}(r,\mu)$.

Remark 10 (ρ_{ln} and u_{logTWR} in finite probability space)**.**

Assume the probability space is finite, e.g., with $\Omega := \{\omega_1,\ldots,\omega_K\}$ *for some fixed* $K \in N$ *and* $p_k := (\{\omega_k\}) > 0$ *for all* $k = 1,\ldots,K$. *Then,* A_{twr} *in Equation* (42) *becomes*

$$A_{\mathrm{twr}} = \left\{ f \in \mathbb{R}^{M+1} \;:\; 1 + T_n(\omega_k)^\top f > 0 \text{ for } n = 1,\ldots,N \text{ and } k = 1,\ldots,K \right\}, \tag{46}$$

where $T_n(\omega k) \in R^{M+1}$ *for each* $n = 1,\ldots,N$ *and* $k = 1,\ldots,K$ *is a vector fixed for a given market (see Equation* (**19**)*). Clearly,* $0 \in A_{twr}$. *Furthermore,* u_{logTWR} *in Equation* (**44**) *becomes*

$$u_{\mathrm{logTWR}}(f) = \sum_{n=1}^{N} \left[\sum_{k=1}^{K} p_k \ln\left(1 + T_n(\omega_k)^\top f\right) \right] \tag{47}$$

for $f \in A_{twr}$. *Then, we obviously get* $dom(u_{logTWR}) = A_{twr}$ *because by definition* $u_{\mathrm{logTWR}}|_{A^c_{\mathrm{twr}}} = -\infty$.

Now, let $(f^m)_{m\in\mathbb{N}} \subset A_{\mathrm{twr}}$ *be a sequence such that* $f^m \to f^* \in \partial A_{\mathrm{twr}}$ *as* $m \to \infty$. *Then, there exist* $n \in \{1,\ldots,N\}$ *and* $k \in \{1,\ldots,K\}$ *such that* $1 + T_n(\omega_k)^\top f^* = 0$. *In this case, we obtain* $u_{\mathrm{logTWR}}(f^m) \to -\infty$ *as* $m \to \infty$. *From this we, can conclude that* A_{twr} *is open and non-empty and, moreover, by Equation* (**47**), $u_{logTWR}|A_{twr}$ *is continuous. In particular, the superlevel sets of* u_{logTWR} *are closed. Consequently, we also must have that* $B_{ulogTWR},A(\mu)$ *is closed for all closed sets A and all* $\mu \in R$.

Analogously, we obtain $(\rho_{\ln}) = A_{\mathrm{twr}} = \mathrm{dom}(u_{\mathrm{logTWR}})$ *where the sublevel sets of* ρ_{ln} *and also* $B_{\rho ln,A}(r)$ *must be closed for all closed sets A and all* $r \in R$. *Then, Proposition 2 (a) and Lemmas 3 and 4 yield that* u_{logTWR} *is closed proper concave and* $\rho_{\ln}$ *is closed proper convex.*

In general, however, when Ω is not finite $u_{\log TWR}$ might not be closed proper concave and $\rho_{\ln}$ might not be closed proper convex. Since we assume these properties in the existence and uniqueness theorem (see Theorem 5 below), we make some more remarks to have a better understanding also in the general situation.

Remark 11 (Notes on $\mathrm{dom}(u_{\log TWR})$ and A_{twr}).

(a) *Clearly* $\mathrm{dom}(u_{\log TWR}) \subset A_{twr}$.

(b) *If* f∈Atwr, *then, using Equation* (**20**), *it follows that* Wn(vtwr(f))=W0∏nk=1(1+TTkf)>0 *a.s. Of course, this is trivial and directly follows from the definition of* Atwr *in Equation* (**42**). *In fact,* Atwr *is defined as the admissible set of* vtwr *(see Example 2).*

(c) *We have*

dom(ulogTWR)

$$= \left\{ f \in \mathbb{R}^{M+1} : \mathrm{E}\left[\ln\left(1 + T_n^\top f\right)\right] > -\infty \textit{ for all } n = 1, \dots, N \right\}$$

$$= \left\{ f \in \mathbb{R}^{M+1} : \int_\Omega \ln\left(1 + T_n(\omega)^\top f\right) \mathrm{d}\,\mathrm{P}(\omega) > -\infty \textit{ for all } n = 1, \dots, N \right\}.$$

Proof: The second equality holds by definition. For the first one, the relation "$\supset$*" is obvious. Let now* f∈$\mathrm{dom}(u_{\log TWR}) \subset$Atwr *be arbitrary. Since* $u_{\log TWR}(f) < \infty$ *by Lemma 3 we have* $u_{\log TWR}(f) \in \mathbb{R}$. *In addition,* $\mathrm{E}\left[\ln\left(1 + T_n^\top f\right)\right] < \infty$ *holds for* n=1,…,N *(cf. Equation* (**45**)*). Hence, it must be* E[ln(1+TTnf)]>−∞ *for* n=1,…,N, *which shows the relation "*$\subset$*" and therefore the equality.*

(d)

Define

A∗twr:={f∈RM+1:thereexistsε>0suchthat1+TTnf>εa.s.foralln=1,…,N}.

$$A^*_{\mathrm{twr}} := \left\{ f \in \mathbb{R}^{M+1} : \right.$$

$$\left. \textit{there exists } \varepsilon > 0 \textit{ such that } 1 + T_n^\top f > \varepsilon \textit{ a.s. for all } n = 1, \dots, N \right\}$$

Then, we obtain $A^*_{\mathrm{twr}} \subset \mathrm{dom}(u_{\log \mathrm{TWR}})$.

Proof: Let $f \in A^*_{\mathrm{twr}}$ *be arbitrary. Then,* $\ln(1 + T_n^\top f) \geq \ln(\varepsilon) > -\infty$ *a.s. This, of course, gives* $\mathrm{E}\left[\ln\left(1 + T_n^\top f\right)\right] \geq \ln(\varepsilon) > -\infty$.

Now, we can show the result for the optimization problems in Equations (MinR) and (MaxU) when using $r:=\rho_{\ln}$ and $u:=u_{\log TWR}$.

Theorem 5 (Existence and uniqueness for $u_{\log TWR}$ and $\rho_{\ln}$).

Assume the multi-period market model S has no nontrivial risk-free trading strategy and $T_n \in \mathcal{L}^1(\Omega, \mathcal{F}_n, P; \mathbb{R}^{M+1})$ *for* n=1,…,N, *where* T_n *is from Equation* (19) *in Example 2. Let the trading strategy generating function be given by* vtwr *(constant weights) from Example 2, with admissible set* A_{twr} *as in Equation* (42). *Assume that* $\rho_{\ln}$ *and* $u_{\log TWR}$ *restricted to some convex and non-empty set* $A \subset \mathrm{dom}(u_{\log TWR}) \cap A_\beta$ *are closed proper convex and closed proper concave, respectively. We define the* minimum log drawdown optimization problem *for fixed* $\mu \in R$ *by*

$$\min_{f \in A} \rho_{\ln}(f) \quad \text{subject to } u_{\log TWR}(f) \geq \mu,\ S_0^{\top}(v_{twr})_1(f) = \beta. \tag{MinDD}$$

We define the maximum log TWR optimization problem *for fixed* $r \in R$ *by*

$$\max_{f \in A} u_{\log TWR}(f) \quad \text{subject to } \rho_{\ln}(f) \leq r,\ S_0^{\top}(v_{twr})_1(f) = \beta. \tag{MaxTWR}$$

The following holds true:

(a) *(Growth optimal trading strategy) The problem in Equation* (MaxTWR) *without risk restriction, i.e.,*

$$\max_{f \in A} u_{\log TWR}(f) \quad \text{subject to } S_0^{\top}(v_{twr})_1(f) = \beta \tag{48}$$

has a unique solution $f^*_{\max} \in A$. *Moreover, we have* $\mu_{\max} = u_{\log TWR}(f^*_{\max}) \in \mathbb{R}$ *and* $r_{\max} = \rho_{\ln}(f^*_{\max}) \in \mathbb{R}_{\geq 0}$, *where* $\mu_{\max}$ *and* $r_{\max}$ *represent the suprema of* $G_{eff}(r,u;A)$ *from Definition 13 for* $r=\rho_{\ln}$ *and* $u=u_{\log TWR}$.

(b) *(Risk minimal trading strategy) The problem in Equation* (MinDD) *without utility restriction, i.e.,*

$$\min_{f \in A} \rho_{\ln}(f) \quad \text{subject to } S_0^{\top}(v_{twr})_1(f) = \beta, \tag{49}$$

has a finite minimum risk value $r_{\min} \in R \geq 0$. *Furthermore, among all* $f \in A$ *which solve Equation (**49**), there is a unique element* $f^*_{\min} \in A$ *with maximal* $u_{\log TWR}$ *value. In particular,* $r_{\min} = \rho_{\ln}(f^*_{\min}) \in R \geq 0$, *but moreover* $\mu_{\min} = u_{\log TWR}(f^*_{\min}) \in \mathbb{R}$ *hold true, where* $r_{\min}$ *and* $\mu_{\min}$ *represent the infima of* $G_{eff(}r,u;A)$ *from Definition 13 for* $r=\rho_{\ln}$ *and* $u=u_{\log TWR}$.

(c). *For each* $\mu \in J=[\mu_{\min}, \mu_{\max}] \neq \emptyset$, *there is exactly one efficient element* $f^*_\mu \in A$ *with* $u_{\log TWR}(f^*_\mu) = \mu$, *which is also the unique solution of Equation* (MinDD). *The mapping* $\tilde{\gamma}: J \to A,\ \mu \mapsto f^*_\mu$ *is continuous.*

(d). For each $r \in I=[r_{min}, r_{max}] \neq \emptyset$*, there is exactly one efficient element* $\tilde{f}_r^* \in A$ *with* $\rho_{\ln}(\tilde{f}_r^*) = r$*, which is also the unique solution of Equation* (MaxTWR). *The mapping* $\tilde{\nu}: I \to A,\ \mu \mapsto \tilde{f}_\mu^*$ *is continuous.*

Proof.

By assumption, ulogTWR and ρln, both restricted to *A*, are closed proper concave and closed proper convex, respectively. Using Lemma 3, we in addition obtain that u_{logTWR} is strictly concave in the set $A \subset dom(u_{logTWR}) \cap A\beta$. Moreover, $B_{ulogTWR,A}(\mu)=B_{ulogTWR}, A_\beta(\mu) \cap A$ is compact for all $\mu \in R$ because of Lemma 3 and Proposition 2 (a). Analogously, $B_{\rho ln,A}(r)$ is compact for all $r \in R$ because of Lemma 4 and Proposition 2 (a). Consequently, Proposition 2 (b) yields that Bρln,ulogTWR,A(r,μ) is compact for all r,μ∈R. Theorem 4 can then be applied, which proves (c) and (d), if we can show that $I=[r_{min}, r_{max}]$, $J=[\mu_{min}, \mu_{max}]$. This is shown in the proofs of (a) and (b).

Proof of (a): Since ul_{ogTWR} is closed proper concave on *A*, we know that u_{logTWR} must be upper semi-continuous (cf. (Rockafellar 1972, Theorem 7.1)). In addition, $Bu_{logTWR,A}(\mu)$ is compact and non-empty for some $\mu \in R$. Hence, there must be a solution of Equation (48) (see (Barbu and Precupanu 2012, Theorem 2.8)). Uniqueness follows from strict concavity of u_{logTWR} restricted to $A \subset dom(u_{logTWR})$. Furthermore, Lemma 2 yields that $\mu_{max} = u_{logTWR}(f^*_{max}) \in \mathbb{R}$ and, since $dom(u_{logTWR})=dom(\rho_{ln})$ by Lemma 4, that $r_{max} = \rho_{ln}(f^*_{max}) \in [0,\infty)$ (also by Lemma 2, note that $B_{ulogTWR,A}$(μmax) contains only f^*_{max}).

Proof of (b): The function ρln is closed proper convex on *A* by assumption and, hence, it is lower semi-continuous (cf. (Rockafellar 1972, Theorem 7.1)). In addition, $B_{\rho_{\ln},A}(r)$ is compact and non-empty for some $r \in R$. Then, there must be a solution of Equation (49) (see (Barbu and Precupanu 2012, Theorem 2.8)). Maximizing u_{logTWR} over all those solutions then, similar to in the proof of (a), gives a unique solution denoted by f^*_{min}. As above, Lemma 2 yields that $r_{min} = \rho_{ln}(f^*_{min}) \in [0,\infty)$. Since $dom(u_{logTWR})=dom(\rho_{ln})$, we get $\mu_{min} = u_{logTWR}(f^*_{min}) \in \mathbb{R}$. Altogether, we obtain that $I=[r_{min}, r_{max}]$ and $J=[\mu_{min}, \mu_{max}]$, which completes the proof.

Note that $A_\beta \cap dom(u_{logTWR}) \neq \emptyset$ because obviously $(\beta, 0, \ldots, 0)^\top \in A_\beta \cap dom(u_{logTWR})$. Hence, there exists such a subset $A \subset A_\beta$ with the above required properties, e.g., $A = A_\beta \cap dom(u_{logTWR})$. Furthermore, the "local" (closed) proper convexity of ρ_{ln} on *A* and the "local" (closed) proper concavity of u_{logTWR} on A, which are relevant according to Setting 1 (because of the domain of definition of both functions) and Theorem 5, can be provided for instance as follows by "global" assumptions.

Lemma 5.

In the situation of Lemma 3 and Lemma 4 assume that ρln *is closed proper convex and* u_{logTWR} *is closed proper concave. For* Aβ *from Equation* (**43**) *with fixed* β>0 *let* $A' \subset A_\beta$ *be closed and convex such that* $A := A' \cap \mathrm{dom}(u_{logTWR})$ *is non-empty. Then,* $A \subset \mathrm{dom}(u_{logTWR}) \cap A_\beta$ *is convex and non-empty. Furthermore,* ρ_{ln} *restricted to A is closed proper convex and* u_{logTWR} *restricted to A is closed proper concave.*

Proof.

First note that u_{logTWR} is closed proper concave and ρ_{ln} is closed proper convex but by definition both on R^{M+1}. Of course, u_{logTWR} is proper concave and ρ_{ln} is proper convex on A as well. Since $\mathcal{B}_{u_{logTWR},A}(\mu) = \{f \in A : u_{logTWR}(f) \geq \mu\} \subset \mathrm{dom}(u_{logTWR})$ we can also write $\mathcal{B}_{u_{logTWR},A}(\mu) = \{f \in \mathbb{R}^{M+1} : u_{logTWR}(f) \geq \mu\} \cap A'$. Hence, $\mathcal{B}_{u_{logTWR},A}(\mu)$ must be closed because {f∈RM+1:ulogTWR(f)≥μ} is closed (cf. Proposition 2 (a) when replacing A by R^{M+1} therein) and A′ is closed by assumption. Proposition 2 (a) then tells us that u_{logTWR} restricted to A is closed proper concave. A similar argumentation yields that ρ_{ln} restricted to A is closed proper convex.

We have seen in Theorem 5 and Lemma 5 that one of the main ingredients to the existence and uniqueness theory for trading off risk and reward with ρ_{ln} and u_{logTWR} is that ρ_{ln} is closed proper convex and that u_{logTWR} is closed proper concave. While for finite probability space Ω this is already derived in Remark 10, in general this is not obvious. Lemma 4 and Lemma 3 just yield proper convex and proper concave, respectively. The following discussion closes this gap under reasonable conditions.

Lemma 6.

For n∈{1,…,N} *fixed let* Tn∈L1(Ω,Fn,P;RM+1). *Define*

hn(f):={E[ln(1+TTnf)],−∞,f∈Dn,f∉Dn,

where Dn:=dom(hn)={f∈RM+1:E[ln(1+TTnf)]>−∞}. *Assume that* 0∈int(Dn), *where* int(Dn) *is the interior of* Dn. *Then,* hn *is closed proper concave.*

Proof.

Note that, according to Lemma 3, the function $h_n\colon \mathbb{R}^{M+1} \to \mathbb{R} \cup \{-\infty\}$ is proper concave and thus D_n is convex. Therefore, h_n is continuous in the interior of D_n (cf. (Rockafellar 1972, Theorem 10.4)).

By assumption, $f_0 := 0 \in \operatorname{int}(D_n)$ and $h_n(f_0) = \ln(1) = 0$. Using (Rockafellar 1972, Theorem 7.5), the closure of h_n is of the form

$$\bar{h}_n(f) = \lim_{\lambda \nearrow 1} h_n((1-\lambda)f_0 + \lambda f) = \lim_{\lambda \nearrow 1} h_n(\lambda f), \qquad f \in \mathbb{R}^{M+1}. \tag{50}$$

The function $\bar{h}_n$ is known to be closed proper concave (see (Rockafellar 1972, Theorem 7.5.1)) with $\bar{h}_n \geq h_n$ and, moreover, $\bar{h}_n$ coincides with h_n everywhere except possibly on ∂D_n (see (Rockafellar 1972, Theorem 7.4)).

If we can show that $\bar{h}_n(f^*) = h_n(f^*)$ for all $f^* \in \partial D_n$, then $\bar{h}_n = h_n$ on $\mathbb{R}^{M+1}$ and thus h_n is closed proper concave as well. To see that, we fix $f^* \in \partial D_n$ and set $\lambda_m := 1 - 1/m \nearrow 1$ (as $m \to \infty$). Since the limit in Equation (**50**) is independent of the sequence realizing $\lambda \nearrow 1$, we have

$$\bar{h}_n(f^*) = \lim_{m\to\infty} h_n(\lambda_m f^*) \in \mathbb{R} \cup \{-\infty\}.$$

Define the random variables $Z_m^+ := \max\{0, \ln(1 + T_n^\top f^* \lambda_m)\}$ and $Z_m^- := \min\{0, \ln(1 + T_n^\top f^* \lambda_m)\}$. By assumption, $T_n \in \mathcal{L}^1(\Omega, \mathcal{F}_n, \mathrm{P}; \mathbb{R}^{M+1})$ and, hence, as in Equation (**45**),

$$Z_m^+ = \max\left\{0, \ln\left(1 + T_n^\top f^* \lambda_m\right)\right\} \leq \max\left\{0, T_n^\top f^* \lambda_m\right\} \leq \max\left\{0, T_n^\top f^*\right\}.$$

Therefore, $0 \leq \mathrm{E}[Z_m^+] \leq \mathrm{E}[\max\{0, T_n^\top f^*\}] =: M_n < \infty$ which gives

$$\bar{h}_n(f^*) = \lim_{m\to\infty} \mathrm{E}\left[\ln\left(1 + T_n^\top f^* \lambda_m\right)\right] = \lim_{m\to\infty} \mathrm{E}\left[Z_m^-\right] + \lim_{m\to\infty} \mathrm{E}\left[Z_m^+\right]$$

Since ln is increasing, Z_m^- is monotonically decreasing in m (i.e., $Z_{m+1}^- \leq Z_m^- \leq 0$ a.s.) and Z_m^+ is monotonically increasing in m (i.e., $0 \leq Z_m^+ \leq Z_{m+1}^+$ a.s.). Hence, the monotone convergence theorem (see (Fristedt and Gray 1997, Section 8.2, Theorem 6)) implies

$$\bar{h}_n(f^*) = \mathrm{E}\left[\lim_{m\to\infty} Z_m^-\right] + \mathrm{E}\left[\lim_{m\to\infty} Z_m^+\right] = \mathrm{E}\left[\ln\left(1 + T_n^\top f^*\right)\right] = h_n(f^*), \tag{51}$$

which completes the proof.

Note that in Equation (**51**) the limit might be finite (i.e., $f^* \in \partial D_n \cap D_n$) or $-\infty$ (i.e., f $f^* \in \partial D_n \cap D_n^c$, where $D_n^c = \mathbb{R}^{M+1} \setminus D_n$). In the latter case the transition of h_n from D_n to D_n^c at the point f∗ is smooth, whereas in the first case h_n jumps at f* (but still maintains upper semi-continuity). Both cases indeed occur as we show in Example 3 below.

Corollary 3.

Let S be a multi-period market model such that $T_n \in \mathcal{L}^1(\Omega, \mathcal{F}_n, P; \mathbb{R}^{M+1})$ *for* n=1,…,N, *where* T_n *is from Equation* **(19)** *in Example 2. Assume that* $0 \in \operatorname{int}(\operatorname{dom}(u_{\log TWR}))$. *Then,* $\rho_{\ln}$ *defined in Lemma 4 is closed proper convex and* $u_{\log TWR}$ *from Lemma 3 is closed proper concave.*

Proof.

Using Lemma 6, hn for n=1,…,N are closed proper concave and thus, in particular, upper semi-continuous (see (Rockafellar 1972, Theorem 7.1)). Hence, $u_{\log TWR}(f) = \sum_{n=1}^{N} h_n(f), f \in \mathbb{R}^{M+1}$, inherits these properties. The proof for ρln is similar. □

We close this section with the already mentioned example.

Example 3 $(\operatorname{dom}(u_{\log TWR}), A_{\text{twr}}$ and $A^*_{\text{twr}})$.

With Remark 11 (a) and (d), we already know that $A^*_{\text{twr}} \subset \operatorname{dom}(u_{\log TWR}) \subset A_{\text{twr}}$. *We want to show at specific examples that* $A^*_{\text{twr}} \subsetneq \operatorname{dom}(u_{\log TWR})$ *as well as* $\operatorname{dom}(u_{\log TWR}) \subsetneq A_{\text{twr}}$ *is possible. In all examples below we use* $\omega = t \in (0,1) =: \Omega$ *with* $P = \lambda_{(0,1)}$ *and* M=N=1 *and, for simplicity, we ignore the risk-free asset.*

(a) *Let* $T_1(t):=\exp(-1/t)-1\in(-1,0)$ *for* t∈(0,1). *Then*

$$u_{\log TWR}(f) = h_1(f) = \int_0^1 \ln(1 + T_1(t)f)\,dt, \quad f \in \operatorname{dom}(h_1) = \operatorname{dom}(u_{\log TWR}).$$

For f∈(−∞,1), *there exists some* M>0 *such that* $1+T_1(t)f \geq M>0$ *for all* t∈(0,1), *but for* f∈(1,∞) *we have* $1+T_1(t)f<0$ *for t with positive measure. Hence,* $A^*_{\text{twr}} = (-\infty, 1)$ *and* $A_{\text{twr}}=(-\infty,1]$. *Calculating*

$$h_1(1) = \int_0^1 \ln(1 + T_1(t))\,dt = \int_0^1 -\frac{1}{t}\,dt = -\infty,$$

we find $f^*:=1\notin\operatorname{dom}(h_1)=\operatorname{dom}(u_{\log TWR})=(-\infty,1)$. *In this example, we thus have* $\operatorname{dom}(u_{\log TWR}) \subsetneq A_{\text{twr}}$. *Moreover, since* $0\in\operatorname{int}(\operatorname{dom}(u_{\log TWR}))$, *by Corollary 3, we obtain that* $u_{\log TWR}$ *is closed proper concave.*

(b). *Let* $\widetilde{T}_1(t) := \exp(-1/\sqrt{t}) - 1 \in (-1,0)$ *for* $t \in (0,1)$. *Reasoning as in (a), we again get* $\widetilde{A}^*_{\text{twr}} = (-\infty,1)$ *and* $\widetilde{A}_{\text{twr}} = (-\infty,1]$. *However, this time*

$$\tilde{h}_1(1) = \int_0^1 \ln(1+\tilde{T}_1(t))\,dt = \int_0^1 -\frac{1}{\sqrt{t}}\,dt = -2.$$

Hence, $f^* := 1 \in \mathrm{dom}(\tilde{h}_1) = \mathrm{dom}(u_{\log TWR}) = (-\infty, 1]$ *and therefore* $\widetilde{A}^*_{twr} \subsetneqq \mathrm{dom}(u_{\log TWR})$. *Again* $u_{\log TWR}$ *is closed proper concave by Corollary 3.*

CONCLUSIONS AND OUTLOOK

In this Part III of our series of papers on a general framework on the portfolio theory, we extend the results from Part I (Maier-Paape and Zhu 2018a) for the one-period financial market to a multi-period market model. We do so by using a modular approach that separates the framework into the four related modules: (a) multi-period market model; (b) trading strategy; (c) risk and utility function; and (d) optimization problem. This work provides an in itself complete general framework for handling the trade-off between competing performance criteria on reward and risk for trading strategies. This framework provides a foundation for implementation which is an interesting direction for further exploration.

Building Block (a) gives a lot of freedom for the market model. The most important assumption on the model should be that there is no nontrivial risk-free trading strategy. Block (b) gives the liberty for choosing a trading strategy. Even more complex trading strategies (besides the buy and hold strategy in Example 1 and fixed fraction strategy in Example 2) are possible, for instance the turtle trading strategy. This allows a more direct link between the portfolio theory and the real implementation of the optimal portfolios/ trading strategies. Since Block (a) allows multi-period market models, the definition of the risk function (and also the utility function) in Block (c) can be path-dependent. This is essential for drawdown risk functions. Although so far we added lots of freedom, Block (d), i.e., the optimization block, is at least formally still very much in the spirit of Markowitz (1952, 1959). As such, this block is fixed in this work. However, different optimization problems might also be possible.

Clearly, with Parts I–III (see Maier-Paape and Zhu (2018a, 2018b)) we by now have a whole zoo of possibilities for how to trade off risk and reward in order to obtain efficient portfolios. Nevertheless, at this point, it is not yet clear what this added variety of possibilities yields when it comes to trading in real financial markets. This last and most important question will be discussed in a subsequent part of our series still to come.

REFERENCES

1. BARBU, VIOREL, AND TEODOR PRECUPANU. 2012. *Convexity and Optimization in Banach Spaces*. Springer Monographs in Mathematics. Heidelberg: Springer.
2. Carr, Peter, and Qiji Jim Zhu. 2018. *Convex Duality and Financial Mathematics*. Springer Briefs in Mathematics. Heidelberg: Springer.
3. Chekhlov, Alexei, Stanislav Uryasev, and Michael Zabarankin. 2003. Portfolio optimization with drawdown constraints. In *Asset and Liability Management Tools*. London: Risk Books, pp. 263–78.
4. Chekhlov, Alexei, Stanislav Uryasev, and Michael Zabarankin. 2005. Drawdown measure in portfolio optimization. *International Journal of Theoretical and Applied Finance* 8: 13–58.
5. Cherny, Vladimir, and Jan Obłój. 2013. Portfolio optimisation under non-linear drawdown constraints in a semimartingale financial model. *Finance and Stochastics* 17: 771–800.
6. Cvitanić, Jaksa, and Ioannis Karatzas. 1995. On portfolio optimization under "drawdown" constraints. In *Mathematical Finance*. Volume 65 of The IMA Volumes in Mathematics and its Applications. Heidelberg: Springer.
7. Fristedt, Bert E., and Lawrence F. Gray. 1997. *A Modern Approach to Probability Theory*. Basel: Birkhäuser.
8. Föllmer, Hans, and Alexander Schied. 2016. *Stochastic Finance: An Introduction in Discrete Time*, 4th ed. Berlin: De Gruyter.
9. Goldberg, Lisa R., and Ola Mahmoud. 2017. Drawdown: From practice to theory and back again. *Mathematics and Financial Economics* 11: 275–97. [
10. Grossman, Sanford, and Zhongquan Zhou. 1993. Optimal investment strategies for controlling drawdowns. *Mathematical Finance* 3: 241–76.
11. Lintner, John. 1965. The valuation of risk assets and the selection of risky investments in stock portfolios and capital budgets. *The Review of Economics and Statistics* 47: 13–37.
12. Maier-Paape, Stanislaus, and Qiji Jim Zhu. 2018a. A general framework for portfolio theory. part I: Theory and various models. *Risks* 6: 53.
13. Maier-Paape, Stanislaus, and Qiji Jim Zhu. 2018b. A general framework for portfolio theory. part II: Drawdown risk measures. *Risks* 6: 76.

14. Markowitz, Harry. 1952. Portfolio selection. *The Journal of Finance* 7: 77–91.
15. Markowitz, Harry. 1959. *Portfolio Selection: Efficient Diversification of Investments*. Number CFM 16. Hoboken: John Wiley & Sons.
16. Mossin, Jan. 1966. Equilibrium in a capital asset market. *Econometrica* 34: 768–83.
17. Platen, Andreas. 2018. Modular Portfolio Theory: A General Framework with Risk and Utility Measures as well as Trading Strategies on Multi-Period Markets. Ph.D. thesis, RWTH Aachen, Aachen, Germany, November 21.
18. Rockafellar, R. Tyrrell. 1972. *Convex Analysis*, 2nd ed. Princeton Landmarks in Mathematics and Physics. Princeton: Princeton University Press.
19. Rockafellar, R. Tyrrell, Stan Uryasev, and Michael Zabarankin. 2006. Master funds in portfolio analysis with general deviation measures. *Journal of Banking & Finance* 30: 743–78.
20. Sharpe, William F. 1964. Capital asset prices: A theory of market equilibrium under conditions of risk. *The Journal of Finance* 19: 425–42. [
21. Vince, Ralph. 2009. *The Leverage Space Trading Model: Reconciling Portfolio Management Strategies and Economic Theory*. Hoboken: John Wiley & Sons.
22. Zabarankin, Michael, Konstantin Pavlikov, and Stan Uryasev. 2014. Capital asset pricing model (capm) with drawdown measure. *European Journal of Operational Research* 234: 508–17.

CHAPTER

11

Research on Regularized Mean–Variance Portfolio Selection strategy with Modified Roy Safetyfirst Principle

Ebenezer Fiifi Emire Atta Mills[1], Dawen Yan[1,2,3], Bo Yu[1] and Xinyu-an Wei[1]

[1]School of Mathematical Sciences, Dalian University of Technology, Dalian, China.

[2]School of Business Management, Dalian University of Technology, Dalian, China.

[3]Bank of Dalian, Dalian, Liaoning, China.

ABSTRACT

We propose a consolidated risk measure based on variance and the safety-first principle in a mean-risk portfolio optimization framework. The safety-

Citation: Atta Mills, E.F.E., Yan, D., Yu, B. et al. "Research on regularized mean–variance portfolio selection strategy with modified Roy safety-first principle". SpringerPlus 5, 919 (2016). https://doi.org/10.1186/s40064-016-2621-7

first principle to financial portfolio selection strategy is modified and improved. Our proposed models are subjected to norm regularization to seek near-optimal stable and sparse portfolios. We compare the cumulative wealth of our preferred proposed model to a benchmark, S&P 500 index for the same period. Our proposed portfolio strategies have better out-of-sample performance than the selected alternative portfolio rules in literature and control the downside risk of the portfolio returns.

Keywords:- Mean-risk portfolio, Safety-first, Stable portfolio, Sparse portfolio

BACKGROUND

The benefits of portfolio optimization are well known to investors and other stakeholders. Strategies in portfolio selection have been well documented and unquestionably explored in finance and other related areas from a theoretical and much more practical perspective. Undoubtedly, a widely used approach is the mean–variance model by Markowitz (1952). He demonstrates how investors can employ risk-return tradeoff for wealth allocation using utility functions. Investors decide where along an efficient frontier a suitable balance between risk and return exist. Markowitz's work has become the bedrock of portfolio selection analysis, thereby earning the name "Modern Portfolio Theory". In a related study, Roy (1952) proposed a safety-first principle, which minimizes the shortfall probability in portfolio selection. For any investor whose general interest is downside risk measure of investment, Roy's (1952) safety-first principle is much more suitable Chiu et al. (2012).

Under the safety-first principle, instead of employing utility functions, which is hard or even impossible to determine, a known amount of the principal is preserved. Thus, an investor predefines a minimum threshold level of return and opts for a portfolio of assets that attains this preservation of principal. Roy's principle has over the years been studied. For instance, Bawa (1978) used the principle to maximize income, given the probability that a predefined threshold is greater than income. Haque et al. (2007) also applied safety-first through extreme value theory to portfolios of Mexican and U.S equities. Rachev (2001) showed how the safety-first approach can be more efficient than the stable Paretian approach in portfolio theory.

Levy and Sarnat (1972) and Nawrocki (1999) show how safety-first principle and mean–variance approach are akin. Levy and Sarnat (1972) found out that in the distinctive case where target rate of return is equal to risk-free return, both Roy's safety-first principle and Markowitz's mean–variance approach lead to the same optimal portfolio selection strategy. However, both models have several restrictions which include but are not limited to the following: Firstly, both Markowitz and Roy employ symmetry about the mean variance as a risk measure. Variance takes into account the situation in which return exceeds mean value and this does not impact risk. Secondly, transaction costs are absent in safety-first principle and mean–variance approach. According to DeMiguel et al. (2014), their loss in the mean–variance approach with the absence of transaction costs was 49.33 %. Borkovec et al. (2010) pointed out that 40 % of financial market participants attribute the fundamental loss in abnormal return to transaction costs. Lastly, both model frameworks are subjected to parameter uncertainty as returns of assets are considered as deterministic parameters being depicted by a single point estimate which leads to estimation risk.

Variance, as a risk measure, penalizes both under-performance and over-performance equally (Markowitz 1968). However, investors are only worried about underperformance. Roman et al. (2007) obtained a better portfolio by employing three indexes i.e. mean–variance-CVaR in comparison to mean–variance and mean-CVaR. The authors demonstrated that the mean-CVaR portfolio policy results in large variance, which leads to a small Sharpe ratio. Again, the CVaR of the portfolio generating from mean–variance model is large. To eliminate these inconsistencies between the strategies of mean–variance and mean-downside risk models, Roman et al. (2007) proposed to merge CVaR and variance in a multi-objective portfolio selection strategy. Inspired by Roman et al. (2007)'s approach of combining a downside risk measure, CVaR, and variance, we consider a portfolio optimization model with multiple risk measures. Recognizing a need to modify and improve Roy's safety first principle, we consider merging our improved Roy-safety first approach with variance as the portfolio risk measure.

The construction of a portfolio of investments is a significant problem faced by investors and institutions. A decision ought to be made to allocate weights to each investment with the intention of striking an appropriate balance between returns and risk. In reality, setting up a new portfolio or rebalancing an existing one requires costs to be incurred and must be inclusive in any realistic analysis. We incorporate proportional transaction costs (Kellerer et al. 2000; Muthuraman and Kumar 2006) which are induced by liquidity costs, tax and brokerage fees (Dumas and Luciano 1991; Kellerer et al. 2000; Lobo et al. 2007) into our portfolio selection model.

In the mean–variance and Roy safety-first models, stock returns are considered deterministic and taken as a single point estimate which results in estimation risk or overfitting (Bawa et al. 1979; Merton 1980). A small variation in the input parameters of the standard Markowitz mean–variance approach and Roy safety-first principle usually lead to changes in the structure of the resulting portfolios (Brandt 2009). To reduce the undesired impact of estimation risk or overfitting, models in the context of robustness (Goldfarb and Iyengar 2003; Tütüncü and Koenig 2004), stochastic programming (Rockafellar and Uryasev 2000), factor models (Green and Hollifield 1992; Nagai 2003; Schultz and Tiedemann 2003) and shrinkage estimators (Jorion 1986; Ledoit and Wolf 2004) have been explored. Another method analogous to our work is the modification of portfolio weights by adding regularizers or additional constraints to the portfolio strategy (Jagannathan and Ma 2003; DeMiguel et al. 2009; Brodie et al. 2009). Correcting undesired portfolio weights are as a result of large estimation risk of unknown parameters leading to the achievement of their desired forms and characteristics. Jagannathan and Ma (2003) employed a short sale constraint in the minimum variance framework and found out the non-negativity constraint performs well as those computed with shrinkage estimators and factor models. In DeMiguel et al. (2009), authors added a convex norm ball constraint to the portfolio weight and observed that the norm ball constrained portfolios had better out of sample performance than the portfolio strategies of the naive $1/N$ diversification (Jagannathan and Ma 2003). The l_2-norm constrained portfolio in general attained higher Sharpe Ratio than l_1-norm constrained portfolios. We study the weight constrained portfolios by specifying the general norm as squared l_2-norm ball.

The optimal portfolio of classical Roy safety-first principle and Markowitz's model in the presence or absence of short sale constraint or stability constraints hold a large number of assets and especially small weights assigned to their proportions. However, holding a large number of assets leads to the investor incurring high transaction costs. Due to this factor and other market and economic frictions, investors often hold only a small number of stocks in their portfolio. This is usually known as a sparse portfolio (a portfolio with few non-zero weights). With sparsity, one pre-sets limit of assets (stocks) with non-zero entries of portfolio allocations.

A method of constructing sparse portfolio termed hard threshold strategy was proposed by Britten-Jones (1999) to statistically test each portfolio weight with a null hypothesis that weight is zero. Using F-test and t-test, Britten-Jones (1999) proposed that if the portfolio weights are statistically not different from zero; assigning zero reduces portfolio risk. Sparse portfolio weights can be derived by cardinality constrained portfolio optimization (CCPO) problem (Chang et al. 2000; Maringer and Kellerer 2003; Ruiz-Torrubiano and Suárez 2015). CCPO takes into account all portfolios of given number of assets and chooses an optimal portfolio. With the inclusion of cardinality constraint, the portfolio selection strategy becomes NP-hard (Moral-Escudero et al. 2006), and standard quadratic program solvers can no longer be adopted to tackle the problem. One resorts to several other methods or relaxations in search of near-optimal solutions at a moderate computational cost. Farrell and Reinhart (1997) suggested classification of assets based on geographical aspect, size, sector, etc. and made a selection of *N* assets from each of these classes. In Pai and Michel (2009), the cardinality constraints were handled via clustering algorithm to reduce the size of the portfolio. More recently, Ruiz-Torrubiano and Suárez (2015) used a memetic approach that combines a genetic algorithm (GA) with an extended set encoding and quadratic programming (QP) in a mean–variance framework to deal with the cardinality constraint. An alternative relaxation method for which we employ in this paper is by constraining the l_1-norm (Tibshirani 1996) which is also in connection with the upper bound of the estimation risk as shown in Fan et al. (2012). The convexity nature of this type of regularization makes it more tractable. The use of norm penalty helps investors limit transaction costs and exposure to risky stocks.

As Roy safety-first principle minimizes the chances that the portfolio's return will fall below the minimum acceptable return, introducing it in the mean–variance model helps control the downside risk of the portfolio return. The modification and improvement of Roy safety-first principle and merging it with variance as a consolidated risk measure in a risk-return framework represents the main novelty of this research paper. In achieving our aim, we minimize our modified and improved Roy safety-first principle and impose the lower constraint on the mean return of the portfolio as in Markowitz's mean–variance model. We also investigate the portfolio strategy with variance and Roy's safety first principle as a consolidated risk measure in a mean-risk framework. To address the problem of estimation risk, we constrain the portfolio weights with squared l_2-norm and proceed to achieve sparsity via l_1-norm heuristic. We will explore the impact of transaction costs on portfolio selection strategies.

The paper is organized as follows. The next section presents our proposed portfolio selection strategies. It's subsections study Roy safety-first principle and its modification, portfolio revision and stable and sparse portfolio. We perform numerical tests and present computational results of our proposed methods in the subsequent section. Concluding remarks are provided in the last section.

PROPOSED PORTFOLIO SELECTION STRATEGIES

Investors allocate proportions of their capital among the assets they invest in. For the purpose of this study, these proportions are allocated to stocks. We denote by N the number of risky assets, R is the required return level, and x^0 is the initial risky assets before rebalancing: x0k is the proportion of capital initially allocated to asset k, $k = 1, 2, 3, ..., N$. Let x, x^b and x^s be N dimensional vectors of controllable variables: x_k is the portfolio invested in risky asset k after rebalancing, xbk is the purchases (proportion used) of risky asset k and xsk is the sales (proportion obtained) of risky asset k. The transaction costs incurred when buying risky assets is c^b and that of selling risky assets is c^s. The financial portfolio is described by a N-dimensional vector of random returns r. The portfolio total random return is $R_p = f(x,r) = \sum_{k=1}^{N} x_k r_k$, portfolio expected return vector $\mu_p = \mathbb{E}_x f(x,r) = \sum_{k=1}^{N} \mathbb{E}_x r_k$, $\sigma_p^2 = \mathbb{E}_x(R_p - \mathbb{E}_x(R_p))^2$ as variance of portfolio return and Q as the variance-covariance matrix of the portfolio return.

Roy safety-first principle and it's modification

Most investors' aim is to maximize returns and minimize risk. The Roy safety-first principle advocates avoiding extreme losses through the minimization of disaster probability. To optimally construct a portfolio strategy, Roy's safety-first principle defines a threshold or a minimum acceptable return R, below which the portfolio wealth is considered to be a disaster. The best portfolio is one that minimizes the chances that the portfolio's return, R_p, will fall below a minimum acceptable return, R. In essence, an investor selects his portfolio by solving this optimization problem:

$$\begin{aligned} &\underset{x\in\mathbb{R}^N}{\text{minimize}} \quad \mathbb{P}(f(x,r) \le R) \\ &\text{subject to} \quad e^T x = 1 \end{aligned} \tag{1}$$

where e is a vector with ones as entries and $\mathbb{P}$ is a probability measure. Roy employed Bienayme–Tchebycheff's inequality as the investor is likely not to know the actual probability function and obtained an approximation

$$\mathbb{P}(R_p \leq R) \leq \frac{\sigma_p^2}{(\mu_p - R)^2}$$

Thus, the optimization problem is reformulated as

$$\begin{aligned} &\underset{x \in \mathbb{R}^N}{\text{minimize}} \quad \frac{\sigma_p^2}{(\mu_p - R)^2} \\ &\text{subject to} \quad e^T x = 1 \end{aligned} \tag{2}$$

The modification to the Roy's approach we adopt is by using a coherent downside risk measure known as Conditional Value-at-Risk or Expected shortfall as it has a set of desirable properties for a risk measure (Platen and Heath 2006) leading to more accurate estimates of probability. For a detailed study on desirable properties of an ideal risk measure in portfolio theory, we refer the reader to Rachev et al. (2008).

A well-known downside risk measure known as Value-at-Risk focuses on the percentiles of loss distributions and measures the predicted maximum loss at a given probability level. Mathematically it is formulated as α-quantile $VaR_\alpha(X) = min\{z|(F_X(z) \geq \alpha\}$, where X is a loss random variable and $\alpha \in (0, 1)$ is the given probability level. Values for α often used are 90 %, 95 % and 99 %. Considering Value-at-Risk (VaR) has undesirable properties such as non-subadditive and non-smooth etc., Rockafellar and Uryasev (2000) introduced a coherent downside risk measure termed Conditional Value-at- Risk (CVaR) and for $\alpha \in (0, 1)$ represented it as

$$CVaR_\alpha(X) = \int_{-\infty}^{+\infty} z dF_X^\alpha(z)$$

where

$$F_X^\alpha(z) = \begin{cases} 0 & \text{when } z < VaR_\alpha(X), \\ \frac{F_X(z)-\alpha}{1-\alpha} & \text{when } z \geq VaR_\alpha(X). \end{cases}$$

Equivalently, for $x \in X \subset \mathbb{R}^N$ and random vector $r \in \mathbb{R}^N$ which represents the actual portfolio return has a continuous density function $p(r)$

$$CVaR_\alpha(x) = \frac{1}{1-\alpha} \int_{f(x,r) \geq VaR_\alpha(x)} f(x,r)p(r)dr$$

First, we will determine the semi deviation of the random return $f(x, r)$ from the α-quantile $VaR_\alpha(x)$. With respect to Bienayme–Tchebycheff's inequality, the following estimate is valid for $VaR_{\alpha)}(x) > R$:

$$\begin{aligned}\mathbb{P}\{f(x,r) \le R\} &= \mathbb{P}\{-f(x,r) \ge -R\} \\ &= \mathbb{P}\{VaR_{\alpha)}(x) - f(x,r) \ge VaR_{\alpha)}(x) - R\} \\ &= \mathbb{P}\{\mid f(x,r) - VaR_{\alpha)}(x) \mid \ge VaR_{\alpha)}(x) - R\} \\ &\le \mathbb{P} \mid \{\min\{f(x,r) - VaR_{\alpha)}(x), 0\} \mid \ge VaR_{\alpha)}(x) - R\} \\ &\le \frac{\mathbb{E}_x \mid \min\{f(x,r) - VaR_{\alpha}(x), 0\} \mid}{VaR_{\alpha)}(x) - R}\end{aligned} \quad (3)$$

Let us consider an α-quantile

$$VaR_{\alpha}(x) = \min\{z \mid \mathbb{P}(f(x,r) \le z) \ge \alpha\} \implies \mathbb{P}\{(f(x,r) \ge VaR_{\alpha}(x)\} = 1 - \alpha$$

and a measure of risk

$$CVaR_{\alpha}(x) = \frac{\mathbb{E}_x \mid \min\{f(x,r) - VaR_{\alpha}(x), 0\} \mid}{\mathbb{P}\{f(x,r) \ge VaR_{\alpha}(x)\}}$$

which is termed expected shortfall from the α-quantile $VaR_{\alpha}(x)$ value. The estimate (3) can be written as

$$\begin{aligned}\mathbb{P}\{f(x,r) \le R\} &\le \frac{\mathbb{P}\{f(x,r) \ge VaR_{\alpha}(x)\}}{VaR_{\alpha}(x) - R} \frac{\mathbb{E}_x \mid \min\{f(x,r) - VaR_{\alpha}(x), 0\} \mid}{\mathbb{P}\{f(x,r) \ge VaR_{\alpha}(x)\}} \\ &\le \frac{(1-\alpha) CVaR_{\alpha}(x)}{VaR_{\alpha}(x) - R}\end{aligned} \quad (4)$$

Therefore (1) can be reformulated by considering the approximation of the right hand side of (4) and obtain the following

$$\begin{aligned}&\underset{x \in \mathbb{R}^N}{\text{minimize}} && \frac{(1-\alpha) CVaR_{\alpha}(x)}{VaR_{\alpha}(x) - R} \\ &\text{subject to} && VaR_{\alpha}(x) > R \\ &&& e^T x = 1\end{aligned} \quad (5)$$

Telser (1955) considered a portfolio strategy by maximizing portfolio returns under the the constraint of Roy safety-first principle. He solved the optimization problem:

$$\begin{aligned}&\underset{x \in \mathbb{R}^N}{\text{maximize}} && \mu^T x \\ &\text{subject to} && \mathbb{P}(f(x,r) \le R) \ge 1 - \epsilon\end{aligned} \quad (6)$$

Inspired by Telser's approach, we constrain (5) with the minimum mean return vector $\mu^T x \ge L$ from below where L is the lower bound of $u^T x$, where $L > R$. Thus we obtain the optimization problem:

$$
\begin{aligned}
P_0: \quad & \underset{x \in \mathbb{R}^N}{\text{minimize}} && \frac{(1-\alpha)CVaR_\alpha(x)}{VaR_\alpha(x) - R} \\
& \text{subject to} && \mu^T x \geq L \\
& && VaR_\alpha(x) > R \\
& && e^T x = 1
\end{aligned} \tag{7}
$$

In another approach, we consider variance and the modified Roy's safety first-principle as a consolidated risk measure in a mean-risk framework. To this end, we propose the optimization problem:

$$
\begin{aligned}
P_1: \quad & \underset{x \in \mathbb{R}^N}{\text{minimize}} && x^T Qx + \frac{(1-\alpha)CVaR_\alpha(x)}{VaR_\alpha(x) - R} \\
& \text{subject to} && \mu^T x \geq L \\
& && VaR_\alpha(x) > R \\
& && e^T x = 1
\end{aligned} \tag{8}
$$

The rest of the modifications is geared towards investigating realistic constraints such as transaction costs, sparsity, and stability to P_0 and P_1 on the financial market.

Portfolio revision

We consider an extension of problems (7) and (8) in which transaction costs are incurred to rebalance or revise the initial portfolio x^0, into an efficient portfolio x. A portfolio of investments may require rebalancing on periodical basis because of updated risk, and return information is generated over time. We make the following assumptions on the transaction cost function c.

Assumption 1

The transaction cost function satisfies the following:

- (i) $c(x)$ is a convex function of x
- (ii) $c(0) = 0$
- (iii) $c(x) \geq 0, \ \forall x$

To achieve portfolio x_k from the previous or initial portfolio x_k^0, we make a payment of transaction costs $c(x - x^0)$. We incorporate proportional transaction costs (Kellerer et al. 2000; Muthuraman and Kumar 2006; Mitchell and Braun 2013) which are induced by liquidity costs, tax, brokerage fees (Dumas and Luciano 1991; Kellerer et al. 2000; Lobo et al. 2007) into our model. Therefore, proportional transaction cost follows this structure:

$$c(x - x^0) = \sum_{k=1}^{N} c_k \left(x_k - x_k^0 \right)$$

where

$$c_k(x_k - x_k^0) = \begin{cases} c_k^b(x_k - x_k^0) & \text{if} \quad x_k \geq x_k^0, \\ c_k^s(x_k^0 - x_k) & \text{if} \quad \text{otherwise.} \end{cases}$$

where cost of buying is $c_k^b 0$ and cost of selling is $c_k^s > 0$.

The P_1 model with proportional transaction costs (P_{1t}) is the optimization problem

$$P_{1t}: \quad \underset{x, x^s, x^b \in \mathbb{R}^N}{\text{minimize}} \quad x^T Qx + \frac{(1-\alpha)CVaR_\alpha(x)}{VaR_\alpha(x) - R} \tag{9}$$

subject to

$$\mu^T x - \sum_{k=1}^{N} \left(c^b x_k^b + c^s x_k^s \right) \geq L \tag{10}$$

$$\sum_{k=1}^{N} x_k + c^b \sum_{k=1}^{N} x_k^b + c^s \sum_{k=1}^{N} x_k^s = 1 \tag{11}$$

$$x_k = x_k^0 + x_k^b - x_k^s, \quad \forall k = 1, \ldots, N \tag{12}$$

$$x_k^b \cdot x_k^s = 0, \quad \forall k = 1, \ldots, N \tag{13}$$

$$VaR_\alpha(x) > R \tag{14}$$

$$x_k^b \geq 0, \quad x_k^s \geq 0, \quad \forall k = 1, \ldots, N \tag{15}$$

The model P_{1t} minimizes the upper bound estimate (5) w.r.t x and superimposes the lower constraint $L \leq \mu' x - \sum_{k=1}^{N} (c^b x_k^b + c^s x_k^s)$ on the average return after the deduction of transaction costs.

Explaining the constraints with respect to transaction costs, the above optimization problem is subjected to a set of linear constraints. Constraint (10) requires the net return of the portfolio after the deduction of transaction costs to be greater or equal to a threshold level L. Constraint (11) is the budget constraint: the capital available to cover transaction costs and shares of stocks. Constraint (12) shows that x_k represents the portfolio position to be chosen explicitly through sold shares x_k^s and purchased shares x_k^b that are rebalanced adjustments to the initial position x_k^0 of stock k. Constraint (13) and constraint (15) are the complementary constraint and non-negative constraint respectively. They prevent any possibility of concurrent purchases and sales (Dybvig 2005). Note that P_0 model with proportional cost (P_{0t}) is defined similarly but without the variance term in the objective function.

Stable and sparse portfolio

In a portfolio selection strategy where the dimensionality of the set of candidate assets is high, sparsity is desired. When the number of assets is large, a non-regularized numerical approach will intensify the effects of estimation risk, leading to an unstable and unreliable estimate of the vector x. Typically, portfolio managers want to set up portfolios with suitable balance between risk and return by investing in a small number of assets, thereby limiting their transaction, management, and monitoring costs.

To obtain meaningful and sparse (zero components) results, a regularization procedure is usually adopted. A standard approach is to augment the objective function of interest with a l_0-norm penalty or adding a cardinality constraint $\|x\|_0 \leq N'$ to optimization problems P_{0t} and P_{1t}, where $\|x\|_0$ is the number of the non-zero entries of x and N' is the upper bound limitation of assets to be managed in the portfolio. However, with the inclusion of cardinality constraint, the portfolio selection strategy becomes NP-hard (Moral-Escudero et al. 2006). We therefore impose its equivalent l_1-norm penalty as employed by Brodie et al. (2009) among others. The l_1-norm is a convex function of x, and such convex relaxation makes portfolio selection strategy more tractable.

To this end, we suggest to evaluate the portfolio weights by

$$\begin{aligned}
S_{1t}: \quad & \underset{x, x^s, x^b \in \mathbb{R}^N}{\text{minimize}} x^T Q x + \frac{(1-\alpha) CVaR_\alpha(x)}{VaR_\alpha(x) - R} + \tau_1 \|x\|_1 \\
& \text{subject to} \quad \mu' x - \sum_{k=1}^{N} \left(c^b x_k^b + c^s x_k^s \right) \geq L \\
& \qquad \sum_{k=1}^{N} x_k + c^b \sum_{k=1}^{N} x_k^b + c^s \sum_{k=1}^{N} x_k^s = 1 \\
& \qquad x_k = x_k^0 + x_k^b - x_k^s, \quad \forall k - 1, \ldots, N \\
& \qquad x_k^b \cdot x_k^c = 0, \quad \forall k = 1, \ldots, N \\
& \qquad VaR_\alpha(x) > R \\
& \qquad x_k^b \geq 0, \quad x_k^s \geq 0, \quad \forall k = 1, \ldots, N
\end{aligned} \tag{16}$$

where the l_1-norm of a vector $x \in \mathbb{R}^N$ is defined by $\|x\|_1 := \sum_{k=1}^{N} | x_k |$ and τ_1 is an adjustable parameter that controls the sparsity of the portfolios. Similarly, S_{0t} can be formulated without the variance term in the objective function.

Optimization models S_{0t} and S_{1t} have estimation risk or overfitting problem. The l_1-norm penalty expedites sparsity of x and leads to a subset of assets receiving zero weights. Such sparsity may result in under-diversification and extreme weights of the portfolio. On the contrary, convex norm ball does not produce sparsity but it can efficiently regularize size of portfolio weight vector. Thus, the norm ball constraint used by DeMiguel et al. (2009) can function as a solution to alleviate the problems of under-diversification and extreme weights of the portfolio aside estimation risk. Following DeMiguel et al.'s (2009) work and specifying the general squared l_2-norm under no short sale constraint, we propose to formulate the portfolio weights by

$$\begin{aligned}
RSMt: \quad & \underset{x,x^b,x^s \in \mathbb{R}^N}{\text{minimize}} \quad \frac{(1-\alpha)CVaR_\alpha(x)}{VaR_\alpha(x)-R} + \tau_1\|x\|_1 + \tau_2\|x\|_2^2 \\
& \text{subject to} \quad \mu^T x - \sum_{k=1}^{N}\left(c^b x_k^b + c^s x_k^s\right) \geq L \\
& \sum_{k=1}^{N} x_k + c^b \sum_{k=1}^{N} x_k^b + c^s \sum_{k=1}^{N} x_k^s = 1 \\
& x_k = x_k^0 + x_k^b - x_k^s, \quad \forall k = 1,\ldots,N \\
& x_k^b \cdot x_k^s = 0, \quad \forall k = 1,\ldots,N \\
& VaR_\alpha(x) > R \\
& x_k^b \geq 0, \quad x_k^s \geq 0, \quad \forall k = 1,\ldots,N
\end{aligned} \tag{17}$$

And

$$\begin{aligned}
RSMVt: \quad & \underset{x,x^b,x^s \in \mathbb{R}^N}{\text{minimize}} \quad x^T Q x + \frac{(1-\alpha)CVaR_\alpha(x)}{VaR_\alpha(x)-R} + \tau_1\|x\|_1 + \tau_2\|x\|_2^2 \\
& \text{subject to} \quad \mu^T x - \sum_{k=1}^{N}\left(c^b x_k^b + c^s x_k^s\right) \geq L \\
& \sum_{k=1}^{N} x_k + c^b \sum_{k=1}^{N} x_k^b + c^s \sum_{k=1}^{N} x_k^s = 1 \\
& x_k = x_k^0 + x_k^b - x_k^s, \quad \forall k = 1,\ldots,N \\
& x_k^b \cdot x_k^s = 0, \quad \forall k = 1,\ldots,N \\
& VaR_\alpha(x) > R \\
& x_k^b \geq 0, \quad x_k^s \geq 0, \quad \forall k = 1,\ldots,N
\end{aligned} \tag{18}$$

where τ_1 and τ_2 are tuning parameters controlling sparsity and stability respectively, with $\|x\|_2^2 = x'x$ as the squared l_2-norm of a vector. We estimate tuning parameters τ_1 and τ_2 by a method of cross-validation. We perform cross-validation for various possible values of the parameters and select the parameter value that produces the minimum cross-validation average error. A combination of the l_2-norm penalty and l_1-norm penalty is referred to as elastic net (Zou and Hastie 2005).

Necessary and sufficient conditions for optimal problems

In this section we would like to identify the necessary and sufficient conditions for optimality of problems (17) and (18). We investigate the Karush-Kuhn-Tucker (KKT) conditions for these problems under the assumption of normality and study whether the constrained problems have optimal solutions.

KKT conditions for optimal problem

The KKT conditions provide necessary conditions for a point to be optimal point for a constrained nonlinear optimal problem. The system

$$
\begin{aligned}
& x_k = x_k^0 + x_k^b - x_k^s, \quad \forall k = 1, \ldots, N \\
& x_k^b \cdot x_k^s = 0, \quad \forall k = 1, \ldots, N \\
& x_k^b \geq 0, \quad x_k^s \geq 0
\end{aligned}
\qquad 19
$$

has a unique solution

$$
x_k^b = \begin{cases} x_k - x_k^0, & \text{if } \; x_k - x_k^0 \geq 0 \\ 0, & \text{else} \end{cases}
$$

$$
x_k^s = \begin{cases} x_k^0 - x_k, & \text{if } \; x_k - x_k^0 < 0 \\ 0, & \text{else} \end{cases}
$$

i.e.

$$
\begin{aligned}
& x_k^b = max\{x_k - x_k^0, 0\} \\
& x_k^s = max\{x_k^0 - x_k, 0\}
\end{aligned}
$$

From the budget constraint $\sum_{k=1}^N x_k + c^b \sum_{k=1}^N x_k^b + c^s \sum_{k=1}^N x_k^s = 1$, we get that $c^b \sum_{k=1}^N x_k^b + c^s \sum_{k=1}^N x_k^s = 1 - \sum_{k=1}^N x_k$. Implementing the first constraint, $\mu^T x - \sum_{k=1}^N (c^b x_k^b + c^s x_k^s) \geq L$ we get that $\mu^T x - (1 - \sum_{k=1}^N x_k) \geq L$. Thus, RSMt can be represented as

$$
\begin{aligned}
& \underset{x \in \mathbb{R}^N}{\text{minimize}} \quad f_1(x) = \frac{(1-\alpha) CVaR_\alpha(x)}{VaR_\alpha(x) - R} + \tau_1 \|x\|_1 + \tau_2 \|x\|_2^2 \\
& \text{subject to} \quad g_1(x) = \mu^T x - \left(1 - \sum_{k=1}^N x_k\right) \geq L \\
& h_1(x) = c^b \sum_{k=1}^N max\left\{x_k - x_k^0, 0\right\} + c^s \sum_{k=1}^N max\left\{x_k^0 - x_k, 0\right\} = 1 - \sum_{k=1}^N x_k \\
& g_2(x) = VaR_\alpha(x) > R
\end{aligned}
\qquad (20)
$$

Similarly, RSMVt can be represented with addition of $x^T Q x$ to the objective function.

Let x be a regular point for the problem RSMt. Then the point x is a local minimum of f subject to constraints (20) if there exists Lagrange multipliers λ_1, λ_2 and λ_3 for the Lagrangian function $L = f_1(x) + \lambda_1 g_1(x) + \lambda_2 h_1(x) + \lambda_3 g_2(x)$ such that the following are true.

1. $\frac{\partial L}{\partial x} = (1-\alpha)CVaR'_\alpha(x)VaR_\alpha(x) - R - \frac{(1-\alpha)CVaR_\alpha(x)VaR'_\alpha(x)}{(VaR_\alpha(x)-R)^2} + \tau_1 c^1 + 2\tau_2 x - \lambda_1 (\mu + I^{N\times 1}) + \lambda_2 c^2 - \lambda_3 VaR'_\alpha(x) = 0$
2. $\lambda_1(L - \mu^T x - (1 - \sum_{k=1}^N x_k) = 0$
3. $\lambda_3(R - VaR_\alpha(x)) = 0$
4. $\lambda_1, \lambda_3 \geq 0$
5. $\mu^T x - (1 - \sum_{k=1}^N x_k) \geq L$
6. $c^b \sum_{k=1}^N max\{x_k - x_k^0, 0\} + c^s \sum_{k=1}^N max\{x_k^0 - x_k, 0\} = 1 - \sum_{k=1}^N x_k$
7. $VaR_\alpha(x) > R$

where

$$c_k^1 = \begin{cases} 1, & \text{if } x_k > 0 \\ -1, & \text{if } x_k < 0 \\ \in [-1,1], & \text{if } x_k = 0 \end{cases}$$

$$c_k^2 = \begin{cases} c^b, & \text{if } x_k > x_k^0 \\ -c^s, & \text{if } x_k < x_k^0 \\ \in [-c^s, c^b], & \text{if } x_k = x_k^0 \end{cases}$$

Remark 1

Since the function h_1 in (20) is linear and the functions g_1 and g_2 are convex, then the feasible region $\Omega = \{x{:}h_1, g_1, \text{ and } g_2\}$ is a convex set. On the other hand, f_1 is a convex function subject to the variable x. We see that any local minimum for problem (20) is a global minimum too and KKT conditions are also sufficient.

Similarly, the above KKT conditions holds for RSMVt for when $f_1(x)$ in (20) is $f_2(x) = x^T Qx + \frac{(1-\alpha)CVaR_\alpha(x)}{VaR_\alpha(x)-R} + \tau_1\|x\|_1 + \tau_2\|x\|_2^2$ and when the following are true

1. $\frac{\partial L}{\partial x} = 2Qx + (1-\alpha)CVaR'_\alpha(x)VaR_\alpha(x) - R - \frac{(1-\alpha)CVaR_\alpha(x)VaR'_\alpha(x)}{(VaR_\alpha(x)-R)^2} + \tau_1 c^1$

   $+2\tau_2 x - \lambda_1(\mu + I^{N\times 1}) + \lambda_2 c^2 - \lambda_3 VaR'_\alpha(x) = 0$
2. $\lambda_1(L - \mu^T x - (1 - \sum_{k=1}^N x_k) = 0$
3. $\lambda_3(R - VaR_\alpha(x)) = 0$
4. $\lambda_1, \lambda_3 \geq 0$

5 $\mu^T x - (1 - \sum_{k=1}^{N} x_k) \geq L$

6. $c^b \sum_{k=1}^{N} max\{x_k - x_k^0, 0\} + c^s \sum_{k=1}^{N} max\{x_k^0 - x_k, 0\} = 1 - \sum_{k=1}^{N} x_k$

7. $VaR_\alpha(x) > R$

where

$$c_k^1 = \begin{cases} 1, & \text{if} \quad x_k > 0 \\ -1, & \text{if} \quad x_k < 0 \\ \in [-1, 1], & \text{if} \quad x_k = 0 \end{cases}$$

$$c_k^2 = \begin{cases} c^b, & \text{if} \quad x_k > x_k^0 \\ -c^s, & \text{if} \quad x_k < x_k^0 \\ \in [-c^s, c^b], & \text{if} \quad x_k = x_k^0 \end{cases}$$

Remark 2

Since the feasible region $\Omega = \{x{:}h_1,\ g_1,\ \text{and}\ g_2\}$ in (20) and the objective function $f_2(x)$ are convex, then the feasible region is a convex set. We can see that the KKT conditions are also sufficient and any local minimum for problem (20) with objective function $f_2(x)$ is a global minimum as well.

EMPIRICAL APPLICATION

Data and models

In this section, we use optimization models RSMt (17) and RSMVt (18) to construct optimal portfolios and evaluate out-of-sample performance using stocks traded on New York Stock Exchange (NYSE). Historical daily returns of 500 randomly selected stocks over the period January 2003 to December 2015 were extracted from Yahoo Finance. The risk-free rate is proxied by the 6-month US Treasury Bill rate. The selection criterion of the random sampling of stocks is based on stocks being traded throughout the evaluation period. We also use S&P 500 index daily stock price data to test the robustness of our results over the same time period even though 21 % of randomly selected stocks are components of S&P 500. We chose a short term period as distribution of stock prices tend to change shape over time. To solve the portfolio strategies, we consider the lower bound of mean return as threshold return level. We compute initial positions $x_k^0, k = 1, \ldots, N$ for constructing portfolios for January 2006 to December 2008 and January 2009 to January 2015 by solving classical Roy safety-first optimization problem:

$$\begin{aligned} \text{minimize} \quad & \frac{\sigma_{\bar{x}}^2}{(\mu_{\bar{x}} - R)^2} \\ \text{subject to} \quad & e^T \bar{x} = 1 \\ & \bar{x} \in \mathbb{R}_N^+, \end{aligned} \qquad (21)$$

and by setting $x_0 = \bar{x}^*$, with $\bar{x}^*$ denoting the optimal solution of the above model.

Two portfolio strategies proposed in this research work, RSMt and RSMVt are compared against these existing ones in literature: (1) sample minimum variance without short sale constraint portfolio (minVu) (2) sample minimum variance short sale constrained portfolio (minVc) (3) sample mean–variance approach (MV) (4) naive equally-weighted (1/*N*) portfolio (5) l_1-penalized mean–variance model (l_1MV) (Brodie et al. 2009) (6) linear combination of sample tangency portfolio, sample minimum variance portfolio and 1/*N* portfolio (TMN) (Tu and Zhou 2011) (7) minimum variance portfolio resulting from using a diagonal covariance matrix (VD) (Kirby and Ostdiek 2012), refer to Table 1.

Table 1. List of proposed and selected alternative portfolio strategies

Symbol		Description
RSMt		l_1- Squared l_2 penalized mean-Roy safety-first portfolio
RSMVt		l_1- Squared l_2 penalized mean–variance-Roy safety-first portfolio
minVu		Sample minimum variance without short sale constraint portfolio
minVc		Sample minimum variance with short sale constraint portfolio
MV		Sample mean–variance portfolio
1/*N*		Equally-weighted portfolio
l_1-MV		l_1 penalized mean–variance portfolio with short sale constraint (Brodie et al. 2009)
TMN		Linear combination of sample tangency portfolio, sample minimum variance portfolio and 1/*N* portfolio (Tu and Zhou 2011)
VD		Minimum variance portfolio resulting from using a diagonal covariance matrix (Kirby and Ostdiek 2012)

We employ a 6-month rolling estimation window for parameter estimations and construct portfolios on a sub-sample periodical basis. To obtain the portfolio for January 2006 to December 2008, we use January 2003 to December 2005 data to construct the initial positions via the classical Roy safety-first optimization problem. For January 2009 to December

2011, we set the portfolio weights from January 2006 to December 2008 as initial positions. We then set the portfolio weights from January 2009 to December 2011 as initial positions to construct portfolios for January 2012 to December 2015. To reflect the true risk of the portfolio, we use January 2006 to December 2008 data as initial positions via the classical Roy safety-first optimization problem to construct portfolios for January 2009 to December 2015. Portfolio revision is made on a monthly basis since in reality they are related to low transaction, management and monitoring costs.

Evaluation criteria

With regards to the return that a portfolio can achieve, we calculate and present the annualized out-of-sample Sharpe ratio with or without transaction cost. Sharpe ratio is defined as the ratio of the expected excess return to standard deviation of portfolio return (Sharpe 1966). The expected excess return is the difference between the return of the portfolio and the return obtained through a risk-free security. Mathematically, we can define Sharpe ratio as $SRatio = \frac{\rho - \rho_f e}{\sigma}$, where ρ_f is the return from the risk-free security (6-month US Treasury Bill rate) and σ is the standard deviation of portfolio return. Regarding portfolio risk, we consider the risk reduction which is defined as the ratio of portfolio risk measure from the portfolio strategies (17) and (18) to that from (7) and (8) respectively. The other portfolio selection strategies in literature's risk reduction are estimated as the ratio of portfolio variance to that from Markowitz's mean–variance framework.

Sparsity may result in under-diversification and extreme weights of the portfolio. We therefore present the average number of a subset of assets with non-zero weights. We consider another performance metric known as portfolio turnover. A portfolio turnover measures the frequency with which assets in this case stocks are bought and sold. We employ the measure used by DeMiguel et al. (2009) and Kourtis et al. (2012) by defining the turnover rate of the portfolio between t to $t + 1$ as

$$Turnover = \frac{1}{\mathring{T} - T - 1} \sum_{t=T}^{\mathring{T}-1} \sum_{k-1}^{N} (| x_{k,t+1} - x_{\bar{k},t+1} |), \tag{22}$$

where $x_{k,t+1}$ is the portfolio weight for stock k at $t + 1$, $x_{\bar{k},t+1}$ is the portfolio weight before revision at $t + 1$, $\mathring{T} - T - 1$ represents the length of the non-zero elements in total portfolio return and N, the number of stocks.

The introduction of transaction costs affects the overall profitability of a portfolio strategy. In practice, they lessen the net returns and diminish capital available for future investments. We assume proportional transaction cost for the purpose of this study (Please refer to Assumption 1 for more details).

We investigate the impact of transaction cost on our portfolio strategy by computing Sharpe ratio with transaction cost. In particular, the optimal portfolios can be obtained by solving problems (17) and (18). To highlight the effect of transaction costs, we consider two situations with transaction costs, $c_k^b = c_k^s = 0$ and $c_k^b = c_k^s = 0.02$.

Computational results

Tables 2 and 3 shows the annualized out-of-sample metrics for different periods of each portfolio considered in this study: RSMt is the regularized mean-safety first portfolio with transaction costs, RSMVt is the regularized mean–variance-safety-first portfolio with transaction costs. The others are selected alternative portfolio strategies in literature.

Table 2. Out-of-sample performance of portfolio strategies on 500 NYSE stocks of sub-sample periods

Portfolio selection strategy	SRatio (transaction costs)	SRatio	Risk reduction	Sparsity	Turnover
Panel A : results for January 2006 to December 2008					
RSMt (17)	0.6891	0.6902	0.2115	165.1612	0.1517
RSMVt (18)	0.7111	0.7209	0.1801	188.6575	0.1783
minVu	−0.1523	0.1902	0.2561	500	0.4013
minVc	0.2042	0.2821	0.2393	500	0.4209
MV	−0.1745	0.1887	1.0000	500	4.1910
$1/N$	0.4084	0.4810	0.2710	500	0.3575
l_1-MV (Brodie et al. 2009)	0.41014	0.5131	0.2604	217.1435	0.3872
TMN (Tu and Zhou 2011)	−0.0158	0.3412	0.2573	500	0.4643
VD (Kirby and Ostdiek 2012)	0.3914	0.4710	0.2104	500	0.4510
Panel B : results for January 2009 to December 2011					
RSMt (17)	1.2319	1.2783	0.0932	151.2341	0.0613
RSMVt (18)	1.3012	1.3114	0.0815	203.2533	0.0791
minVu	0.0985	0.4612	0.1044	500	0.2002

minVc	0.4202	0.5319	0.1012	500	0.2125
MV	0.0812	0.4104	1.0000	500	2.1234
1/*N*	0.7264	0.8041	0.1395	500	0.1453
l_1-MV (Brodie et al. 2009)	0.9011	0.9511	0.1391	326.5532	0.1877
TMN (Tu and Zhou 2011)	0.2216	0.6013	0.1091	500	0.2563
VD (Kirby and Ost-diek 2012)	0.8153	0.8921	0.0995	500	0.2418
Panel C : results for January 2012 to December 2015					
RSMt (17)	1.3112	1.3205	0.0901	195.0186	0.0717
RSMVt (18)	1.3813	1.3984	0.0737	245.1227	0.0801
minVu	0.1120	0.4824	0.1018	500	0.2011
minVc	0.4413	0.5302	0.1006	500	0.2289
MV	0.1091	0.4591	1.0000	500	2.3914
1/*N*	0.7641	0.8321	0.1391	500	0.1465
l_1-MV (Brodie et al. 2009)	0.9403	0.9821	0.1294	267.8271	0.1903
TMN (Tu and Zhou 2011)	0.2426	0.63841	0.1067	500	0.2613
VD (Kirby and Ost-diek 2012)	0.8534	0.9010	0.0945	500	0.2485

Performance measures and metrics of portfolio selection strategies

Table 3. Out-of-sample performance of portfolio strategies for 2009–2015

Portfolio selection strategy	**SRatio (transaction costs)**	**SRatio**	**Risk reduction**	**Sparsity**	**Turnover**
Results for January 2009 to December 2015					
RSMt (17)	0.8912	0.9111	0.1521	175.5103	0.9859
RSMVt (18)	0.9577	0.9610	0.1112	228.3181	0.8113
minVu	0.0698	0.2413	0.1742	500	1.1612
minVc	0.2408	0.2958	0.1663	500	1.6209
MV	0.0451	0.1959	1.0000	500	7.3224
1/*N*	0.5514	0.5911	0.1977	500	0.9418
l_1-MV (Brodie et al. 2009)	0.6899	0.7094	0.1790	254.8511	0.9958
TMN (Tu and Zhou 2011)	0.1354	0.2985	0.1551	500	1.4705
VD (Kirby and Ost-diek 2012)	0.4544	0.5871	0.1605	500	1.2403

Performance measures and metrics of portfolio selection strategies

Analyzing portfolio risk, the regularized portfolio rules have lower risk than other portfolio selection strategies. This can be attributed to the downside risk measure, Roy safety-first principle considered in the regularized strategies. The variance and the safety-first principle as a consolidated risk measure in RSMVt have lower risk reduction compared to RSMt.

The portfolio turnover indicates how frequently assets in a portfolio are bought and sold. This performance measure is preferred to be small. In terms of turnover, the naive 1/*N* portfolio has the lowest turnover. The two portfolio selection strategies considered in this paper have relatively low turnover rates with RSMVt the preferred choice. The portfolio rules in selected from literature have a high turnover rate as compared to the regularized models in this paper. The highest turnover comes from TMN, MV and minVu portfolios leading to smaller Sharpe ratios. We observe that with large turnovers, feasible transaction costs lowers the monetary gains of many selected portfolios strategies in literature, as seen by Sharpe ratio deductions after transaction costs have been considered.

The Sharpe ratio allows investors to analyse risk-adjusted returns in exchange for the level of risk they are assuming. The higher the Sharpe ratio, the more returns the investor gets per unit of risk. The lower the Sharpe ratio, the more risk the investor bears to get more returns. Comparing all strategies in this study, the regularized portfolios have the highest Sharpe ratios with RSMVt leading the way. With regards to sparsity, more than 30 % of stocks are selected by RSMt and RSMVt in all the evaluation periods. The l_1-penalized mean–variance model selects at least 50 % of the stocks across the evaluation periods. The smaller set of sparse portfolio optimizes the budget allocation by focusing on stocks believed to foster diversification.

From Table 2, the sample-based portfolios i.e. minVu, minVc, MV and TMN perform worse due to a large number of stocks that increases the degree of estimation risk or over-fitting. Apart from VD, all other non-regularized portfolios selected from literature (minVu, minVc, MV, TMN) perform worse than 1/*N* regarding both Sharpe ratio and turnover. We observe an increment in Sharpe ratio, lower risk reduction, lower turnover for post financial crisis sub-sample periods. Among the selected alternative portfolio strategies considered in this paper, l_1MV has a better out-of-sample performance than minVu, minVc, MV, 1/*N*, TMN and VD. In comparing periods 2006–2008 to 2009–2015, we observe that during the financial crisis period, the Sharpe ratio is lower and the risk reduction is much higher with turnover ratio also higher. This amounts to lower returns during 2006–2008 as compared to the other periods. The Sharpe ratio is much higher and the risk reduction is lower in periods 2009–2011 and 2012–2015 as compared to periods 2006–2008 and 2009–2015. In all the sub-sample periods 2006–2008, 2009–2011 and 2012–2015 including the period 2009–2015, our proposed models RSMt and RSMVt have better out-of-sample performance than the selected alternative models.To further gain financial insights, we use S&P 500 index data from January 2009 to December 2015 as a benchmark portfolio. With a starting wealth value of $1, we compare the cumulative wealth of portfolio strategies RSMt, RSMVt and l_1MV to that of S&P 500 index. To provide evidential proof, Fig. 1 plots the cumulative wealth of the portfolios strategies relative to S&P 500 benchmark.

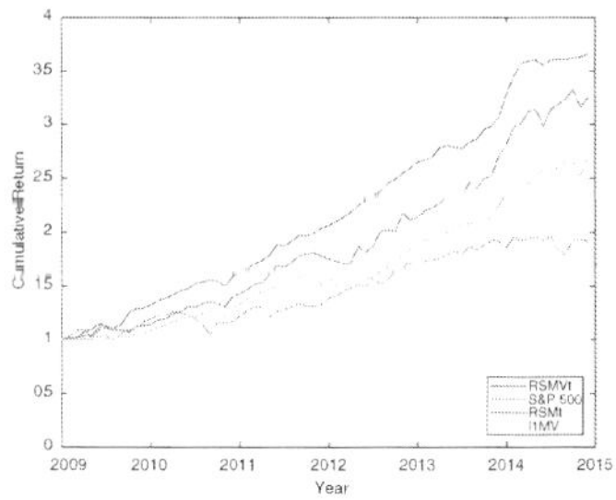

Figure1. Cumulative wealth of portfolio strategy RSMt, RSMVt and l_1MV relative to S&P 500 portfolio

In Fig. 1, $1 is used as initial wealth and it grows at a monthly return of the portfolio selection strategies considered. The figure shows distinctly the higher performance of the regularized portfolio over S&P 500 benchmark.

CONCLUSION

In this paper, we seek near-optimal sparse and stable portfolios to reduce the difficulty of portfolio management. Theoretical results are established to guarantee the stability and sparsity of our novel portfolio strategies. Computational evidence indicates that l_1- squared l_2 penalized mean-Roy safety-first portfolio and l_1- squared l_2 penalized mean–variance-Roy safety-first portfolio are able to choose optimal sparse and stable portfolios while maintaining satisfactory out-of-sample performance We compare the performance of our proposed models (l_1- squared l_2 penalized mean-Roy safety-first portfolio and l_1- squared l_2 penalized mean–variance-Roy safety-first portfolio) of optimal asset allocation relative to selected alternative portfolio strategies in literature (minVu, minVc, MV, 1/*N*, l_1MV, TMN and VD). Our results show that our regularized proposed models have a better out-of-sample performance with high Sharpe ratios and relatively low turnover rates. Except for VD, the Sharpe ratio of 1/*N* portfolio when compared to other selected non-regularized portfolio rules considered in this paper is higher, which shows that estimation errors in returns shrink gains from other selected classical portfolio strategies in literature. The norm penalty improves Sharpe ratio and turnover. As a result, l_1MV has a better out-of-sample performance than minVu, minVc, MV, 1/*N*, TMN and VD.

To gain more financial acumen, we compare our proposed models and the best performing portfolio strategy among the selected models from literature considered in this paper, to a benchmark, S&P 500 index. The results indicate that given an initial wealth of $1, the excess returns from l_1- squared l_2 penalized mean–variance-Roy safety-first portfolio is the highest. Our proposed models for optimal asset allocation are favourable since they overcome unsteady and extreme portfolio weights induced by estimation error due to parameter uncertainty.

ACKNOWLEDGEMENTS

We thank the reviewers for taking time to add their comments and suggestions to this paper. We also thank SpringerPlus for giving us the platform to contribute to literature. We acknowledge support by the National Nature Science Foundation of China (11571061, 71301017).

REFERENCES

1. Bawa VS, Brown SJ, Klein RW (1979) Estimation risk and optimal portfolio choice. North-Holland Publ Co, New York
2. Bawa VS. Safety-first, stochastic dominance, and optimal portfolio choice. J Financ Quant Anal. 1978;13(02):255–271. doi: 10.2307/2330386.
3. Borkovec M, Domowitz I, Kiernan B, Serbin V. Portfolio optimization and the cost of trading. J Invest. 2010;19(2):63–76. doi: 10.3905/joi.2010.19.2.063.
4. Brandt M. Portfolio choice problems. Handbook Financ Econom. 2009;1:269–336.
5. Britten-Jones M. The sampling error in estimates of mean-variance efficient portfolio weights. J Finance. 1999;54(2):655–671. doi: 10.1111/0022-1082.00120.
6. Brodie J, Daubechies I, De Mol C, Giannone D, Loris I. Sparse and stable Markowitz portfolios. Proc Natl Acad Sci. 2009;106(30):12267–12272. doi: 10.1073/pnas.0904287106.
7. Chang TJ, Meade N, Beasley JE, Sharaiha YM. Heuristics for cardinality constrained portfolio optimisation. Comput Oper Res. 2000;27(13):1271–1302. doi: 10.1016/S0305-0548(99)00074-X.
8. Chiu MC, Wong HY, Li D. Roys safety-first portfolio principle in financial risk management of disastrous events. Risk Anal. 2012;32(11):1856–1872. doi: 10.1111/j.1539-6924.2011.01751.x.
9. DeMiguel V, Garlappi L, Nogales FJ, Uppal R. A generalized approach to portfolio optimization: improving performance by constraining portfolio norms. Manag Sci. 2009;55(5):798–812. doi: 10.1287/mnsc.1080.0986.
10. DeMiguel V, Mei X, Nogales FJ (2014) Multiperiod portfolio optimization with many risky assets and general transaction costs. Available at SSRN 2295345
11. Dumas B, Luciano E. An exact solution to a dynamic portfolio choice problem under transactions costs. J Finance. 1991;46(2):577–595. doi: 10.1111/j.1540-6261.1991.tb02675.x.
12. Dybvig PH (2005) Mean-variance portfolio rebalancing with transaction costs. Working paper, Washington University in Saint Louis
13. Fan J, Zhang J, Yu K. Vast portfolio selection with gross-exposure

constraints. J Am Stat Assoc. 2012;107(498):592–606. doi: 10.1080/01621459.2012.682825.

14. Farrell JL, Reinhart WJ. Portfolio management: theory and application. New York: McGraw-Hill; 1997.
15. Goldfarb D, Iyengar G. Robust portfolio selection problems. Math Oper Res. 2003;28(1):1–38. doi: 10.1287/moor.28.1.1.14260.
16. Green RC, Hollifield B. When will mean-variance efficient portfolios be well diversified? J Finance. 1992;47(5):1785–1809. doi: 10.1111/j.1540-6261.1992.tb04683.x.
17. Haque M, Varela O, Hassan MK. Safety-first and extreme value bilateral US–Mexican portfolio optimization around the peso crisis and NAFTA in 1994. Q Rev Econ Finance. 2007;47(3):449–469. doi: 10.1016/j.qref.2006.01.005.
18. Jagannathan R, Ma T. Risk reduction in large portfolios: why imposing the wrong constraints helps. J Finance. 2003;58(4):1651–1684. doi: 10.1111/1540-6261.00580.
19. Jorion P. Bayes–Stein estimation for portfolio analysis. J Financ Quant Anal. 1986;21(03):279–292. doi: 10.2307/2331042.
20. Kellerer H, Mansini R, Speranza MG. Selecting portfolios with fixed costs and minimum transaction lots. Ann Oper Res. 2000;99(1–4):287–304. doi: 10.1023/A:1019279918596.
21. Kirby C, Ostdiek B. It's all in the timing: simple active portfolio strategies that outperform naive diversification. J Financ Quant Anal. 2012;47(02):437–467. doi: 10.1017/S0022109012000117.
22. Kourtis A, Dotsis G, Markellos RN. Parameter uncertainty in portfolio selection: shrinking the inverse covariance matrix. J Bank Finance. 2012;36(9):2522–2531. doi: 10.1016/j.jbankfin.2012.05.005.
23. Ledoit O, Wolf M. A well-conditioned estimator for large-dimensional covariance matrices. J Multivar Anal. 2004;88(2):365–411. doi: 10.1016/S0047-259X(03)00096-4.
24. Levy H, Sarnat M. Safety firstan expected utility principle. J Financ Quant Anal. 1972;7(03):1829–1834. doi: 10.2307/2329805.
25. Lobo MS, Fazel M, Boyd S. Portfolio optimization with linear and fixed transaction costs. Ann Oper Res. 2007;152(1):341–365. doi: 10.1007/s10479-006-0145-1.

26. MaringerD,KellererH.Optimizationofcardinalityconstrainedportfolios with a hybrid local search algorithm. OR Spectrum. 2003;25(4):481–495. doi: 10.1007/s00291-003-0139-1.
27. Markowitz H. Portfolio selection. J Finance. 1952;7(1):77–91.
28. Markowitz HM. Portfolio selection: efficient diversification of investments. New Haven: Yale University Press; 1968.
29. Merton RC. On estimating the expected return on the market: an exploratory investigation. J Financ Econ. 1980;8(4):323–361. doi: 10.1016/0304-405X(80)90007-0.
30. Mitchell JE, Braun S (2013) Rebalancing an investment portfolio in the presence of convex transaction costs, including market impact costs. Optim Methods Softw 28(3):523–542
31. Moral-Escudero R, Ruiz-Torrubiano R, Suárez A (2006) Selection of optimal investment portfolios with cardinality constraints. In: IEEE congress on evolutionary computation, CEC 2006, IEEE; pp 2382–2388
32. Muthuraman K, Kumar S. Multidimensional portfolio optimization with proportional transaction costs. Math Finance. 2006;16(2):301–335. doi: 10.1111/j.1467-9965.2006.00273.x.
33. Nagai H. Optimal strategies for risk-sensitive portfolio optimization problems for general factor models. SIAM J Control Optim. 2003;41(6):1779–1800. doi: 10.1137/S0363012901399337.
34. Nawrocki DN. A brief history of downside risk measures. J Invest. 1999;8(3):9–25. doi: 10.3905/joi.1999.319365.
35. Pai G, Michel T. Evolutionary optimization of constrained-means clustered assets for diversification in small portfolios. IEEE Trans Evol Comput. 2009;13(5):1030–1053. doi: 10.1109/TEVC.2009.2014360.
36. Platen E, Heath D. A benchmark approach to quantitative finance. New York: Springer Science & Business Media; 2006.
37. Rachev ST. Safety-first analysis and stable paretian approach to portfolio choice theory. Math Comput Model. 2001;34(9):1037–1072.
38. Rachev S, Ortobelli S, Stoyanov S, Fabozzi FJ, Biglova A. Desirable properties of an ideal risk measure in portfolio theory. Int J Theor Appl Finance. 2008;11(1):19–54. doi: 10.1142/S0219024908004713.
39. Rockafellar RT, Uryasev S. Optimization of conditional value-at-risk. J Risk. 2000;2:21–42. [Google Scholar]

40. Roman D, Darby-Dowman K, Mitra G. Mean-risk models using two risk measures: a multi-objective approach. Quant Finance. 2007;7(4):443–458. doi: 10.1080/14697680701448456.
41. Roy AD (1952) Safety first and the holding of assets. Econom J Econom Soc 20:431–449
42. Ruiz-Torrubiano R, Suárez A. A memetic algorithm for cardinality-constrained portfolio optimization with transaction costs. Appl Soft Comput. 2015;36:125–142. doi: 10.1016/j.asoc.2015.06.053.
43. Schultz R, Tiedemann S. Risk aversion via excess probabilities in stochastic programs with mixed-integer recourse. SIAM J Optim. 2003;14(1):115–138. doi: 10.1137/S1052623402410855.
44. Sharpe WF. Mutual fund performance. J Bus. 1966;39(1):119–138. doi: 10.1086/294846.
45. Telser LG. Safety first and hedging. Rev Econ Stud. 1955;23(1):1–16. doi: 10.2307/2296146.
46. Tibshirani R (1996) Regression shrinkage and selection via the lasso. J R Stat Soc Ser B (Methodol) 1:267–288
47. Tütüncü RH, Koenig M. Robust asset allocation. Ann Oper Res. 2004;132(1–4):157–187. doi: 10.1023/B:ANOR.0000045281.41041.ed.
48. Tu J, Zhou G. Markowitz meets Talmud: a combination of sophisticated and naive diversification strategies. J Financ Econ. 2011;99(1):204–215. doi: 10.1016/j.jfineco.2010.08.013.
49. Zou H, Hastie T. Regularization and variable selection via the elastic net. J R Stat Soc Ser B (Stat Methodol) 2005;67(2):301–320. doi: 10.1111/j.1467-9868.2005.00503.x.

CHAPTER 12

Systemic Risk from Investment Similarities

Danilo Delpini[1,2], Stefano Battiston[3], Guido Caldarelli[2,4], and Massimo Riccaboni[2,5]

[1]Dept. of Economics and Business, Università degli Studi di Sassari, Sassari, Italy,

[2]IMT Institute for Advanced Studies, Lucca, Italy,

[3]Dept. of Banking and Finance, University of Zurich, Zurich, Switzerland,

[4]ISC-CNR Uos "Sapienza", Rome, Italy,

[5]Dept. of Managerial Economics, Strategy and Innovation, KU Leuven, Leuven, Belgium

Citation: Delpini D, Battiston S, Caldarelli G, Riccaboni M (2019) "Systemic risk from investment similarities". PLoS ONE 14(5): e0217141. https://doi.org/10.1371/journal.pone.0217141

ABSTRACT

Network theory proved recently to be useful in the quantification of many properties of financial systems. The analysis of the structure of investment portfolios is a major application since their eventual correlation and overlap impact the actual risk by individual investors. We investigate the bipartite network of US mutual fund portfolios and their assets. We follow its evolution during the Global Financial Crisis and study the diversification, as understood in modern portfolio theory, and the similarity of the investments of different funds. We show that, on average, portfolios have become more diversified and less similar during the crisis. However, we also find that large overlap is far more likely than expected from benchmark models of random allocation of investments. This indicates the existence of strong correlations between fund investment strategies. We exploit a deliberately simplified model of shock propagation to identify a systemic risk component stemming from the similarity of portfolios. The network is still partially vulnerable after the crisis because of this effect, despite the increase in the diversification of multi asset portfolios. Diversification and similarity should be taken into account jointly to properly assess systemic risk.

INTRODUCTION

The Global Financial Crisis (GFC) of 2007-2008 has highlighted the systemic risk stemming from the increasing interdependence both between large institutional investors and among global assets [1–3]. On the one hand, large institutional investors allow for a better diversification of individual risk: the larger the number of different assets in a portfolio, the smaller the fraction of an idiosyncratic shock an investor has to bear. On the other hand, the GFC has shown that cross-sectional dependencies between assets can cause idiosyncratic shocks (i.e. related to the distress/bankruptcy of a single specified asset) to spread, ultimately threatening the stability of the entire financial system. According to modern portfolio theory, risk depends on the share of individual stock holdings and the variance–covariance matrix among its holdings [4]. Hence, theoretical models imply that a portfolio should be properly diversified to reduce risks (unsystematic risks), but how to construct a well-diversified portfolio when multiple investors are simultaneously active on the market still remains not fully understood. There is consensus in the literature that a strong risk reduction of holdings can be realized by increasing the number of assets in a portfolio [5–8]. Indeed, between 1997 and 2012, assets in the equity, balanced, and fixed income mutual funds have increased by more than 400 percent [9]. The efficacy of diversification strategies depends, however, on market conditions and systemic risk that may moderate the relationship between diversification, fund performance, and risk. Recently, it has been shown that the benefit of diversification increases in high market volatility conditions, such as the GFC of 2007-2008, meaning that the number of stocks needed to achieve a well-diversified portfolio increases under those market conditions [10]. Unfortunately, the implications of such an increase of portfolio diversification during crises is still unclear. Moreover, the empirical evidence accumulated during the GFC has raised legitimate doubts on the effectiveness of portfolio diversification strategies to reduce risk [11].

Recently, systemic risk in financial systems has been increasingly investigated through the lens of network theory [1–3, 12–21]. Most of the work so far has focused on the network of interbank loans, even though there is a paucity of data about the real-world structure of financial networks. Interestingly, it has been found that, when the magnitude of negative shocks is large and the network is scale-free, a more densely connected financial network (corresponding to a more diversified pattern of investment) serves as a mechanism for the propagation of shocks, leading to a more fragile financial system, thus increasing systemic risk [14].

Even though a better diversification may reduce risk for an individual portfolio, the structure of the similarities across mutual funds could be a key element of systemic risk [22]. Depending on that structure, on the way diversification is pursued and the level of interdependence between investment strategies, increasing portfolio diversification during a crisis might possibly increase the cross-correlation among assets thus amplifying systemic risk. Therefore, the exact role played by the evolution of the financial network during crisis in potentially creating systemic risk remains, at best, imperfectly understood. The same holds for the role of global fund managers which appears to have been studied mostly in simulated scenarios [23]. We contribute to the literature by analyzing, through the lens of network theory, the bipartite network of US mutual funds over time and throughout the GFC as well as in simulated scenarios.

The fund holding network has been considered previously in [24] where the correlation between changes in a firm's position in the network and future stock market performance is considered, and in [25] where the CRSP database was studied to detect possible conflicts of interest in the strategies of multi-fund managers. In [26] the size distribution of funds has been investigated in detail, while in [27] a size-dependent model of fund growth has been proposed to explain its shape. The existence of a positive flow-performance relationship and fire-sales have been documented for mutual funds [11]: investors tend to redeem their investments in response to negative shocks. In some circumstances, overlapping portfolios and asset liquidations can force mutual funds to sell additional assets potentially triggering cascading effects in the market [22, 28].

In this paper we consider bipartite networks of portfolio holdings [1, 15, 16] with the intent to provide a better understanding of the relationship between the similarity of individual investment strategies and systemic riskiness. We use the network of US equity mutual funds as a test case. We exploit a deliberately simple model of distress propagation and stress test the network to study how the systemic fragility of the system depends on the overlap between portfolios due to correlated investment strategies. This is done through simulations and comparison with null models of random versus correlated investments. Finally we make an assessment of how much network vulnerability has changed across the Global Financial Crisis.

MATERIALS AND METHODS

Dataset

The US Mutual Fund market is the largest in the world: with 15$ trillion assets under management at year-end 2013, it accounts for about half the total value in mutual fund assets worldwide. In this study, we analyse data from the Survivor-Bias-Free Mutual Fund Database provided by Chicago Booth Center for Research in Security Prices (CRSP).

It includes open-ended mutual funds registered with the Securities and Exchange Commission and provides detailed information about the composition of managed portfolios. A mapping of funds to the portfolios of assets they manage is provided and detailed information about portfolio holdings, including the market value of each holding, is delivered on a quarterly basis.

In the following, we provide a formal construction of the bipartite network of holdings. We also present a schematic description of the model of distress propagation considered in the main text. Access to the data requires a valid (non-free) subscription that can be acquired issuing a subscription enquiry to the CRSP center.

Bipartite graph of portfolio holdings

We represent portfolio holdings in terms of a bipartite graph [16]. The two vertex classes are the set of mutual funds $i = 1, \ldots, N_f$, and the set of the different assets $\alpha = 1, \ldots, N_a$ in their portfolios. The degree k_i of vertex *i* is exactly the number of distinct assets held in the portfolio of fund *i*. Edges incident at vertex *i* can be assigned weights $W_{i\alpha}$ equal to the total market value of the shares of asset α and the graph is also conveniently represented by a $N_f \times N_a$ matrix *B* with elements $B_{i\alpha} = W_{i\alpha}$. We indicate as $G(N_f, N_a)$, or simply *G*, the undirected bipartite network of portfolio holdings. If we wanted to retain just the topological information of which asset is owned by which fund, we could define an unweighted graph $G^0(N_f, N_a)$. This would correspond [29, 30] to the incidence matrix B^0 whose elements B0iα are 1 if asset α is in the portfolio of fund *i* and 0 otherwise. A schematic picture of the network of holdings is provided in Fig 1.

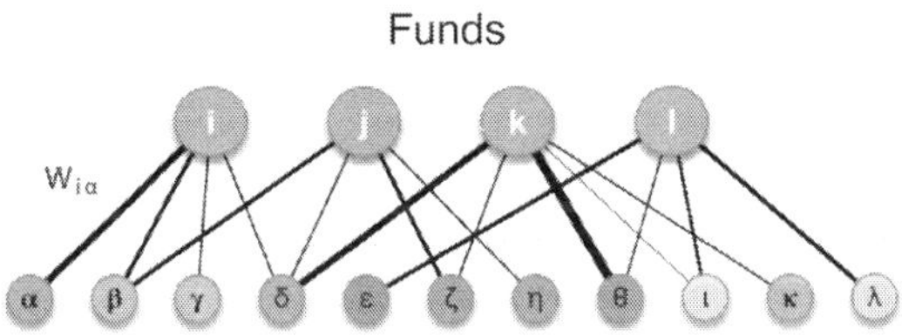

Figure 1. Graph of portfolio holdings.

The network of portfolio holdings can be represented as a bipartite graph. The two vertex classes are the *funds* $\{i, j, k, \ldots\}$ and the *assets* $\{\alpha, \beta, \gamma, \ldots\}$ in their portfolios. Each edge (i, α) represents a specific holding relationship. The edge weight $W_{i\alpha}$ is equal to the total market value of security α owned by fund *i* in its portfolio.

The *size* of fund *i* is given by

$$S_i = \sum_{\alpha=1}^{N_a} W_{i\alpha},$$

where it is understood that $W_{i\alpha} = 0$ if asset α is not in the portfolio. We clearly see that, in network terms, fund size corresponds to *node strength* [31, 32]. The sum $S_{tot} = \sum_i S_i$ represents the *total value* of the system.

The quantities S_i and $W_{i\alpha}$ are expressed in currency units. We also define the *portfolio weights* $w_{i\alpha} = W_{i\alpha}/S_i$, that represent the fraction of portfolio wealth corresponding to each asset. The indices of diversification and similarity discussed in the main text are all expressed in terms of the $w_{i\alpha}$s.

Due to portfolio reallocations, the set of assets in a portfolio and the edge weights can change in time; also the number of indexed funds in the dataset depends on time. Reporting of portfolio reallocations by funds are not synchronous and some choices are due in order to aggregate the information about holdings over a given time frame and construct the graph representing the system. Since portfolio composition is reported on a quarterly basis, in our analysis we choose a three-month time frame for aggregation. We create quarterly snapshots of the bipartite network by means of the following procedure. For each quarter, we consider the set of the funds $i = 1, \ldots, N_f$ for which a mapping to a portfolio exists. For the given quarter and for each fund i, we retrieve the holding information at the most recent report date t_i. Basically, this information is given by the set of assets in the portfolio and the corresponding market values. Fund i is inserted into the graph for the current quarter and so are the links (i, α), with $\alpha \in [1, N_a]$, for all assets held by the fund. Each link is assigned weight $W_{i\alpha}$ equal to the market value of the holding it represents.

When parsing the holding relationships in the dataset, attention has to be paid to the issue of *fund classes*. As a matter of fact, a fund may issue different types of shares all corresponding to the same underlying portfolio. In the database, different classes of a funds are associated to different unique identifiers, as if they were distinct funds. For the purpose of our analysis, we consolidated information about fund classes, to avoid including the same fund multiple times.

A basic model of distress propagation

Let us suppose that during period t the prices of some stocks undergo a negative shock $\delta_\alpha(t) = [v_\alpha(t) - v_\alpha(t-1)]/v_\alpha(t-1) < 0$, where $v_\alpha(t)$ is the market price of α. These downward moves produce a negative variation in the values of the portfolios that hold those stocks. As a consequence, we expect individual fund investors to redeem their shares, eventually forcing the asset manger to sell assets in order to meet the requests of redemption. Generally, we expect a larger (relative) drop Δ_i of the total net assets of fund i to correspond to a higher probability for an investor to redeem her shares of the fund. Let V_α be the total market value of α disinvested in the process of asset selling. The instantaneous increase in the offer of α will determine a negative impact in its price on the market and a new drop $\delta_\alpha(t+1) = \lambda[V_\alpha(t+1)]$, where the function λ is called the *price impact* (deterministic or stochastic). We expect the structure of similarities between real portfolios to be reflected also in the actions taken by agents (i.e. asset managers) in response to a shock. Accordingly, we enrich the above dynamics including a heterogeneous probability for agents to mimic the activity of funds that have already liquidated part of their portfolio. More precisely, a fund *i* that reacts to a negative shock may imitate the behavior of the most similar fund *j* that has already reacted. This occurs with probability given by the value of their cosine similarity. Imitation consists in the fact that *i* would liquidate only the stocks liquidated by *j* that are in both portfolios.

In the absence of large random effects and portfolio reallocations, such a recursive dynamics will bring a progressive reduction of the total value of the system. In the literature, the price impact of asset liquidations has been modeled as a function of average daily volume traded and asset's price volatility [16]. For the sake of simplicity, in our analysis we assume the impact on the price of security α to be a linear function of the traded value V_α. More precisely we take

$$\lambda(V_\alpha) = \frac{V_\alpha}{V_{\alpha,tot}},$$

where $V_{\alpha,tot}$ is the total value of stock α owned by funds at the beginning of each trading period. The values $V_{\alpha,tot}$ can be considered as proxies of stock liquidity [33, 34]. In this sense, the previous choice for λ accomodate for heterogeneity in stock liquidity characteristics. The simulations discussed in the next section also assume that fund managers liquidate a fraction of portfolio holdings equal to the relative drop of total net assets. When a fund's trading occurs through imitation of another fund's liquidations, we also assume that the imitating fund can liquidate only common assets. In this case, it is not allowed to trade more than the value accounted for by such assets. Results refer to a negative shock of −30% to ten most common assets in fund portfolios. This shock is propagated for a number of periods and the percentage total loss of value of the system is computed for each one.

Such a model provides an admittedly simplified representation of shock propagation within a bipartite network of portfolios and their holdings, which incorporates *contagion effects* due to common exposures of overlapping portfolios. It should be considered a very crude approximation of the dynamics in the time span that divides subsequent reallocations executed by fund managers, wherein the network can be regarded as static.

It is worth noting that we model a system where agents correspond to asset managers. Redemption requests by individual fund investors act as a driver in transmitting shocks but their behavior is not modeled explicitly. In future work a fully-fledged agent based model of asset management can be develop to accommodate more sophisticated behavioral assumptions and market features.

RESULTS AND DISCUSSION

As discussed in the previous section, a network of portfolio holdings is conveniently represented as a bipartite graph where the two vertex classes are the set of portfolios and the set of the assets they hold, respectively. For our purposes, we make the identification of funds with the portfolios they manage.

Within this framework, we consider the usual notion of diversification of an individual portfolio. This depends on the number of different assets in the portfolio, as well as on their weights and on the correlation between asset price variations. Diversification reduces the average loss that a portfolio undergoes when a random idiosyncratic shock hits one of its holdings. It is a widely accepted and employed strategy to mitigate the effects of market risk at the portfolio level. Then we add a second dimension to the discussion, by studying the *similarity* of two portfolios, also known as asset overlap [11]. The opposite of similarity is a measure of the difference between two portfolios and we may name it "differentiation", to distinguish such notion from the notion of portfolio diversification.

The analysis is performed with an attention to the evolution of the network throughout the GFC. One event that has marked the escalation of the crisis was the UK investors queuing to take their money out of Northern Rock in September 2007 and, as it is well known, the peak of the crisis has been reached in September 2008, culminating in the collapse of Lehman Brothers. To make comparisons of the network at different times, we focus on three reference quarters, 2006Q3, 2007Q3, 2008Q3.

Diversification and similarity

It is widely understood that diversification can reduce unsystematic risk, this wisdom dating back to the studies of Markowitz on portfolio diversification. However, less well understood is the optimal diversification strategy when multiple investors are simultaneously active on the market with possibly similar, or otherwise correlated, investment strategies [35]. In this paper we consider diversification and overlap between portfolios as related notions. We make the hypothesis that, while diversification reduces risk for an individual portfolio, the similarity of portfolios can be a source of systemic risk. There can be multiple ways to attain the same level of diversification for a given portfolio. And this is true independently of how the other portfolios invest their money. It cannot be excluded that, depending on the topological properties of the holding network, on the criteria of selection of new holdings and how the decisions by different portfolio managers depend on each other, a raise in diversification may increase portfolio overlap and translate into a raise of systemic riskiness.

We support these conjectures by first considering three illustrative cases for a stylized network of holdings. They are built on top of the graphs depicted in Fig 2, with two funds i, j investing in five different assets $\alpha, \beta, \gamma, \delta, \epsilon$. We are interested in the systemic effect of a negative external shock impacting the market value of an asset owned by funds. To this aim we use a deliberately simplified toy model of dynamics to keep the discussion as schematic as possible and consider the effects of similarity and portfolios diversification separately. We will make the following assumptions:

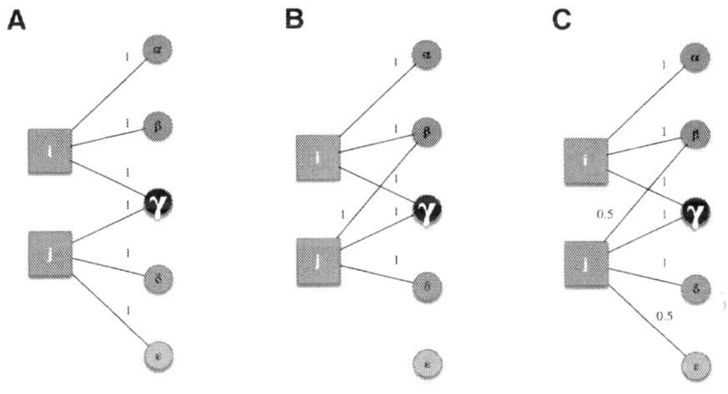

Figure 2. Three case studies, with the same values of the portfolios ($S_i = S_j = 3$) **but different values of the inverse** Herfindahl-Hirschmann index and of the cosine similarity between them.

1. All assets have unit value at first;
2. A portfolio manager liquidates the whole position as a response to the negative shock, such as in a scenario of extreme flow-performance relationship;
3. When assets are disinvested by a fund, this determines a proportional decrease in the market value of the asset.

We measure diversification by the *inverse of the Herfindahl–Hirschmann index*. This is defined as

$$h_i = [\sum w_{i\alpha}^2]^{-1}.$$

where $w_{i\alpha}$ is the weight of asset α in the portfolio of fund i. The Herfindahl–Hirschman index is a typical measure of concentration [36–38] and its inverse can be regarded as the number of leading assets in the portfolio. The value of h_i will be close to one for a fund that invests primarily in a single asset; the opposite case is represented by a uniform investment, with the same fraction of portfolio wealth invested in each asset, when we have that h_i is equal to the degree of fund i in the network.

To measure the overlap between two portfolios, we adopt *cosine similarity* [11]

$$s_{ij} = \frac{\sum w_{i\alpha} w_{j\alpha}}{\|\vec{w}_i\| \ \|\vec{w}_j\|} .$$

where $\|\vec{w}_i\|$ stands for the Euclidean norm of the vector $\vec{w}_i$ of the portfolio's weights. Portfolio similarity depends indeed on two factors: the number of assets they have in common and the similarity of the weights attached to common assets. This index is equal to one for two portfolios that contain exactly the same assets in exactly the same proportions. It will be smaller than one otherwise, and equal to zero for two portfolios that do not overlap at all.

For the illustrative examples of Fig 2, we assume that the value of asset γ goes to zero due to an external shock and either one of the two funds reacts by liquidating the whole portfolio and quitting the market. The portfolios have equal value (or "size"), $S_i = S_j = 3$, and the total value of the system is the sum of the sizes. Here, we define systemic damage as the percentage variation in total value after portfolio liquidation and subsequent reduction in asset values.

Case A

The two portfolios are equally diversified, with $h_i = h_j = 3$, and their similarity is $s_{ij} = 1/3$. Fund i reacts to the shock liquidating its portfolio and quits the market. Asset γ is the only one shared with fund j and the systemic damage amounts to $\Delta_A = -2/3 \simeq -67\%$.

Case B

In this scenario j prefers to invest one unit of value in β instead of ϵ. In doing so, it attains the same diversification as in case A but mimics i's investment strategy more closely. This results in double the similarity of case A, $s_{ij} = 2/3$. When i liquidates its portfolio, it sells one unit of β and, as a consequence, the value of β decreases by one half. This provokes a corresponding reduction in the value of j and the systemic damage is now $\Delta_B = \frac{(1/2+1)-6}{6} = -\frac{3}{4} \simeq -75\% < \Delta_A$.

Case C

With respect to case A, j attains higher diversification with a holding of 0.5 of ϵ and investing the other 0.5 on β: $h_j = \frac{18}{5} \simeq 3.6$. In this particular case, the increase in diversification causes an increase of similarity as well, $s_{ij} = \frac{1/2+1}{\sqrt{3}\ \sqrt{5/2}} \simeq 0.55$. After portfolio liquidation by i the value of β reduces to 1/3 and we get $\Delta_C = \frac{1}{6}\left(\frac{1}{2} \times \frac{1}{3} + 1 + \frac{1}{2}\right) - 1 = -\frac{13}{18} \simeq -72\% < \Delta_A$.

To summarize, the above toy examples highlight the following aspects:

1. Case B exemplifies that for a fixed level of diversification a higher level of similarity between portfolios can lead to a riskier system in the presence of feedback effects;
2. Case C exemplifies that higher diversification may be obtained to the cost of an increase in portfolio similarity, again making the system riskier.

In cases B and C the increase of overlap is direct consequence of the fact that, in network terms, the *degree* of β has increased by one. That makes the system more fragile also because an initial shock to β would have a wider outreach that it had in case A. However, in our example a shock is given only to γ. So the degree of the common asset β plays a role only in the second stage, where it acts as the medium that transmits to fund j the effects of the liquidation of the other fund's portfolio.

The network of U.S. equity funds

In this section we investigate diversification and similarity of real portfolios, considering the case of US mutual fund portfolios. Holding information was aggregated as detailed in section Materials and methods. Summary statistics for this network over the considered years are reported in Table 1.

Table 1. Summary statistics of the bipartite network of holdings.

Quarter	N_f	N_a	$\tilde{h}$	S_{tot} ($B)
2005Q3	1875	6963	42.4	1950.9
2005Q4	1749	6739	44.1	2248.0
2006Q1	1976	7272	44.3	2457.5

2006Q2	2437	9537	43.2	2654.2
2006Q3	2458	9602	42.9	2684.7
2006Q4	2552	10796	43.2	2900.9
2007Q1	2448	11405	42.8	2712.8
2007Q2	2813	11048	43.3	3331.8
2007Q3	2698	14340	43.1	3332.7
2007Q4	2608	20142	45.6	3638.1
2008Q1	2902	26319	51.1	4174.4
2008Q2	3555	27135	49.7	4332.6
2008Q3	3854	26366	45.9	3299.1
2008Q4	3991	29917	46.7	2621.1
2009Q1	3984	30554	51.0	3314.6
2009Q2	4056	31702	53.3	3832.7
2009Q3	3992	30963	53.5	4377.0
2009Q4	4061	29989	52.8	4515.8
2010Q1	3990	30761	50.1	4594.3
2010Q2	3814	30437	53.7	4365.8

Number of funds N_f and assets N_a, median of the inverse Herfindahl-Hirschmann index h˜ and total value of the funds in billion dollars.

The network of US funds holdings is heterogeneous in several respects. For instance, the distribution of portfolio sizes is known to be highly skewed and fat tailed. Evidences are presented in [27] in favor of a log-Normal distribution. Of major interest to our analysis are the distributions of diversification, asset degree and portfolio similarity, which we show in Fig 3 for the three reference quarters. In network terms, the *degree* of an asset represents the number of funds that holds it. We see that the popularity of assets in fund portfolios varies over a broad range: many assets are in the portfolios of few funds, but there are also assets that are owned by hundreds or thousands of funds. The distribution of h (first panel of Fig 3) decays slowly and extends over several orders of magnitude. Most funds manage portfolios with just few leading assets, the median value h˜ stays between 40 and 50 (see Table 1), but still we see that some funds invest in thousands of different stocks. Between the two extremes, a wide spectrum of intermediate investment strategies exists. The larger the value of h, the more the portfolio can be considered diversified. As discussed previously, the definition of h is independent on the likeness of different portfolios, while the latter aspect is of major interest in a systemic perspective. The probability density functions of the cosine similarity across portfolios are shown in the third panel of Fig

3. The network is extremely heterogeneous in this respect, with a probability density that spans more than 10 orders of magnitude. Interestingly, the existence of a non negligible probability for similarities close to one points to correlations between investment strategies.

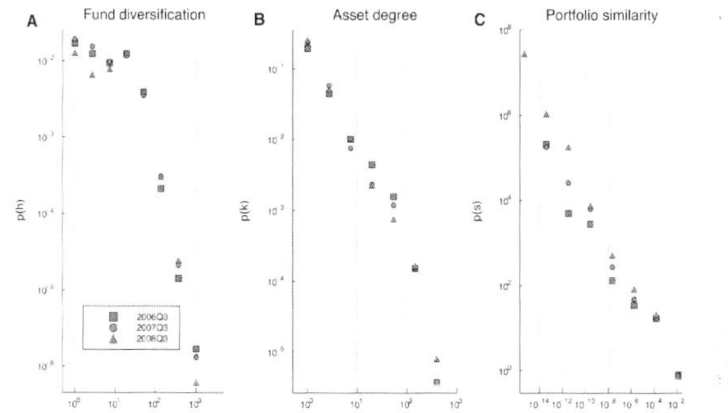

Figure 3. Probability distributions.

Probability density functions of the inverse Herfindahl–Hirschmann index h (A), asset degree k (B) and cosine similarity s (C) for three reference quarters. The network is heterogeneous and all the distributions are broad and fat tailed. Most portfolios have few dominant assets but there exist some that have thousands; most assets belong to few funds, but some are extremely popular. Similarities extend over more than 10 orders of magnitude and there exist pairs of almost identical portfolios. Only values $s > 0$ have been considered to compute the histogram.

Comparison with benchmark models of random investments

The real network is heterogeneous with respect to portfolio diversification and asset popularity and exhibits extreme values of both quantities. It is therefore important to assess if large values of portfolio similarities may originate simply from finite size effects (limited number of possible different holdings) or from the presence of very popular assets. The latter would correspond to vertexes with anomalously large connectivity ("hubs" in network terms). We analyze the level of heterogeneity and information content in the network of holdings by comparing the structure of the real network with benchmark models obtained through randomization of the original investments by means of two alternative strategies. Both schemes preserve the number and values of the original holdings of each portfolio but holdings are reassigned to potentially different assets in a random fashion. See also [22] for a similar approach.

In the first case, the new assets are taken randomly with uniform probabilities. We refer to this case as a "random holdings" (RH) model. For holdings of unitary value, it would be equivalent to a bipartite version of an Erdös–Renyi random graph with a constrained degree sequence for funds. In the second case, assets are reassigned through degree-preserving randomization (DPR). We notice that the funds' degree sequence is preserved by both schemes, while the assets' is preserved only by the DPR model.

In the left panel of Fig 4 we show the complementary cumulative distribution function *CCDF*(s) of the similarity for a snapshot of the real network corresponding to quarter 2006Q3 (the plot does not show the probability mass at $s = 0$). This is compared with its counterparts from the random benchmarks. The *CCDF*(s) gives the probability of finding a value of the cosine similarity that is greater or equal to s and allows to evaluate differences in the probabilities of very large similarities. In the RH model, the probability decays most rapidly and we see that values of $s > 0.05$ are virtually absent. In the DPR case, the probability of intermediate values of s is much larger than in the RH case. Hubs within the assets are not present in the RH models but they are in the DPR case, because there are some in the real case and asset degree is preserved by the DPR scheme. Still, we find that high similarity is unlikely to occur even in the DPR model. In particular, we can find pairs of almost identical portfolios only in the real case. It can be noticed that the probability of low to intermediate values is much larger in the DPR model than in the real network. This suggests that the real network is highly clustered, with groups of similar portfolios and small similarities across different clusters. In this case, we can expect that the rewiring procedure used in the DPR case strongly alters the clustered structure, rearranging holdings in a way that makes original clusters more similar to each other at the expanses of a smaller overlap within each one.

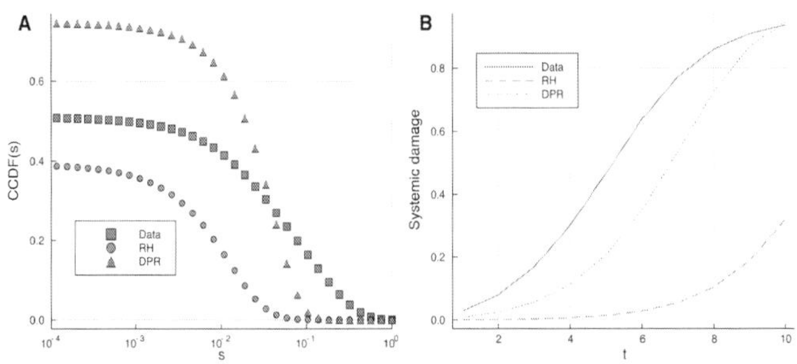

Figure 4. Similarity and systemic risk: Comparison to the null models.

Complementary cumulative distribution functions of the similarity for a snapshot of the real network (2006Q3) and the random benchmarks RH and DPR (A); the same comparison is performed for the system's riskiness measured by the total percentage loss ensuing from the propagation of a negative shock of -30% in the value of the 10 most common stocks in the network (B). Large values of the similarity between portfolios are more likely in the real world than in the randomized networks even when the degree sequence of assets is preserved. The real network appears as the most fragile. Such fragility is not explained by the role of very popular assets, as shown by the comparison with the DPR case.

We conclude that the similarities observed in the real network are much larger than can be expected by chance and, more interestingly, they cannot be explained with the existence of "hubs" among the assets. Many different mechanisms have been suggested in the literature to account for such a high degree of similarity across portfolios including connections between mutual fund managers and corporate board members, herding behavior and imitation of successful diversification strategies.

The original values of *h* are preserved by the randomization procedures and we can exploit the benchmark models RH and DPR to also test the hypothesis that portfolio overlap is a major source of systemic risk. To this aim, we consider the basic model of distress propagation discussed in Materials and methods, which describes the transmission through the network of an idiosyncratic shock due to common exposures of portfolios. We assume that a negative shock to the value of some assets spread across the network because individual fund investors redeem their portfolio shares in response to a drop in the value of their fund. This produces a negative feedback on the market value of the assets being sold, which in turn triggers a new round of losses for the portfolios owning those assets. It is, of course, a simplified view of the real network dynamics and it does not consider portfolio reallocations. However, reallocation of mutual fund investments is a slow process which might only partially ameliorate the negative impact of the asset sell-off process. In particular, we expect portfolio reallocations by fund managers to take place on a time scale larger than the one of asset price movements on the market and of buying/selling decisions by an individual investor in the fund.

In this context, we measure systemic fragility by the total percentage "loss of value" of the network. We define the initial shock as a relative drop of -30% in the value of the 10 most common stocks in the network.

In the right panel of Fig 4 we show the total loss as a function of time for the real network and the benchmark models. The comparison supports the hypothesis that large portfolio similarities contribute to a large extent to the system's vulnerability to financial shocks.

The less risky case is a model of random holdings. This is expected as it corresponds to a system that is homogeneous on the asset side. Hubs, which are well known to play as accelerators for shock propagation, are absent here and assets are likely to have similar popularity, close in value to the mean of their degree distribution. The real network is most fragile and such riskiness cannot be reproduced by the DPR model either. The damage ensuing from a shock to the most common stocks is large and propagates rapidly. Hubs are present in the DPR case. They correspond to the same assets as in the real network and the targets of the initial shock are the same in the two cases. We conclude again that it is not possible to explain the observed fragility as a pure effect of very strongly connected assets and it must depend on the similarity structure of the real portfolios. This latter fact is a major result of our analysis.

Systemic risk and similarity across the crisis

To further support the previous result about the role of portfolio overlap, we compare the level of similarity and vulnerability in the network of US equity funds over the years, see Fig 5. The first is measured by the median of the values of s (conditionally to s being larger than zero), while riskiness results from simulations of the dynamics introduced earlier over $T = 4$ trading periods.

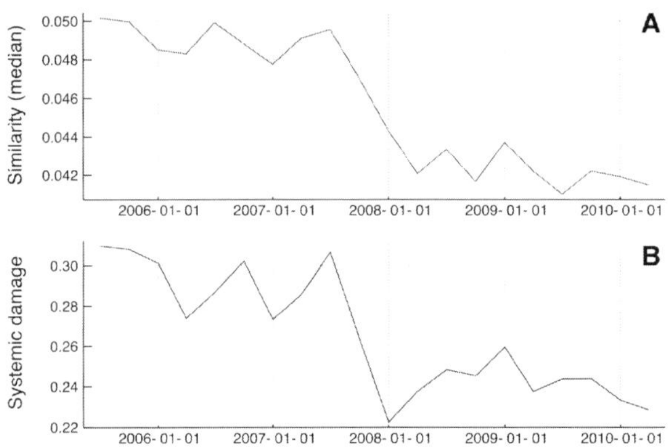

Figure 5. Similarity and systemic damage over time.

Median of the similarity s (A) and systemic damage after $T = 4$ periods (B). The similarity across portfolios and the systemic riskiness in the real network are strongly correlated. Both quantities have reduced and the network is less risky after the crisis.

The similarity of portfolios has decreased during crisis. This agrees with the fast growth in the number of holdings and the holdings/funds ratio over years. This is shown by the summary statistics of Table 1 and it has been recently documented also by using other data sources, such as the Pensions and Investments survey [9]. It is also interesting to note that the median of h, also reported in Table 1, has increased as expected during crises, but the number of holdings has increased much more rapidly.

The bottom panel of Fig 5, that shows the evolution of systemic risk with time, exhibits a noticeable resemblance with the evolution of the similarity. We find a value of 0.91 for the correlation coefficient between the two time series and 0.59 for the correlation between their variations. This supports the idea that the level of similarity could be considered as a proxy of vulnerability. The simulated systemic damage has decreased by 22% from 2007Q2 to 2008Q1; we can therefore conclude that the network is more robust after the crisis and this is correlated to modifications of the similarity structure of portfolios.

CONCLUSION

We perform an extensive study of the structure and the evolution of the US mutual fund network throughout the Global Financial Crisis of 2007–2008. Even though in normal times households rarely rebalance their retirement saving portfolios [39], it has been found that 21% of them changed investment strategy between February and November 2009 [40]. Such a dramatic recomposition of investment portfolios during crises could have severe consequences on financial stability, even though the role of the asset management industry during the crisis is still questionable. The size of the "ecosystem" of different fund investments has grown steeply over time [9] and, as an average, in the aftermath of the crisis mutual funds have become better diversified and more differentiated. However, simple summary statistics do not tell the whole story. Inspection of the probability density functions shows evidence of a heterogeneous system, with few largely diversified hubs and many specialized funds. Moreover, the probability of the similarity between portfolios decays slowly and large similarities are far more likely that can be expected from benchmark models of a random network of investments. We conclude that a high degree of correlation exists between investment decisions of different funds. This correlation limits the effectiveness of fund of funds diversification strategies.

One of the leading forces behind the emergence of such correlation can be found in the social network of relationships between fund managers [41] and the effects of managerial sharing. Other reasons may be due to herding behavior and the fact that professional investors with similar targets and risk profiles are likely to adopt similar investment strategies. Portfolio managers try to maximize profits and the strategies of many of them will likely include those assets that have proved to be profitable or that can be selected by shared quantitative analysis techniques. Extreme market uncertainty can act as a driver for fund investments during a crisis, when an important fraction of their invested capital is moved from equity mutual funds to fixed-income mutual funds. During the crisis, defined contribution equity mutual funds experienced a large outflow of more that -15%, while flow into the fixed income mutual funds reached a historical peak of +20% [9]. Similarly, many funds might be damaged during crisis at the same time and trigger a second-order effect by which other funds get hit in a failure cascade. In our stylized representation of distress propagation, such second-order effect is induced by fund managers that liquidate portfolios in response to a potentially large number of individual investors that simultaneously redeem their fund shares. However, a complete representation would also consider changes in the network topology as a result of the fact that portfolio managers will try to rebalance their portfolios. Massive co-movements in fund allocations as a response to crisis may have an even higher impact on the market value of securities. This, in turn, may result in significant effects back to the mutual fund network and possibly lead to higher levels of overlap between funds.

Differentiation of portfolios provides a different notion of diversification of investments. Exploiting a stylized model of shock propagation on a simple and static bipartite network, we have shown how strongly portfolio overlap can impact the fragility of the network. We find that the systemic damage in the network is much larger than the damage that can be procured to a network of random investments. By comparison with a degree-preserving random model, we also find that such higher riskiness cannot be ascribed solely to the presence of hubs among the assets. In the random benchmarks the diversification of original portfolios is preserved and thus the comparison with the real case provides a way to quantify the systemic risk induced by the similarity of portfolios. We conclude that the Global Financial Crisis has stimulated an increase of diversification but a systemic risk component still exists because of the similarity of investments.

We believe that the evidences presented in our study have implications for both the modeling and the regulation of financial networks. In particular, we show that similarity is correlated with systemic vulnerability. Thus the degree of overlap between the portfolios of large institutional investors should be taken into account for the purpose of assessing systemic risk in holding networks and for devising effective policy actions. In this respect, future research should integrate into our network model more realistic assumptions about the behavior of agents (i.e. asset managers) as well as more detailed information about asset specific market conditions.

ACKNOWLEDGMENTS

Guido Caldarelli acknowledges support from EU projects Multiplex 317532, Simpol 610704, Dolfins 640772, CoeGSS 676547 and SoBigdata 654024. The authors thank the anonymous referees for their valuable comments and suggestions.

REFERENCES

1. Huang X, Vodenska I, Havlin S, Stanley HE. Cascading Failures in Bi-partite Graphs: Model for Systemic Risk Propagation. Scientific Reports. 2013;3:1219 10.1038/srep01219
2. Caccioli F, Shrestha M, Moore C, Farmer JD. Stability analysis of financial contagion due to overlapping portfolios. Journal of Banking & Finance. 2014;46:233–245. 10.1016/j.jbankfin.2014.05.021
3. Gualdi S, Cimini G, Primicerio K, Di Clemente R, Challet D. Statistically validated network of portfolio overlaps and systemic risk. Scientific Reports. 2016;6:39467 10.1038/srep39467
4. Statman M. How Many Stocks Make a Diversified Portfolio? Journal of Financial and Quantitative Analysis. 1987;22(3):353–363. 10.2307/2330969
5. Chen Y, Mabu S, Ohkawa E, Hirasawa K. Constructing Portfolio Investment Strategy Based on Time Adapting Genetic Network Programming. In: Proceedings of the Eleventh Conference on Congress on Evolutionary Computation. CEC'09. IEEE Press; 2009. p. 2379–2386.
6. Domian DL, Louton DA, Racine MD. PORTFOLIO DIVERSIFICATION FOR LONG HOLDING PERIODS: HOW MANY STOCKS DO INVESTORS NEED? Studies in Economics and Finance. 2003;21(2):40–64. 10.1108/eb028774
7. Domian DL, Louton DA, Racine MD. Diversification in Portfolios of Individual Stocks: 100 Stocks Are Not Enough. The Financial Review. 2007;42(4):557–570. 10.1111/j.1540-6288.2007.00183.x
8. Statman M. The Diversification Puzzle. Financial Analysts Journal. 2004;60(4):44–53. 10.2469/faj.v60.n4.2636
9. Sialm C, Starks L, Zhang H. Defined Contribution Pension Plans: Mutual Fund Asset Allocation Changes. American Economic Review. 2015;105(5):432–36. 10.1257/aer.p20151006
10. Hu JL, Chang TP, Chou RY. Market conditions and the effect of diversification on mutual fund performance: should funds be more concentrative under crisis? Journal of Productivity Analysis. 2014;41(1):141–151. 10.1007/s11123-012-0331-x
11. Fricke C, Fricke D. Vulnerable asset management? The case of mutual funds. Discussion Papers 32/2017, Deutsche Bundesbank.
12. Battiston S, Farmer JD, Flache A, Garlaschelli D, Haldane AG,

Heesterbeek H, et al. Complexity theory and financial regulation. Science. 2016;351(6275):818–819. 10.1126/science.aad0299

13. Corsi F, Marmi S, Lillo F. When Micro Prudence Increases Macro Risk: The Destabilizing Effects of Financial Innovation, Leverage, and Diversification. Operations Research. 2016;64:1073–1088. 10.1287/opre.2015.1464
14. Acemoglu D, Ozdaglar A, Tahbaz-Salehi A. Systemic Risk and Stability in Financial Networks. American Economic Review. 2015;105(2):564–608. 10.1257/aer.20130456
15. Guo W, Minca A, Wang L. The Topology of Overlapping Portfolio Networks; 2015. Available at SSRN: https://ssrn.com/abstract=2619514 or 10.2139/ssrn.2619514.
16. Braverman A, Minca A. Networks of Common Asset Holdings: Aggregation and Measures of Vulnerability; 2014. Available at SSRN: https://ssrn.com/abstract=2379669 or 10.2139/ssrn.2379669.
17. Battiston S, Delli Gatti D, Gallegati M, Greenwald B, Stiglitz JE. Liaisons dangereuses: Increasing connectivity, risk sharing, and systemic risk. Journal of Economic Dynamics and Control. 2012;36:1121–1141. 10.1016/j.jedc.2012.04.001
18. Battiston S, Gatti DD, Gallegati M, Greenwald B, Stiglitz JE. Default cascades: When does risk diversification increase stability? Journal of Financial Stability. 2012;8:138–149. 10.1016/j.jfs.2012.01.002
19. Elliott M, Golub B, Jackson MO. Financial Networks and Contagion. American Economic Review. 2014;104(10):3115–53. 10.1257/aer.104.10.3115
20. Galbiati M, Delpini D, Battiston S. The power to control. Nature Physics. 2013;9:126–128. 10.1038/nphys2581
21. Delpini D, Battiston S, Riccaboni M, Gabbi G, Pammolli F, Caldarelli G. Evolution of Controllability in Interbank Networks. Scientific Reports. 2013;3:1626 10.1038/srep01626
22. Fricke D. Are specialist funds "special"? Financial Management. 2018; p. 1–32.
23. Feldman T. Portfolio manager behavior and global financial crises. Journal of Economic Behavior & Organization. 2010;75:192–202. 10.1016/j.jebo.2010.03.014
24. Wool PS. Essays Concerning the Network Structure of Mutual Fund Holdings and the Behavior of Institutional Investors; 2013. Ph.D.

Thesis. University of California.

25. Abdesaken G. Conflicts of Interest in Multi-Fund Management. Journal of Asset Management. 2019;20:54–71. 10.1057/s41260-018-00104-2
26. Schwarzkopf Y, Farmer JD. Empirical study of the tails of mutual fund size. Physical Review E. 2010;81:066113 10.1103/PhysRevE.81.066113
27. Schwarzkopf Y, Farmer JD. What drives mutual fund asset concentration?; 2010. Available at SSRN: https://ssrn.com/abstract=1173046 or 10.2139/ssrn.1173046.
28. Greenwood R, Landierc A, Thesmarde D. Vulnerable banks. Journal of Financial Economics. 2015;115:471–485. 10.1016/j.jfineco.2014.11.006
29. Caldarelli G. Scale-free networks Complex webs in nature and technology. Oxford University Press; 2007.
30. Newman MEJ. Networks: an introduction. Oxford University Press; 2010.
31. Barrat A, Barthélemy M, Pastor-Satorras R, Vespignani A. The architecture of complex weighted networks. Proceedings of the National Academy of Sciences of the United States of America. 2004;101:3747–3752. 10.1073/pnas.0400087101
32. Yook SH, Jeong H, Barabasi AL. Modeling the Internet's large-scale topology. Proceedings of the National Academy of Sciences of the United States of America. 2002;99:13382–13386. 10.1073/pnas.172501399
33. Gompers PA, Metrick A. Institutional Investors and Equity Prices. The Quarterly Journal of Economics. 2001;116(1):229–259. 10.1162/003355301556392
34. Lou D. A Flow-Based Explanation for Return Predictability. Review of Financial Studies. 2012;25(12):3457–3489. 10.1093/rfs/hhs103
35. Ibragimov R, Jaffee D, Walden J. Diversification disasters. Journal of financial economics. 2011;99(2):333–348. 10.1016/j.jfineco.2010.08.015
36. Battiston S. Inner structure of capital control networks. Physica A: Statistical Mechanics and its Applications. 2004;338:107–112.
37. Glattfelder J, Battiston S. Backbone of complex networks of corporations: The flow of control. Physical Review E. 2009;80:036104–036104. 10.1103/PhysRevE.80.036104

38. Woerheide W. An index of portfolio diversification. Financial Services Review. 1993;2:73–85. 10.1016/1057-0810(92)90003-U
39. Agnew J, Balduzzi P, Sundén A. Portfolio Choice and Trading in a Large 401(k) Plan. American Economic Review. 2003;93(1):193–215. 10.1257/000282803321455223
40. Hurd MD, Rohwedder S. Effects of the Financial Crisis and Great Recession on American Households. National Bureau of Economic Research; 2010. 16407.
41. Augustiani C, Casavecchia L, Gray J. Managerial Sharing, Mutual Fund Connections, and Performance. International Review of Finance. 2015;15(3):427–455. 10.1111/irfi.12054

SECTION 4
VALUATION OF DERIVATIVES OPTIONS

CHAPTER 13

The Value of Monte Carlo Model-Based Variance Reduction Technology in the Pricing of Financial Derivatives

Yunyu Zhang

School of Law, Xiamen University, Xiamen, China

ABSTRACT

The purpose of the study is to reduce the error in the pricing process of financial derivatives, as well as to obtain more accurate product values, thereby reducing transaction costs, accelerating transaction speed, establishing a larger investment scale, and enabling investors to obtain excellent returns under market conditions as much as possible. Based on the variance reduction technology, a Monte Carlo model that can effectively analyze financial prices is added to analyze price fluctuations and find the

Citation: Zhang Y (2020) "The value of Monte Carlo model-based variance reduction technology in the pricing of financial derivatives". PLoS ONE 15(2): e0229737. https://doi.org/10.1371/journal.pone.0229737

optimal holding time for users of financial derivatives, thereby reducing the risk of holding the financial derivatives. The results show that the Monte Carlo model-based variance reduction technology can significantly improve the simulation efficiency of financial derivatives pricing. In addition, the importance sampling method is used to optimize the selection, thereby making it closer to the theoretical values. The proposed method is easy to implement and has higher computational efficiency, which can ensure the financial benefits of users holding financial derivatives during the holding period. It can be seen that the Monte Carlo model-based variance reduction technology has high application value in the pricing of financial derivatives, and it is of great significance for the pricing of other products.

INTRODUCTION

With the continuous development of the world economy, financial derivatives become increasingly various. The financial derivatives are derived from basic financial instruments, such as currency, interest rate stocks, and commodities. Financial derivatives cannot exist independently, and their values largely depend on basic financial instruments. The global financial crisis triggered by the US subprime mortgage crisis in 2008 caused a huge blow to the major economic entities worldwide, making the world economy be into a recession. The financial markets of many developed countries also suffered heavy losses during the financial crisis. However, the most serious losses were the outbreak of the financial crisis and the United States, the most developed financial derivatives market. After research, the most important factors causing the crisis is considered to be the globalization of financial derivatives and the global proliferation of financial derivatives caused by lack of regulation. At present, there are more companies and users interested in financial derivatives. However, considering the value and risk of these products, there will be more considerations for the holding period of these derivatives [1].

The Monte Carlo method has a wide range of applications in financial engineering. It can measure the number of occurrences of events and thereby the probability of occurrence of events. According to the law of large numbers and the central limit theorem in probability theory, the large number theorem can ensure that the estimated value obtained by the Monte Carlo method can gradually converge to the actual value according to the increase of the number of simulations, and the central limit theorem can give the error estimation range of the Monte Carlo model [2]. At the same time, since the Monte Carlo model is very adaptable, the conditional constraints are small when solving the problems. Financial products are usually priced to maximize the attractiveness of new and existing customers and to obtain the most theoretically priced prices; however, such pricing usually takes into account market share, high returns, and reduced competitors [3]. These financial derivatives are the major businesses of the bank, and the method of pricing the bank products should be based on the cost of the product and be determined flexibly in combination with the states of supply and demand, as well as the degree to which the customer is willing to pay.

Previous research has focused on the prediction of derivatives prices in various industries, while Monte Carlo models and variance reduction techniques can better predict the value of derivatives, which is vital for better judging the holding time. Therefore, considering the application of the Monte Carlo model in financial pricing and the related situation of financial derivatives pricing, the Monte Carlo model-based variance reduction technology is proposed to analyze the pricing problem of financial derivatives, which has a certain significance for the holding situations of users. It is expected that through the application of the Monte Carlo model, price fluctuations can be analyzed in the financial market to find the optimal holding time for users of financial derivatives, thereby reducing the risk of holding financial derivatives.

LITERATURE REVIEW

With the collapse of the global fixed exchange rate regime and the continuous innovation in the financial sector, in addition to the traditional financial products, financial derivatives such as options, futures, and exchanges have developed rapidly. Monte Carlo is a commonly used statistical analysis method that can obtain an approximate solution to a problem based on economic indicators.

Pricing of financial derivatives

Wang Yulin (2017) combed and summarized the relevant pricing methods of real estate derivatives to further understand the research process and achievements of real estate derivatives, thereby providing ideas for the subsequent research [4]. Rigatos and Zervos (2017) used stochastic differential equations and diffusion-type partial differential equations to describe financial derivatives and options, pricing models. Based on the descriptions, a method based on decomposing nonlinear partial differential equations that were grouped into a set of ordinary differential equations with respect to time was proposed. The accuracy of the model was verified by state estimation. The results showed that when the Black-Scholes PDE model was verified, the parameters in the Black-Scholes PDE model were inconsistent [5]. Hefter and Jentzen (2019) used the Cox-Ingersoll-Ross (CIR) process extensively in the latest financial derivatives pricing model and analyzed the convergence speed of the discretization method. The results showed that the discrete method can achieve up to one strong convergence order [6]. Jang and Lee (2019) proposed a Bayesian learning model in which the knowledge of the risk-neutral pricing structure was integrated, which provided a fairer price for the deep ITM and deep OTM options of few transactions. The research results showed that the established estimation model was compared with the machine learning option model. The machine learning prediction model had better prediction ability and showed better overall prediction performance [7].

Monte Carlo model

Spurzem and Giersz (2018) introduced a new method to analyze the gas models with different standards and performed dynamic analysis by Monte Carlo technology. The results showed that the standard theoretical model and the direct N-body model simulation results were more consistent, which also indicated that the method was an efficient and realistic model [8]. Wang et al. (2017) used the Monte Carlo model to analyze the impact of classification accuracy index and the variant on sample size, number of latent categories, category distance, and number of indicators, and their combinations on latent profile analysis. The results showed that there was a high correlation between Entropy value and classification accuracy, but it also changed with the sample size and index number. When other conditions were unchanged, the larger the sample size was, the smaller the Entropy value was, and the worse the accuracy of the classification was. If the sample size was small, the more the number of indicators was, the Entropy value also showed a better effect [9]. Zhang (2017) considered the real strategy of options trading and analyzed the trend of option value. In the process of using Monte Carlo in simulation, the value trend was analyzed from the perspective of intrinsic value, time value, and external volatility. The results showed that there was a certain correlation between the intrinsic value of the option and the time value. In addition, as the option time was shortened, especially on the option expiration date, the intrinsic value and time value of the option would decrease, while the external fluctuation rate and leverage would increase. Therefore, users who held the options were reminded to hold the options as early as possible before the expiration date [10].

Variance reduction technology

Zhu and Chen (2018) used the Monte Carlo method to estimate the option price error and convergence rate. In addition, they utilized the least-squares method, the storage reduction technique, and the variance reduction technique to price the options of derivative securities. The research results showed that the standard deviation reduction effects of the control variable method and the dual variable method were obvious, and the dual variable method had a relatively average performance under different parameters. In addition, the effect of the control variable method depended on the correlation coefficient between the target assets, which was determined by the selected control variable. The convergence of the Monte Carlo method was improved by using a deterministic low-deviation sequence instead of a random point sequence [11]. Gao and Chen (2017) pointed out that under the Vasicek interest rate model, the pricing equations of ordinary bonds and European call options were deduced, and the discrete barrier options were priced based on them while the variances such as conditional expectation and importance sampling were used to reduce the variance and optimize the experimental analysis. The results showed that the Monte Carlo simulation method of conditional expectation and importance sampling and the two-variance reduction technology could be used to analyze the price of discrete obstacles, which could obtain more stable price data [12].

In summary, previous studies have focused on the prediction of derivatives prices in various industries, while the Monte Carlo model and variance reduction technology can better predict the value of derivatives. Therefore, in this paper, based on the Monte Carlo model of variance reduction technology, the value of financial derivatives pricing can make a good judgment on the holding time of financial derivatives holders, which is very important to judge the better time of holding.

METHODS

Monte Carlo method

The Monte Carlo method, also known as statistical simulation and random sampling, is a stochastic simulation method based on probability and statistical theory, which uses a random number (or more common pseudo-random number) to solve a variety of computational problems. Whether it is used on the computer is taken as an important sign. Therefore, although it belongs to the calculation method, it is quite different from the general calculation method. It links the solved problem to a fixed probability model and uses a computer to implement random sampling or statistical simulation, thereby obtaining an approximate solution to the problem [13].

With the continuous development of computer technology, the application of the Monte Carlo method becomes increasingly extensive. In addition, because of the excellent adaptability of the method, the calculation of integral, the solution of equations, and the calculation of eigenvalues can be easily obtained. Currently, the Monte Carlo method is mainly used to solve two types of problems, i.e., the randomness and certainty. Among these problems, the deterministic problem needs to establish a probability model. In such a process, the model is randomly observed, and its arithmetic mean is used to solve the approximate estimate of the model [14]. The random problem is that it is affected by random factors and needs to be solved by random sampling. The Monte Carlo algorithm is mainly used to study the probability density function, random number generator, sampling rule, simulation result record, error estimation, variance reduction technology, and parallel and vectorization [15–16]. The probability density function mainly generates a set of probability density functions describing a problem or system. Then, it generates a uniformly distributed random number of [0, 1] through a random number generator and randomly extracts obedience from these random numbers. For a given random variable, the corresponding simulation results are recorded, the errors of these results are estimated, and the variance reduction technology is used to reduce the number of calculations in the simulation process. Eventually, the method is operated [17–19].

When analyzing its convergence and error estimation, it is necessary to analyze its arithmetic mean value and expected value. The number of samples is denoted as N, the arithmetic mean of the samples is assumed as X_1, X_2, X_3, … X_N, as shown in Eq (1):

$$\bar{X}_N = \frac{1}{N}\sum_{i=1}^{N} X_i \qquad (1)$$

As an approximation of the solution, the sequence of variables in the sample is independent and identically distributed, and the probability value [20] can be obtained according to its expected value, as shown in Eq (2):

$$P\{\lim_{N\to\infty} \bar{X}_N - E(X)\} = 1 \tag{2}$$

However, since there is a certain error between the Monte Carlo approximation and actual value, and its variance value is shown in Eq (3):

$$0 < \sigma^2 = \int (x - E(X))^2 f(x)dx < \infty \tag{3}$$

In the above equation, f(x) is the distribution density function of X, then:

$$\lim_{N\to\infty} P\left(\frac{\sqrt{N}}{\sigma}|\bar{X}_N - E(X)| < x\right) = \frac{1}{\sqrt{2\pi}}\int_{-x}^{x} e^{-\frac{t^2}{2}dt} \tag{4}$$

If N tends to infinity, Eq (4) can be approximated as Eq (5):

$$P\left(|\bar{X}_N - E(X)| < \frac{\lambda_\alpha \sigma}{\sqrt{N}}\right) \approx \frac{2}{\sqrt{2\pi}}\int_{0}^{\lambda_\alpha} e^{-\frac{t^2}{2}dt} = 1 - \alpha \tag{5}$$

Therefore, the error of the Monte Carlo method can be obtained as shown in Eq (6):

$$\varepsilon = \frac{\lambda_\alpha \sigma}{\sqrt{N}} \tag{6)}$$

In the above equation, σ indicates the fluctuation rate of the market, N indicates the times of simulation, α is the confidence degree, $1-\alpha$ indicates the confidence level, the order of the error convergence speed is $O(N^{-1/2})$, and the corresponding estimated value is as shown in Eq (7):

$$\hat{\sigma} = \sqrt{\frac{1}{N}\sum_{i=1}^{N} X_i^2 - \left(\frac{1}{N}\sum_{i=1}^{N} X_i\right)^2} \tag{7}$$

As shown in Eq (6), the convergence speed is not related to the dimension of the problem sought, and its multiple integrals are proportional to the power of the dimension.

Variance reduction technology-importance sampling

There are many methods for variance reduction. At present, the more common methods are controlled variable method, dual variable method, conditional expectation method, stratified sampling method, importance sampling method, related sampling method, and mixed-method [21].

Importance sampling increases the sampling probability of a given sample space and increases the sampling weights of important regions so that the events that have an important effect on the simulation results are more likely to occur, thereby reducing the variance and improving the sampling efficiency and estimation accuracy [22]. The importance of the original weight sampling principles is to select a density function g (x) and make it close to the shape of f (x). Herein, the density function g (x) is the envelope function [23]. Then, the expectation can be expressed by Eq (8):

$$\theta = E[h(X)] = \int h(x)f(x)dx = \int h(x)\frac{f(x)}{g(x)}dx \tag{8}$$

Sample estimation is performed on the data in g(x), and Eq (9) is obtained:

$$\hat{\theta} = \frac{1}{N}\sum_{i=1}^{N} h(X_i)w(X_i) \tag{9}$$

In the above equation, $w(X_i)$ is the original weight, i.e., f(x)/g(x).

The standard weight can be expressed by Eq (10):

$$w^*(X_i) = \frac{w(X_i)}{\sum_{i=1}^{N} w(X_i)} \tag{10}$$

Thus, the corresponding standard expected value is obtained, as shown in Eq (11):

$$\theta = E[h(X)] = \frac{\int h(x)f(x)dx}{\int f(x)dx} = \frac{\int h(x)\frac{f(x)}{g(x)}g(x)dx}{\int \frac{f(x)}{g(x)}g(x)dx} \tag{11}$$

The operation of the estimation of g(x) is repeated. After repeating m times, the sample variance becomes Eq (12):

$$\theta' = \frac{1}{m}\sum_{i=1}^{m}(\hat{\theta}_i - \bar{\hat{\theta}}_i)^2 \tag{12}$$

Asian option pricing

The risk-return characteristics of financial derivatives are different from those of general financial products. For example, an option can not only restrict market risk but also have a reservation effect on the market to retain profitable space. It is another characteristic that is different from the risk-return of basic assets. It is precise because of these advantages of financial derivatives that financial derivatives have won the favor of many investment enthusiasts, which makes financial derivatives gradually become an integral part of the basic asset portfolio.

As a financial derivative, options have a certain effect on risk avoidance and can avoid undesired consequences. In addition, it can largely benefit from good results. Currently, options have become one of the most versatile investments in financial instruments. In stock investment, continuous compounding is usually adopted. Also, the rise and fall of stock prices are generally irregular and will be affected by many factors. However, the randomness of the stock price also allows users to treat it as the value of a variable during the research process, which varies in an uncertain way over time. When the future prediction value of the variable in the stochastic process is only related to the current value of the variable and is independent of the past value, the stochastic process can be called as the Wiener process. The standard basic Wiener process is shown in Fig 1.

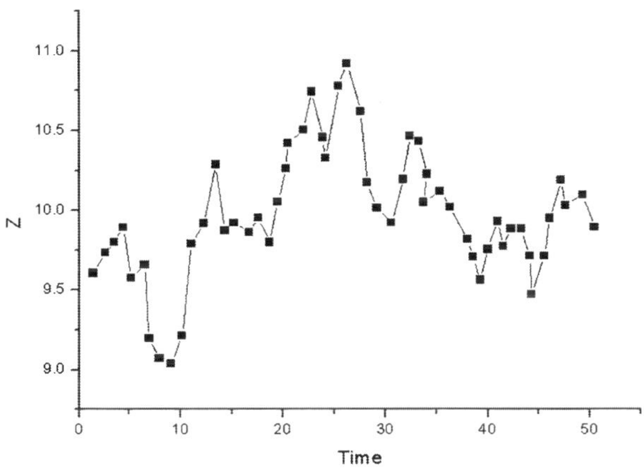

Figure 1. Basic Wiener process.

As shown in Fig 1, the behavior of the variable Z can be seen by the basic Wiener process generated by R, thereby predicting that the short time interval is a change of the variable Z. Asian options, also known as average price options, are derivatives of stock options and are based on lessons learned from the implementation of options, such as real options, virtual options, and stock options. Asian options are one of the most actively traded options in the current financial derivatives market. The difference between stock options and stock options in the usual sense is the limit on the execution price, which is the average price of the stock price for a period of time before the execution date. The difference from the standard option is that when the options income is determined on the maturity date, instead of using the market price of the underlying asset at the time, the average value of the asset price marked for a certain period of the contract period is used. Such a period is called the average period. When averaging the prices, the arithmetic mean or geometric mean is used.

Assuming that an Asian option is priced at K and the expiration time is T, *P* represents the fee payable when purchasing an option, and its calculation can be expressed as Eq (13):

$$P = \left[\exp\left(\frac{1}{T}\int_0^T \ln(S_t)dt\right) - K\right]^+ \tag{13}$$

In the above equation, S_t represents the stock price at time t. However, since the currency is time-worthy, *P*' represents the current value of payment fees, and its calculation can be expressed as Eq (14):

$$P' = \exp(-rT)\left[\exp\left(\frac{1}{T}\int_0^T \ln(S_t)dt\right) - K\right]^+ \tag{14}$$

Assuming that the option is bought at time t, *P** represents the current value of payment fees, and its calculation can be expressed as Eq (15):

$$P^* = -f_c + \exp(-rT)\left[\exp\left(\frac{1}{T}\int_0^T \ln(S_t)dt\right) - K\right]^+ \tag{15}$$

The return of the lookback option has a certain relationship with the maximum value of the underlying asset before the maturity date. The simulation results are shown in Fig 2:

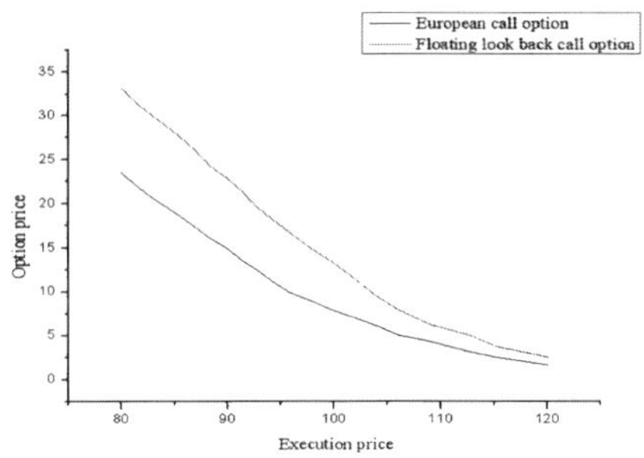

Figure 2. Simulation results of option pricing.

The pricing of fixed-looking call options and European call options is related to the strike price, while the price of the fixed-back call option is higher than the corresponding European call option. If the risk does not pose a threat, the expected yield of the stock is the risk-free rate. Therefore, the expected stock price on the maturity date will be higher than the average stock price during the effective period. At this time, the price of the floating look-up call option will be higher than the price of the corresponding Asian call option.

The value of the lookback option mainly depends on the highest or lowest price of the underlying asset that can be observed during the life of the option. In particular, there is an analytical solution to the pricing of the option under the continuous model, i.e., the lowest price or the highest price of the stock can be obtained when the continuous observation is performed. However, since the price of the underlying asset is discrete, the price under the continuous model can only be considered as an approximation of the actual price. As the frequency of observation increases, the probability of observing the extreme value increases. At this time, the price of the lookback option also shows an upward trend with the increase of the observation frequency. The Monte Carlo simulation of the European option is shown in Fig 3:

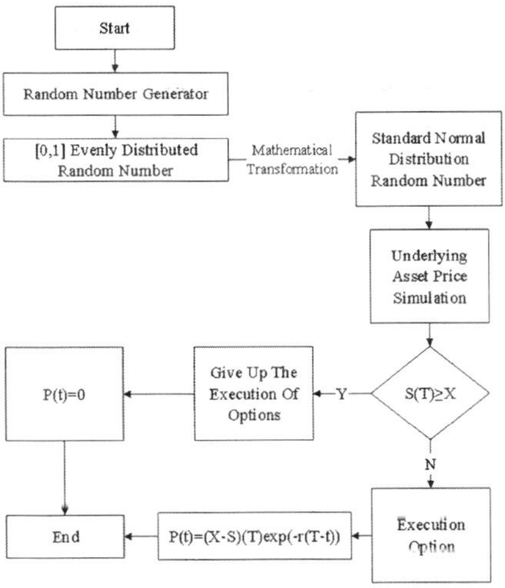

Figure 3. Monte Carlo simulation flow of European options.

The Asian option differs from the standard option. In such a case, when the options income is determined on the maturity date, the average value of the asset price of a short time target within the short time of the option is used to determine the price average operation. The method is mainly to avoid malicious manipulation of stock prices, and it can also reduce the insider trading of internal employees and damage the interests of the company to a certain extent.

RESULTS AND DISCUSSIONS

It is supposed that the current price of a stock is 50, the market volatility is σ, which is set at 40%, the risk-free rate is 10%, and the expiration is 3 months. According to the European lookback option, the value of the call option is 8.03. The Monte Carlo model-based variance reduction technology is used to calculate the price of the option, and the possible risks in the 3 months are also compared to simulate the stock path, thereby setting the price observation point and finding the lowest stock price as the exercise price. Then, the profit and loss of the option are calculated. The trends of log prices of ten-year Treasury bill futures and spots in 2015–2017 are collected, and the data in Fig 4 are obtained:

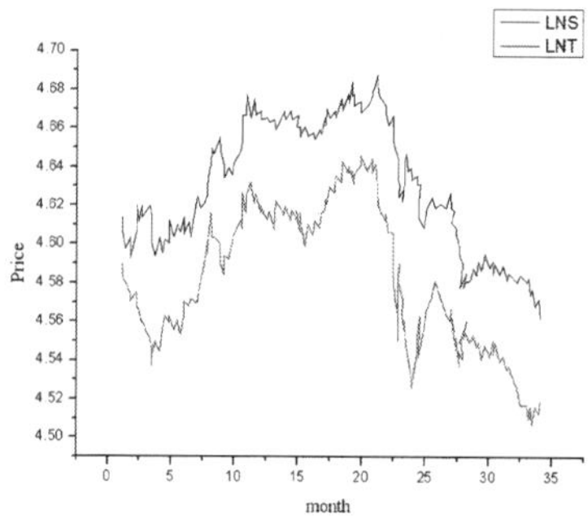

Figure 4. Trends of log prices of ten-year Treasury bill futures and spots in 2015–2017.

As shown in Fig 4, the price has decreased in recent years, and it is necessary to predict future trends.

Therefore, the data are assumed to be solved in Excel VBA, and the results of Table 1 are obtained:

Table 1. Simulation results.

Number of simulations	Equity value when the price observation point is 100 times	The value of the option when the price observation point is 1000 times	Number of simulations	Equity value when the price observation point is 100 times	The value of the option when the price observation point is 1000 times
1000	7.4724	7.6051	6000	7.5791	7.9262
2000	7.5616	7.7757	7000	7.5204	7.9254
3000	7.3442	7.7219	8000	7.6163	7.8825

4000	7.4721	7.7763	9000	7.6014	7.9051
5000	7.4746	7.7768	10000	7.4794	7.8861

In the hypothesis that the stock price obeys several Brownian movements, the stock price movement path can be further stimulated. When the static comparison is made, the time value of the currency is not considered, and the calculated theoretical value and the option value are increased in the simulation process. The error will become increasingly smaller. When performing dynamic comparison, according to the big number theorem, if the number of simulations increases, the simulated value will be closer to the true value. According to the central limit theorem, the deviation in the simulation process will also be concentrated in a certain range.

It requires to transform the number of simulations while controlling other conditions. From the results, the relative error between the analog and real values is simulated. Afterward, it is necessary to use the Monte Carlo model-based variance reduction technology to carry out simulation experiments to ensure that the dimension is floating in a small range, which can avoid the adverse effects of random scatter on the experimental results.

By analyzing the simulation data results, the average absolute error data of different simulation times under important sampling is obtained, as shown in Fig 5:

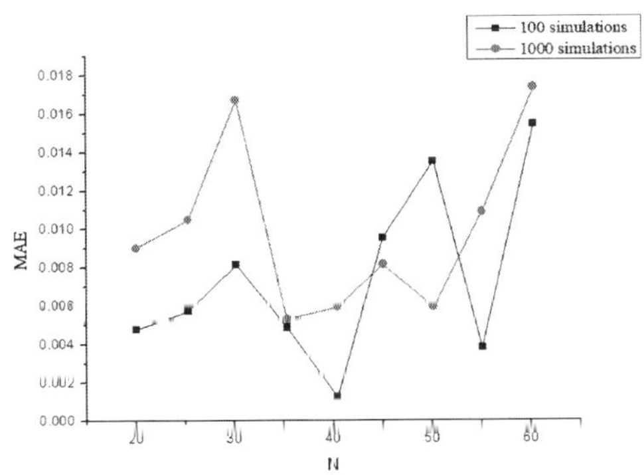

Figure 5. Average absolute error of different simulation times under importance sampling.

In addition, according to the above simulation calculation results, the convergence of Fig 6 is obtained:

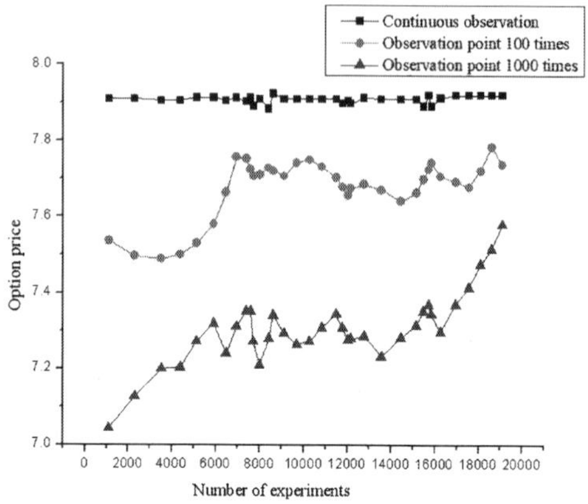

Figure 6. Convergence process of the option value.

As shown in Fig 6, during the continuous observation, the price change is small, and the lowest price and the highest price of the option can be accurately understood.

In summary, based on the importance sampling method in the variance reduction technology, the Monte Carlo model is used to simulate the hypothesis data. It can be seen that as the observations increase, the prediction results are better. In addition, with the increase with frequency, i.e., the observation time is shortened, the possibility of observing the highest price and the lowest price is greater. Since it can solve this problem well when the input is significantly uncertain during the simulation, the Monte Carlo model has been accepted and adopted by many scholars. The Monte Carlo model is performed under certain assumptions, which is different from the real trading environment and has a certain impact on the empirical results. The accuracy of the Monte Carlo's calculation results is high. When a certain number of simulations are reached, the accuracy of the results is extremely high. The disadvantage of the Monte Carlo method is that the computational workload is huge, and it takes more time. However, technology is constantly improving, and the performance of all aspects of computers has been greatly improved. The computing efficiency of the Monte Carlo model has been correspondingly improved.

CONCLUSIONS

Based on the variance reduction technology, the Monte Carlo model that can effectively analyze the financial price is added. Through the simulation of the test and the analysis of the value of the stock price at 100 and 1000, the estimated value is reduced. The actual value error helps investment enthusiasts find the best holding time for financial derivatives holdings, as well as reducing the holding risks. The application of variance reduction based on the Monte Carlo model in the pricing process of financial derivatives has a certain guiding role in the pricing of financial derivatives, which can help related companies adopt better strategies in pricing and attract customers while expanding the market share. In addition, at the same time, the market risk can be adjusted to varying degrees to avoid bringing greater losses to financial derivatives, thereby providing certain ideas for investment enthusiasts in selecting financial derivatives. This study has preliminarily researched and applied the Monte Carlo model; however, it lacks comparative analysis with other prediction models to find the shortcomings of the Monte Carlo model in data prediction. In the subsequent study, other commonly used price analysis models will be analyzed to find out how to minimize the error between the simulated value and the actual value, thereby reducing the transaction costs, accelerating the transaction speed, and increasing the returns on investment.

REFERENCES

1. 1. Ye X., & Wang X. (2017). Enhancing quasi-monte carlo simulation by minimizing effective dimension for derivative pricing. *Computational Economics*, (1), pp. 1–24.
2. 2. Spiegeleer J. D., Madan D. B., Reyners S., & Schoutens W. (2018). Machine learning for quantitative finance: fast derivative pricing, hedging and fitting. Quantitative Finance, pp. 1–9.
3. 3. Huang J., Yang S. S., & Chang C. (2018). Modeling temperature behaviors: application to weather derivative valuation. *Journal of Futures Markets*, 38(9), pp. 1152–1175.
4. 4. Yulin Wang. (2017). Summary of research on pricing methods of real estate financial derivatives. *Cooperative economy and technology*, (8), pp. 98–100.
5. 5. Rigatos G., & Zervos N. (2017). Detection of mispricing in the black–scholes pde using the derivative-free nonlinear kalman filter. *Computational Economics*, 50(1), pp. 1–20.
6. 6. Hefter M., & Jentzen A. (2019). On arbitrarily slow convergence rates for strong numerical approximations of cox–ingersoll–ross processes and squared bessel processes. *Finance and Stochastics*, 23(1), pp. 139–172.
7. 7. Jang H., & Lee J. (2019). Generative bayesian neural network model for risk-neutral pricing of american index options. *Quantitative Finance*, 19(1), pp. 1–17.
8. 8. Spurzem R., & Giersz M. (2018). A stochastic monte carlo approach to modelling of real star cluster evolution—i. the model. *Monthly Notices of the Royal Astronomical Society*, 283(3), pp. 805–810.
9. 9. Mengcheng Wang, Qiaowen Deng, Xiangyang Bi, Haosheng Ye, Wendeng Yang, & Meng-Cheng WANG et al. (2017). The performance of classification accuracy index entropy in latent profile analysis: a Monte Carlo simulation study. *Psychology Journal*, 49(11), pp. 1473–1482.
10. 10. Wei Zhang, Fu Li, & Fengyi Wu. (2017). Dynamic Risk Management Based on Monte Carlo Simulation. *Statistics and Decisions* (14), pp.164–166.
11. 11. Changpeng Zhu, & Ping Chen. (2018). Application of Importance Sampling in Discrete Barrier Option Pricing. *Economic Research Guide*, (6). pp. 87–90.

12. 12. Gao S., & Chen C. (2017). Transnationalism and financial regulation change: a case of derivative markets. *European Business Organization Law Review*, 18(1), pp. 193–223.

13. 13. Hoberg G., & Moon S. K. (2017). Offshore activities and financial vs operational hedging . *Journal of Financial Economics*, 125, pp. 217–244.

14. 14. Roy R. P., & Roy S. S. (2017). Financial contagion and volatility spillover: an exploration into indian commodity derivative market . *Economic Modelling*, 67, pp. 368–380.

15. 15. Mccrickerd R., & Pakkanen M. S. (2017). Turbocharging monte carlo pricing for the rough bergomi model. *Papers*, 18(11), pp. 1877–1886.

16. 16. Gerhard K. (2017). Augmenting propensity score equations to avoid misspecification bias–Evidence from a Monte Carlo simulation. *AStA Wirtschafts- und Sozialstatistisches Archiv*, 11(3–4), pp. 205–231.

17. 17. Ahmed M. F., Yasar S., & Sang H. C. (2018). A monte carlo model of a benchtop x-ray fluorescence computed tomography system and its application to validate a deconvolution-based x-ray fluorescence signal extraction method. *IEEE Transactions on Medical Imaging*, (99), pp. 1–1.

18. 18. Borges C. L. T., & Dias J. A. S. (2017). A model to represent correlated time series in reliability evaluation by non-sequential monte carlo simulation. *IEEE Transactions on Power Systems*, 32(2), pp. 1511–1519.

19. 19. Shang F., Zhou K., Liu H., Cheng J., Tsang I., & Zhang L., et al. (2018). Vr-sgd: a simple stochastic variance reduction method for machine learning. *IEEE Transactions on Knowledge & Data Engineering*, (99), pp. 1–1.

20. 20. Andrei C., Christoph R. (2017). A mixed Monte Carlo and partial differential equation variance reduction method for foreign exchange options under the Heston-Cox-Ingersoll-Ross model. *Journal of Computational Finance*, 20(3), pp. 109–149.

21. 21. Shen X., & Hartarska V. (2018). Winners and losers from financial derivatives use: evidence from community banks. *Applied Economics*, 50, pp. 1–16.

22. 22. Kim H. T., Papanastassiou M., & Nguyen Q. (2017). Multinationals and the impact of corruption on derivatives use and firm value: evidence

from east asia. *Journal of Multinational Financial Management*, 39, pp. 39–59.

23. 23. Riccardo P., Giulio D. G., Francesco C. (2017). A Monte Carlo evolution of the Functional Resonance Analysis Method (FRAM) to assess performance variability in complex systems. *Safety Science*, 91, pp. 49–60.

CHAPTER
14

Modelling and Computation in the Valuation of Carbon Derivatives with Stochastic Convenience Yields

Shuhua Chang [1, 2] **and Xinyu Wang** [1]

[1] Research Center for Mathematics and Economics, Tianjin University of Finance and Economics, Tianjin 300222, China,

[2] Institute of Policy and Management, Chinese Academy of Sciences, Beijing 100190, China,

ABSTRACT

The anthropogenic greenhouse gas (GHG) emission has risen dramatically during the last few decades, which mainstream researchers believe to be the main cause of climate change, especially the global warming. The mechanism of market-based carbon emission trading is regarded as a policy

Citation: Chang S, Wang X (2015) "Modelling and Computation in the Valuation of Carbon Derivatives with Stochastic Convenience Yields". PLoS ONE 10(5): e0125679. https://doi.org/10.1371/journal.pone.0125679

instrument to deal with global climate change. Although several empirical researches about the carbon allowance and its derivatives price have been made, theoretical results seem to be sparse. In this paper, we theoretically develop a mathematical model to price the CO_2 emission allowance derivatives with stochastic convenience yields by the principle of absence of arbitrage opportunities. In the case of American options, we formulate the pricing problem to a linear parabolic variational inequality (VI) in two spatial dimensions and develop a power penalty method to solve it. Then, a fitted finite volume method is designed to solve the nonlinear partial differential equation (PDE) resulting from the power penalty method and governing the futures, European and American option valuation. Moreover, some numerical results are performed to illustrate the efficiency and usefulness of this method. We find that the stochastic convenience yield does effect the valuation of carbon emission derivatives. In addition, some sensitivity analyses are also made to examine the effects of some parameters on the valuation results.

INTRODUCTION

As we all have known, climate change is one of the main environmental problems. Over the last decades, numerous scientific studies have proved that greenhouse gases such as carbon dioxide undoubtedly contribute to the climate change a lot. The accumulation of greenhouse gases, particularly CO_2, contributes to the high concentration of solar energy in the air, and the greenhouse effects can be reflected by the increase of the number of extreme weather events, such as tsunamis, floods and droughts. The greenhouse gas emission increases so fast that the temperature in the atmosphere is rising, which is now a big threat to all the species on the earth, and leads to the glacier melting and the sea level rising. Thus, one of the tasks for the environmental management is to mitigate the above changes, that is to find a way to reduce the CO_2 emission effectively.

In response to climate changes, the Kyoto Protocol proposed a carbon emission trading scheme to establish a market and reduce the concentration of GHG in the atmosphere. Several national and regional emission markets such as Chicago Climate Exchange (CCX) have been established. The European Union Emission Trading Scheme (EU ETS), which is the world's largest single market for CO_2 emission allowances, was also formed under the Kyoto Protocol in 2005 and is referred to as a cap-and-trade system. The cap means the pre-allocated amounts of greenhouse gas that the emitter can release freely. It started with a period, 2005–2007, followed by the Kyoto commitment period 2008–2012. The third period will last from 2013 to 2020.

Emission permits trade of carbon is an environmental management method specifical for carbon dioxide emission, and it is an effective way of reducing emissions. In fact, it is estimated that the carbon emission permits trading implemented and to be planned will be able to reduce 330 million tons of carbon dioxide equivalent emissions each year at least, which accounts for about 7% of global annual emissions. America is the first to carry out the study on the theory of the emission trading. Croker [1] proposed a property method to control the air pollution, which has laid a theoretical foundation for emission permits trading. Montgomery [2] studies the joint cost minimization in a market equilibrium of competition, and proves theoretically that the market-based emission permits trading is superior to the traditional environmental governance policy obviously. Hahn [3] believes that the initial allocation of emission permits may influence the monopolistic behavior of sewage plants with the monopoly power, and influence further the market efficiency. In fact, the pollutant emission permits trading is always a means of local governments to implement environmental management. However, the carbon emission reduction problem is a global environmental problem, and the environmental effect is obviously cross-regional, so the carbon emission permits trading has become a global environmental problem. In this case, the old theories are no longer enough, and a series of new studies is required.

The basic idea of carbon emission permits trading, which is regarded as a key environmental management method, is to limit the total amount of emission by creating rights to emit a certain amount of carbon, which is called the cap, and to make these rights tradable. Because of the difference in the efficiency of using energy and emitting CO_2 for different countries and regions, the emission permits have become a scarce resource under the quota system. Due to the scarcity of emission permits, the emitters have to bear higher costs for production, which should encourage them to reduce emissions as soon as possible.

The management for emission permits trading is to analyze efficiently the costs of emission reduction, and utilize the markets to reach the goal of management. In the emission permits trading system, which has been established for the purpose of environmental management, the emitters can trade their own excessive permits.

In the studies of carbon trading based on the global environmental management in the past decades, Edwards and Hutton [4] simulate the allocation of carbon permits in the United Kingdom, and pointed out that the auctioning of emission permits can allow much of the potential 'double dividend' to be realized. Cramton [5] proposed that auction is a better way to allocate the cap of emission permits, as it distributes the costs flexibly, incents innovations, and reduces the contentious arguments among members. Klaassen [6] conducts three experiments for the six largest carbon emission regions by taking advantage of experimental economics, of which the first is a single bid auction and the second is a Walrasian, and the third relies on the bilateral trading. Moreover, the three measures can all capture a significant part of the potential cost savings of emission trading. Many disputes exist and some theoretical issues still need to be completed further by the scholars from all countries, of which the valuation of carbon derivatives is the most important.

The growing CO_2 derivatives market attracts a wide range of industries and financial institutions. There are two meanings about such a market. On one hand, CO_2 derivatives contracts meet the primary need of risk transfer from those who are risk averse to those who are willing to accept the risk of permits shortage situation. By hedging the price risk through carbon derivatives, buyers and sellers can make better plans about their businesses. On the other hand, the project investments, i.e. investments committed under the so-called Clean Development Mechanism (CDM) and Joint Implementation Mechanism (JIM), which return the CO_2 emission reduction payoff that depends on the CO_2 permit market price at a fixed period, can be considered as real option contracts. It is reasonable to regard these projects as contracts whose values depend on the CO_2 permit spot prices. Similarly, the saved costs or revenue from the emission permits market may encourage the emitters to increase the technological abatement investments.

CO_2 emission allowances markets have become more and more liquid, and may grow up to the most prosperous commodities markets around the world in the near future. Moreover, the literature is available for examining the CO_2 allowance prices from the econometric or risk management angle. For instance, see the survey paper by Mnif and Davison [7]. Also, in [8] and [9] the dynamics of short-term price of CO_2 emission allowances in EU ETS is studied by Benz and Truck, and some statistical properties is obtained. In [10], an empirical framework is proposed for the pricing and hedging of intra-phase and inter-phase futures and options on futures, in which it has been pointed out that for inter-phase futures, the cost-of-carry model is still applicable, but a stochastic and mean reverting convenience yield is needed for accurate pricing. In addition, the ADF, ECM-GARCH, and ECM-TGARCH models are used to probe the mean-reversion properties and volatility features of stochastic convenience yields for CO_2 emissions allowances in [11]. Furthermore, see [12–14] for the empirical studies of risk premia in CO_2 allowance spots and futures prices and the forecasting ability of volatility, respectively, and the evaluation of the progress of this market from the trial phase to the next commitment period (Phase II). Specially, it has been suggested that the marginal costs of abatement are equal for each company [15], and the equilibrium price for European Union Allowances (EUAs) equals the marginal costs of abatement [16].

However, to the best of our knowledge, there are very few theoretical studies in the previous literature for the valuation of carbon emission permits derivatives. From [11] and [10] we can see that the convenience yields for CO_2 emission allowances show a mean-reverting process. Convenience yields are defined as the benefits or premiums associated with holding an underlying product or physical good, rather than the contract or derivatives product [17]. Spot holders can achieve potential benefits due to price volatility, but the holders of futures contracts cannot attain such benefits. The pricing problem, which contains a stochastic convenience yield, cannot be done as the one in a complete market. As we all know, the derivative can be replicated by a portfolio of existing assets such as stocks and bonds, and the risk exposure can be eliminated in a complete market, while it is not valid in an incomplete market. Schwartz et al. [18, 19] and Hilliard [20] have studied the valuation of oil contingent claims and other derivatives with stochastic convenience yield. Besides, from the viewpoint of game theory, many researches have been done on how people make decisions to adapt climate change. See, for example, [21–28].

In order to bridge the gap between theory and practice, in this paper we theoretically derive the pricing partial differential equations of CO_2 emission allowance derivatives, futures and options, with stochastic convenience yields by using the no-arbitrage principle and risk premium. Since the value of an American option is determined by a linear complementarity problem, a power penalty approach is proposed for the linear complementarity problem. That is, we will approximate the linear complementarity problem by a nonlinear parabolic PDE in two spatial dimensions with an l_k penalty term. In addition, a so-called finite volume method is presented for the numerical solution of the two-dimensional nonlinear partial differential equation. The innovation of this method is in that it combines a finite volume formulation with a fitted approximation. The finite volume method possesses a special feature of the local conservativity of the numerical flux, and is becoming more and more popular. See, for instance, Wang [29] for degenerate parabolic problems, Leveque [30] for hyperbolic problems, and Liu [31] for elliptic problems. Here we aim to present the theoretical valuation of carbon emission derivatives using partial differential equations coupled with numerical methods.

The paper is organized as follows. In Section 2, the pricing PDEs of the CO_2 emission allowance derivatives are obtained, and the final and boundary conditions are prescribed for different contingent claims. Then, a power penalty method is introduced for the American option in Section 3. In Section 4, a so-called fitted volume method is proposed for the discretization of the pricing PDE. Some numerical experiments are performed to illustrate the efficiency and usefulness of the numerical method in Section 5. In addition, some sensitivity analyses are also made to examine the effects of some parameters on the valuation results in Section 6. Finally, some comments are given in Section 7.

THE PRICING MODEL OF CO_2 CONTINGENT CLAIMS

In order to achieve the pricing equation, which applies to any contingent claims for CO_2 emission allowance spot prices, we assume that the spot price of CO_2 emission allowance S_t at time t and the convenience yield δ follow a joint diffusion process specified by (see, for example, [10, 32])

$$dS_t = \left(k_s\left(\mu(t) - \ln(S_t) + \frac{1}{k_s}\left(\frac{1}{2}\sigma_s^2 + \frac{d\mu}{dt}\right)\right) - \delta\right)S_t dt + \sigma_s S_t dW_s^P, \tag{1}$$

$$d\delta = k_c(\theta_c - \delta)dt + \sigma_c dW_c^P, \tag{2}$$

where k_s and k_c denote the speeds of mean-reversion of spot price and convenience yield, σ_s and σ_c denote the constant volatilities of spot price and convenience yield, respectively, θ_c is the long-run mean yield, and $\mu(t)$ = ln (the marginal abatement cost) + ξt is the logarithm of the marginal abatement cost C_s growing at rate ξ. In addition, dW^P_s and dW^P_c are correlated increments to standard Brownian process under the historical measure $\mathbb{P}$, and they have the correlation coefficient $_{sc}$. From the concept of absence of arbitrage opportunities, if the market prices of spot and convenience yield are given, then we know

$$W_s^Q = W_s^P + \int_0^t \lambda_s(\tau)d\tau \quad \text{and } W_c^Q = W_c^P + \int_0^t \lambda_c(\tau)d\tau$$

are Brownian motions under the risk-neutral measure $\mathbb{Q}$, where λ_s and λ_c are the market prices of spot and convenience yield, respectively.

Note that Eq (1) is based on the relationship between the CO_2 emission spot price and the marginal abatement cost given by

$$\ln(S_t) = \mu(t) + X_t, \tag{3}$$

where X_t is governed by an Ornstein-Uhlenbeck process with a zero long run mean, a speed of adjustment k_s, and a constant volatility σ_s:

$$dX_t = -k_s X_t dt + \sigma_s dW_s^P.$$

Eq (3) implies that the natural logarithm of the CO_2 emission allowance spot price deviates around a mean-reversion level, and this level grows with the discount rate ξ and is determined by the marginal cost of abatement C_s. Moreover, it is expected to be mean-reversed for convenience yields because of the strong tendency, and short-term random convenience yields converge to their mean values in the long run.

The partial differential equation for the price of carbon derivatives

Assume that the price of carbon derivative $F(S, \delta, t)$ is a twice continuously differentiable function of S and δ, where we omit the subscript t of S for brevity. Then, we can use Itô's lemma to derive the instantaneous price change as follows:

$$\begin{aligned} dF &= \frac{\partial F}{\partial t}dt + \frac{\partial F}{\partial S}dS + \frac{\partial F}{\partial \delta}d\delta + \frac{1}{2}\frac{\partial^2 F}{\partial S^2}(dS)^2 + \frac{1}{2}\frac{\partial^2 F}{\partial \delta^2}(d\delta)^2 + \frac{\partial^2 F}{\partial S \partial \delta}dS d\delta \\ &\left(\frac{\partial F}{\partial t} + (k_s(n(t) - \ln S) - \delta)S\frac{\partial F}{\partial S} + k_c(\theta_c - \delta)\frac{\partial F}{\partial \delta} + \frac{1}{2}\sigma_s^2 S^2\frac{\partial^2 F}{\partial S^2} + \frac{1}{2}\sigma_c^2\frac{\partial^2 F}{\partial \delta^2}\right. \\ &\left. + \rho_{sc}\sigma_s\sigma_c S\frac{\partial^2 F}{\partial S \partial \delta}\right)dt + \sigma_s S\frac{\partial F}{\partial S}dW_s^P + \sigma_c\frac{\partial F}{\partial \delta}dW_c^P. \end{aligned}$$

from which the relative change can be obtained via dividing both sides by F

$$\begin{aligned} \frac{dF}{F} &= \left(\left(\frac{\partial F}{\partial t} + (k_s(n(t) - \ln S) - \delta)S\frac{\partial F}{\partial S} + k_c(\theta_c - \delta)\frac{\partial F}{\partial \delta} + \frac{1}{2}\sigma_s^2 S^2\frac{\partial^2 F}{\partial S^2} + \frac{1}{2}\sigma_c^2\frac{\partial^2 F}{\partial \delta^2}\right.\right. \\ &\left.\left. + \rho_{sc}\sigma_s\sigma_c S\frac{\partial^2 F}{\partial S \partial \delta}\right)/F\right)dt + \left(\sigma_s S\frac{\partial F}{\partial S}/F\right)dW_s^P + \left(\sigma_c\frac{\partial F}{\partial \delta}/F\right)dW_c^P,\end{aligned}$$

where

$$n(t) = \mu(t) + \frac{1}{k_s}\left(\frac{1}{2}\sigma_s^2 + \frac{d\mu}{dt}\right).$$

For simplicity, we define

$$k = \left(\frac{\partial F}{\partial t} + (k_s(n(t) - \ln S) - \delta)S\frac{\partial F}{\partial S} + k_c(\theta_c - \delta)\frac{\partial F}{\partial \delta} + \frac{1}{2}\sigma_s^2 S^2 \frac{\partial^2 F}{\partial S^2} + \frac{1}{2}\sigma_c^2 \frac{\partial^2 F}{\partial \delta^2} \right.$$
$$\left. + \rho_{sc}\sigma_s\sigma_c S \frac{\partial^2 F}{\partial S \partial \delta}\right)/F, \tag{4}$$

$$L_1 = \left(\sigma_s S \frac{\partial F}{\partial S}\right)/F, \quad L_2 = \left(\sigma_c \frac{\partial F}{\partial \delta}\right)/F. \tag{5}$$

Following Brennan and Schwartz [33], we construct a portfolio P by investing amounts of x_1, x_2, and x_3 in three options of maturities $_1$, $_2$, and $_3$, respectively. Then, the rate of return on this portfolio is given by

$$\frac{dP}{P} = [x_1 k + x_2 k + x_3 k]dt + [x_1 L_1 + x_2 L_1 + x_3 L_1]dW_s^P + [x_1 L_2 + x_2 L_2 + x_3 L_2]dW_c^P, \tag{6}$$

which is non-stochastic if the portfolio proportions are chosen so that the coefficients of dW^P_s and dW^P_c in Eq (6) are zero. That is,

$$x_1 L_1 + x_2 L_1 + x_3 L_1 = 0, \quad x_1 L_2 + x_2 L_2 + x_3 L_2 = 0. \tag{7}$$

To eliminate the arbitrage opportunity, the above return on the portfolio must be risk-free over short time intervals, this is, the return rate is equal to r, the instantaneous risk-free interest rate. As a consequence, the portfolio risk premium is zero:

$$k - r = \lambda_s L_1 + \lambda_c L_2, \tag{8}$$

The arbitrage free condition Eq (8) and the two zero risk conditions Eq (7) have a solution only if

$$k - r = \lambda_s L_1 + \lambda_c L_2, \tag{9}$$

where $_s$ and $_c$ are the market prices of per unit spot and convenience yield, respectively.

Subsequently, substituting Eqs (4) and (5) into Eq (9) results in

$$\left(\frac{\partial F}{\partial t} + (k_s(n(t) - \ln S) - \delta)S\frac{\partial F}{\partial S} + k_c(\theta_c - \delta)\frac{\partial F}{\partial \delta} + \frac{1}{2}\sigma_s^2 S^2 \frac{\partial^2 F}{\partial S^2} + \frac{1}{2}\sigma_c^2 \frac{\partial^2 F}{\partial \delta^2} \right.$$
$$\left. + \rho_{sc}\sigma_s\sigma_c S \frac{\partial^2 F}{\partial S \partial \delta}\right)/F - r - \lambda_s\left(\sigma_s S \frac{\partial F}{\partial S}\right)/F + \lambda_c\left(\sigma_c \frac{\partial F}{\partial \delta}\right)/F.$$

That is,

$$\frac{\partial F}{\partial t} + (k_s(n(t) - \ln S) - \delta - \lambda_s \sigma_s) S \frac{\partial F}{\partial S} + (k_c(\theta_c - \delta) - \lambda_c \sigma_c) \frac{\partial F}{\partial \delta} + \frac{1}{2}\sigma_s^2 S^2 \frac{\partial^2 F}{\partial S^2} + \frac{1}{2}\sigma_c^2 \frac{\partial^2 F}{\partial \delta^2} + \rho_{sc}\sigma_s\sigma_c S \frac{\partial^2 F}{\partial S \partial \delta} - rF = 0. \tag{10}$$

which is the partial differential equation for the contingent claims depending on the evolution of the spot price S and the convenience yield δ.

Remark 1. *In the case of futures contract, the return rate on the portfolio* Eq (6) *should be zero [34], such that the* Eq (8) *changes into* $x_1 k(_1)+x_2 k(_2)+x_3 k(_3) = 0$. *Then, the pricing partial differential equation becomes*

$$\frac{\partial F}{\partial t} + (k_s(n(t) - \ln S) - \delta - \lambda_s \sigma_s) S \frac{\partial F}{\partial S} + (k_c(\theta_c - \delta) - \lambda_c \sigma_c) \frac{\partial F}{\partial \delta} + \frac{1}{2}\sigma_s^2 S^2 \frac{\partial^2 F}{\partial S^2} + \frac{1}{2}\sigma_c^2 \frac{\partial^2 F}{\partial \delta^2} + \rho_{sc}\sigma_s\sigma_c S \frac{\partial^2 F}{\partial S \partial \delta} = 0. \tag{11}$$

Remark 2. *The same pricing partial differential equation can be also obtained by using two-dimensional version of the Feynman-Kac formula.*

Boundary and final conditions for different carbon derivatives

It stands to reason that we should prescribe the final conditions for different derivatives to fully define the problem. In addition, to solve the pricing PDE numerically, we also need the boundary conditions. In this paper, we propose three kinds of derivatives, European and American options as well as futures, which are the popular derivatives in any financial market.

Options

We only take the call option's final and boundary conditions into consideration, since the put option is similar. The most common final condition is the vanilla option's payoff given by

$$F(S, \delta, T) = \max\ (0, S - K), \quad S \in (0, S_{\max}), \tag{12}$$

where K denotes the exercise price of the option satisfying $0 < K < S_{\max}$. A second choice is the cash-or-nothing payoff given by

$$F(S, \delta, T) = B\mathbb{H}(S - K), \quad S \in (0, S_{\max}), \tag{13}$$

where $B > 0$ is a constant and $\mathbb{H}$ denotes the Heaviside function. Obviously, this final condition is a step function, which is zero if $S < K$ and B if $S \geq K$.

The price of any derivatives except the futures contract will satisfy Eq

(10), and the boundary conditions have to be adjusted according to the specific exercise features of derivatives. In the case of European options, there are four boundaries in the solution domain: $S = 0$, $S = S_{max}$, $\delta = \delta_{min}$, and $\delta = \delta_{max}$. Obviously, the boundary conditions at $S = 0$ and $S = S_{max}$ are simply taken to be the extension of the final condition:

$$F(0, \delta, t) = F(0, \delta, T) = 0 \quad \text{and} \quad F(S_{\max}, \delta, t) = F(S_{\max}, \delta, T). \tag{14}$$

In the case of American options, since they can be exercised at any point in time before maturity, part of the valuation problem involves identifying the optimal exercise policy, or the exercise time that maximizes the option value. The boundary condition at $S = S_{max}$ should be replaced by the following two classic value-matching and smooth-pasting boundary conditions [35]:

$$F(S^{\star}(t), \delta, t) = S^{\star}(t) - K, \tag{15}$$

$$\frac{\partial F(S^{\star}(t), \delta, t)}{\partial S} = 1, \tag{16}$$

for the American call option.

To determine the boundary conditions at $\delta = \delta_{min}$ and $\delta = \delta_{max}$, we need to model the pricing Eq (10) again for two particular values $\delta = \delta_{min}$ and $\delta = \delta_{max}$, and solve the two resulting one-dimensional equations. These two approximate boundary conditions will be used to implement our numerical methods.

Futures

In the case of futures, the way to determine the boundary conditions is similar to that for European options. We ignore the discussion about boundary conditions and only propose the final condition in this subsection. As mentioned in [36], when the delivery period is being reached, the futures price is very close or even equal to the spot price, which means that the final condition for futures should be

$$F(S, \delta, T) = S. \tag{17}$$

THE POWER PENALTY APPROACH

Since the valuation of American option is a free boundary problem, its exact solution is not available analytically. Therefore, numerical approximation to the solution is normally sought in practice. In fact, we formulate the free boundary problem as a linear complementarity problem, and then develop a power penalty method to solve it. That is, a nonlinear parabolic PDE in two spatial dimensions with an l_k penalty term is utilized to approximate the linear complementarity problem. Moreover, in the next section, a fitted finite volume method is proposed to solve the penalized nonlinear equation.

Formulation of the problem into a complementarity problem

Denote the value of the American option with expiry date T by $F(S, \delta, t)$ and define

$$LF = -\frac{\partial F}{\partial t} - \frac{1}{2}\left(\sigma_s^2 S^2 \frac{\partial^2 F}{\partial S^2} + 2\rho_{sc}\sigma_s\sigma_c S \frac{\partial^2 F}{\partial S \partial \delta} + \sigma_c^2 \frac{\partial^2 F}{\partial \delta^2}\right) - \left(h(S,\delta,t)\frac{\partial F}{\partial S} + g(\delta,t)\frac{\partial F}{\partial \delta}\right) + rF, \tag{18}$$

where $h(S, \delta, t) = S(k_s(n(t) - \ln S) - \lambda_s \sigma_s - \delta)$ and $g(\delta, t) = k_c(\theta_c - \delta) - \lambda_c \sigma_c$. Here we define $h(0, \delta, t) = 0$ to keep the continuity of $h(S, \delta, t)$ at $S = 0$. The free boundary $S^\star(t)$ divides the domain $\Omega = (0, S_{max}) \times (\delta_{min}, \delta_{max})$ into the continuation region Σ_1 and the stopping region Σ_2. In the continuation region Σ_1, $F > F(S, \delta, T)$, $LF = 0$; in the stopping region Σ_2, note that $S > K$, $F = S - K$, then $LF > 0$. In a word, the American option value F satisfies the following partial differential complementarity problem:

$$\begin{cases} LF \geq 0, \\ F - F^\star \geq 0, \\ LF \cdot (F - F^\star) = 0, \end{cases} \tag{19}$$

for $(S, \delta, t) \in \Omega \times [0, T)$ with the boundary conditions:

$$\begin{cases} F(0,\delta,t) = 0, F(S_{max},\delta,t) = S_{max} - K, \\ F(S,\delta_{min},t) = g_1(S,t), F(S,\delta_{max},t) = g_2(S,t), \end{cases} \tag{20}$$

and the terminal condition

$$F(S, \delta, T) = F^\star(S, \delta), \tag{21}$$

where

$$F^\star(S, \delta) = \max(S-K, 0)$$

is the payoff function, g_1, and g_2 are the boundary conditions to be determined in the next section.

For the ease of theoretical analysis, we rewrite Eq (18) as the following divergent form:

$$LF = -\frac{\partial F}{\partial t} - \nabla \cdot (A\nabla F + \underline{b}F) + \bar{c}F, \tag{22}$$

where

$$A = \begin{pmatrix} a_{11} & a_{12} \\ a_{21} & a_{22} \end{pmatrix} = \begin{pmatrix} \frac{1}{2}\sigma_s^2 S^2 & \frac{1}{2}\rho_{sc}\sigma_s\sigma_c S \\ \frac{1}{2}\rho_{sc}\sigma_s\sigma_c S & \frac{1}{2}\sigma_c^2 \end{pmatrix},$$

$$\underline{b} = \begin{pmatrix} b_1 \\ b_2 \end{pmatrix} = \begin{pmatrix} k_s(n(t) - \ln S)S - \lambda_s\sigma_s S - \delta S - \sigma_s^2 S \\ k_c(\theta_c - \delta) - \lambda_c\sigma_c - \frac{1}{2}\rho_{sc}\sigma_s\sigma_c \end{pmatrix}, \tag{23}$$

$$\bar{c} = r + k_s(n(t) - \ln S) - \delta - \lambda_s\sigma_s - \sigma_s^2 - k_s - k_c.$$

Since the homogeneous Dirichlet boundary conditions are convenient for theoretical discussion, we transform Eqs (19)–(21) to be homogeneous, which is not necessary in computations. To this purpose, we introduce a new function

$$u(S, \delta, t) = e^{\beta t}(F_0 - F), \tag{24}$$

where $\beta = \frac{1}{2}(\sigma_s^2 + \sigma_c^2)$ and $F_0(S, \delta)$ is a twice differentiable function satisfying the boundary conditions in Eq (20). Transforming F in Eq (19) into the new function u, we have

$$\begin{cases} \mathcal{L}u \le f, \\ u - u^* \le 0, \\ (\mathcal{L}u - f) \cdot (u - u^*) = 0, \end{cases} \tag{25}$$

where

$$\begin{cases} \mathcal{L}u = -u_t - \nabla \cdot (A\nabla u + \underline{b}u) + \underline{c}u, \\ \underline{c} = \bar{c} + \beta, u^* = e^{\beta t}(F_0 - F^*), f(S, \delta, t) = e^{\beta t}LF_0. \end{cases} \tag{26}$$

At the same time, the boundary and the terminal conditions in Eqs (20)–(21) are also transformed into

$$u(0, \delta, t) = 0 = u(S_{\max}, \delta, t), \quad t \in [0, T] \text{ and } \delta \in [\delta_{\min}, \delta_{\max}],$$
$$u(S, \delta_{\min}, t) = 0 = u(S, \delta_{\max}, t), \quad t \in [0, T] \text{ and } S \in [0, S_{\max}],$$

and $u(S, \delta, T) = u^*(S, \delta)$, respectively.

Formulation of the complementarity problem into a variational inequality problem

Next we reformulate Eq (25) as a variational inequality problem in an appropriate functional setting. To this end, we first introduce some standard notations. We define $\Omega = (0, S_{max}) \times (\delta_{min}, \delta_{max})$ and Γ as the computational domain and its boundary, respectively. Let $L^p(\Omega)$ be the space of all p-integrable functions for any $1 \leq p \leq \infty$ on Ω with the norm $\|\cdot\|_{L^p(\Omega)}$. For the standard Sobolev space $H^{m,p}(\Omega)$, for $p = 2$ we use $H^m(\Omega)$ and $\|\cdot\|_{m,\Omega}$ to stand for $H^{m,2}(\Omega)$ and $\|\cdot\|_{m,2,\Omega}$, respectively. Moreover, a weighted Sobolev space H1$\omega(\Omega)$ and its norm $\|\cdot\|_{1,\omega}$ are defined as follows:

$$H^1_\omega(\Omega) = \{v \cdot v, Sv_S, v_\delta \in L^2(\Omega)\}, \qquad \| v \|^2_{1,\omega} = \int_\Omega (S^2 v_S^2 + v_\delta^2 + v^2) d\Omega.$$

Also, we define

$$H^1_{0,\omega}(\Omega) = \{v : v \in H^1_\omega(\Omega), \quad v|_\Gamma = 0\},$$

$$\mathcal{K} = \{v(t) : v(t) \in H^1_{0,\omega}(\Omega), v(t) \leq u^*(t), \quad \text{a.e.} \quad \text{in } (0, T)\},$$

where $u^\star(t)$ is defined by Eq (26). Clearly, $\mathcal{K}$ is a convex and closed subset of H10,$\omega(\Omega)$. We also define the norm of $L^p(0, T; H(\Omega))$ for any Hilbert space $H(\Omega)$ as follows:

$$\| v(\cdot, \cdot, t) \|_{L^p(0,T;H(\Omega))} = \left(\int_0^T \| v(\cdot, \cdot, t) \|^p_H \, dt \right)^{1/p}.$$

Hereinafter, $v(\cdot, \cdot, t)$ will be written simply as $v(t)$ when there is no confusion.

Now, we are in the position to define our variational inequality problem.

Problem 1. *Find $u \in \mathcal{K}$ such that for all $v \in \mathcal{K}$,*

$$\left(-\frac{\partial u}{\partial t}, v - u \right) + B(u, v - u, t) \geq (f, v \quad u) \text{ a.e. in } (0, T), \tag{27}$$

where $B(\cdot, \cdot; t)$ is a bilinear form defined by

$$B(u, v; t) = (A\nabla u + \underline{b}u, \nabla v) + (\underline{c}u, v), \quad u, v \in H^1_{0,\omega}(\Omega). \tag{28}$$

For this variational inequality problem, we have the following theorem.

Theorem 1. *Problem 1 is the variational form of the complementarity problem* Eq (25).

Proof. For any $w \in \mathcal{K}$, it follows from the definition of $\mathcal{K}$ that

$$w - u^* \leq 0 \text{ a.e. on } \Theta = \Omega \times (0, T).$$

Multiplying both sides of the first inequality of Eq (25) by $w{-}u^\star$, we obtain by integration that

$$\left(-\frac{\partial u}{\partial t}, w - u^*\right) - (\nabla \cdot (A\nabla u + \underline{b}u) - \underline{c}u, w - u^*) \geq (f, w - u^*), \text{ a.e. in } (0, T).$$

Using the Gauss-divergence theory, we obtain

$$\left(-\frac{\partial u}{\partial t}, w - u^*\right) + B(u, w - u^*; t) \geq (f, w - u^*), \text{ a.e. in } (0, T). \tag{29}$$

Since $\mathcal{K}$ is a convex and closed subset of $H^1_{0,\omega}(\Omega)$, we may write w as $w = \theta v + (1 - \theta)u$, where $u, v \in \mathcal{K}$ and $\theta \in [0, 1]$. Therefore, Eq (29) becomes

$$\left(-\frac{\partial u}{\partial t}, \theta(v - u)\right) + B(u, \theta(v - u); t)$$
$$\geq (f, \theta(v - u)) - \left(\left(-\frac{\partial u}{\partial t}, u - u^*\right) + B(u, u - u^*; t) - (f, u - u^*)\right). \tag{30}$$

On the other hand, from the third relationship of Eq (25) we know that

$(\mathcal{L}u - f, u - u^\star) = 0$,

i.e.

$$\left(-\frac{\partial u}{\partial t}, u - u^*\right) + B(u, u - u^*; t) - (f, u - u^*) = 0.$$

Therefore, Eq (30) reduces to

$$\left(-\frac{\partial u}{\partial t}, \theta(v - u)\right) + B(u, \theta(v - u); t) \geq (f, \theta(v - u)),$$

from which we can obtain

$$\left(-\frac{\partial u}{\partial t}, v - u\right) + B(u, v - u; t) \geq (f, v - u), \text{ a.e. in } (0, T).$$

Lemma 1. *There exist positive constants C and M, such that for any* v,w∈H10, $\omega(\Omega)$, *there hold*

$$B(v, v; t) \geq C \| v \|_{1,\omega}^2 \quad \text{and} \quad |B(v, w; t)| \leq M \| v \|_{1,\omega} \| w \|_{1,\omega}.$$

Proof. For any $v \in H^1_{0,\omega}(\Omega)$, we have via integration by parts that

$$\int_\Omega \underline{b}v \cdot \nabla v d\Omega = \int_{\partial\Omega} v^2 \underline{b} \cdot n ds - \int_\Omega v \nabla \cdot (\underline{b}v) d\Omega = -\int_\Omega v\underline{b} \cdot \nabla v d\Omega - \int_\Omega v^2 \nabla \cdot \underline{b} d\Omega, \tag{31}$$

which leads to

$$\int_\Omega \underline{b}v \cdot \nabla v d\Omega = -\frac{1}{2}\int_\Omega \nabla \cdot \underline{b}v^2 d\Omega.$$

From Eqs (28) and (31) we obtain

$$\begin{aligned} B(v, v; t) &= (A\nabla v + \underline{b}v, \nabla v) + \underline{c}(v, v) \\ &= (A\nabla v, \nabla v) + (\underline{b}v, \nabla v) + \underline{c}(v, v) \\ &= \int_\Omega (\sigma_s^2 S^2 v_S^2 + \sigma_c^2 v_\delta^2) d\Omega + \left(\underline{c} - \frac{1}{2}\nabla \cdot \underline{b}\right)(v, v) \\ &= \int_\Omega (\sigma_s^2 S^2 v_S^2 + \sigma_c^2 v_\delta^2) d\Omega + \left(\frac{1}{2}(k_s(n(t) - \ln S) - \lambda_s \sigma_s - y - \sigma_s^2 - k_s - k_c) + r + \beta\right) \| v \|_0^2 \\ &\geq C \| v \|_{1,\omega}^2 . \end{aligned}$$

Now, let us show the continuity of B. For any $v, w \in H^1_{0,\omega}(\Omega)$, we have

$$|B(v, w; t)| = |(A\nabla v + \underline{b}v, \nabla w) + \underline{c}(v, w)| \leq |(A\nabla v, \nabla w)| + |(\underline{b}v, \nabla w)| + |\underline{c}(v, w)|. \tag{32}$$

For $|(A\nabla v, \nabla w)|$ in Eq (32), it follows from the Cauchy-Schwartz inequality that

$$\begin{aligned} |(A\nabla v, \nabla w)| &= \left| \int_\Omega \left(\frac{1}{2}\sigma_s^2 S^2 v_S w_S + \frac{1}{2}\sigma_c^2 v_\delta w_\delta \right) d\Omega \right| \\ &\leq \frac{1}{2}\left(\int_\Omega \sigma_s^2 S^2 v_S^2 d\Omega \right)^{\frac{1}{2}} \left(\int_\Omega \sigma_s^2 S^2 w_S^2 d\Omega \right)^{\frac{1}{2}} \\ &\quad + \frac{1}{2}\left(\int_\Omega \sigma_c^2 v_\delta^2 d\Omega \right)^{\frac{1}{2}} \left(\int_\Omega \sigma_c^2 w_\delta^2 d\Omega \right)^{\frac{1}{2}} \\ &\leq M \left(\int_\Omega (\sigma_s^2 S^2 v_S^2 + \sigma_c^2 v_\delta^2) d\Omega \right)^{\frac{1}{2}} \left(\int_\Omega (\sigma_s^2 S^2 w_S^2 + \sigma_c^2 w_\delta^2) d\Omega \right)^{\frac{1}{2}} \\ &\leq M \| v \|_{1,\omega} \| w \|_{1,\omega}. \end{aligned}$$

For $|(b-v, \nabla w)|$ in Eq (32), it follows from the expression of b− in Eq (23) and the Cauchy-Schwarz inequality that

$$\begin{aligned} |(\underline{b}v, \nabla w)| &= \left|\int_\Omega \underline{b}v \cdot \nabla w d\Omega\right| = \left|\int_{\partial\Omega} vw\underline{b} \cdot n ds - \int_\Omega w\nabla \cdot (\underline{b}v) d\Omega\right| \\ &= \left| - \int_\Omega w\underline{b} \cdot \nabla v d\Omega - \int_\Omega vw\nabla \cdot \underline{b} d\Omega\right| \\ &\le \left|\int_\Omega w\underline{b} \cdot \nabla v d\Omega\right| + \left|\int_\Omega vw\nabla \cdot \underline{b} d\Omega\right| \\ &\le \left|\int_\Omega w(b_1 v_S + b_2 v_\delta) d\Omega\right| + \left|\int_\Omega vw\nabla \cdot \underline{b} d\Omega\right| \\ &\le M\| w \|_0 \| v \|_{1,\omega} + M\| v \|_0 \| w \|_0. \end{aligned}$$

For $|\underline{c}(v, w)|$ in Eq (32), it is easy to see

$$|\underline{c}(v, w)| \le M\| v \|_0 \| w \|_0 .$$

Summarizing the above, the continuity of B is obtained as follows:

$$B(v, w; t) \le M(\| v \|_{1,\omega} \| w \|_{1,\omega} + \| w \|_0 \| v \|_{1,\omega} + \| v \|_0 \| w \|_0) \le M\| v \|_{1,\omega} \| w \|_{1,\omega}.$$

Remark 3. *From Lemma 1 we know that* $\| \cdot \|_\theta = \sqrt{B(\cdot,\cdot)}$ *is a norm.*

Lemma 1, together with the theory of abstract variational inequalities, yields the following theorem.

Theorem 2. *Problem 1 has a unique solution.*

The power penalty approach

First of all, we introduce the following nonlinear variational inequality to obtain the power penalty method:

Seek $u_\lambda \in H^1_{0,\omega}(\Omega)$ to satisfy that for all $v \in H^1_{0,\omega}(\Omega)$,

$$\left(-\frac{\partial u_\lambda}{\partial t}, v - u_\lambda\right) + B(u_\lambda, v - u_\lambda; t) + j(v) - j(u_\lambda) \ge (f, v - u_\lambda) \quad \text{a.e. in } (0, T), \tag{33}$$

where

$$j(v) = \frac{\lambda k}{k+1}[v - u^*]_+^{\frac{k+1}{k}}, \quad k > 0, \quad \lambda > 1, \tag{34}$$

and $[z]_+ = \max\{0, z\}$. According to Lemma 1 and the lower semi-continuity of j, the problem has a unique solution. In addition, $j(v)$ is differentiable, which means that Eq (33) is equivalent to the following problem.

Problem 2. *Seek* $u_\lambda \in H^1_{0,\omega}(\Omega)$ *to satisfy that for all* $v \in H^1_{0,\omega}(\Omega)$,

$$\left(-\frac{\partial u_\lambda}{\partial t}, v\right) + B(u_\lambda, v; t) + (j'(u_\lambda), v) = (f, v) \quad \text{a.e.} \quad in \quad (0, T), \tag{35}$$

where

$$j'(v) = \lambda[v - u^*]_+^{\frac{1}{k}}. \tag{36}$$

Note that the penalized variational equation of variational inequality Eq (27) is just Eqs (35)–(36), and the penalized equation, which approximates Eq (25), can be written as:

$$\mathcal{L}u_\lambda + \lambda[u_\lambda - u^*]_+^{\frac{1}{k}} = f, \quad (S, \delta, t) \in \Omega \times [0, T), \tag{37}$$

with the following boundary and final conditions:

$$u_\lambda(S, \delta, t)|_\Gamma = 0 \quad \text{and} \quad u_\lambda(S, \delta, T) = u^*(S, \delta). \tag{38}$$

It has been proved in several papers [37–40] that the solution of Eq (37) converges to that of Eq (19) at the rate ($\lambda^{-k/2}$) in a Sobolev norm as $\lambda \to +\infty$.

THE FITTED FINITE VOLUME METHOD

As the valuation models for futures and European options are the special forms of Eq (37), the so-called fitted finite volume method is presented only for the American option model here, and it can be also applied to the other two models. The innovation of this method is that it combines two existing techniques, a finite volume method and a fitted approximation, together, in which the flux of a given function is approximated by a constant locally, yielding a locally nonlinear approximation to the function. In what follows, we will develop the method for our two dimensional nonlinear partial differential equation with a penalty term.

Boundary conditions

First of all, how to determine the boundary condition functions $g_1(S, t)$ and $g_2(S, t)$ is discussed below.

1. The boundary condition $g_1(S, t)$ on the boundary $\delta = \delta_{min}$ is determined by solving the following parabolic partial differential equation:

$$\begin{cases} -\dfrac{\partial F}{\partial t} - \dfrac{1}{2}\sigma_s^2 S^2 \dfrac{\partial^2 F}{\partial S^2} - h(S, \delta_{\min}, t)\dfrac{\partial F}{\partial S} + rF - \lambda[F^* - F]_+^{\frac{1}{k}} = 0, \\ F(0,t) = 0, F(S_{\max}, t) = S_{\max} - K, \\ F(S,T) = \max(S-K, 0). \end{cases} \tag{39}$$

1. The boundary condition $g_2(S, t)$ on the boundary $\delta = \delta_{\max}$ is determined by solving the following initial-boundary problem:

$$\begin{cases} -\dfrac{\partial F}{\partial t} - \dfrac{1}{2}\sigma_s^2 S^2 \dfrac{\partial^2 F}{\partial S^2} - h(S, \delta_{\max}, t)\dfrac{\partial F}{\partial S} + rF - \lambda[F^* - F]_+^{\frac{1}{k}} = 0, \\ F(0,t) = 0, F(S_{\max}, t) = S_{\max} - K, \\ F(S,T) = \max(S-K, 0). \end{cases} \tag{40}$$

The fitted finite volume method

From Eq (24) we can easily find that the problem Eqs (37)–(38) can be rewritten as

$$-F_t - \nabla \cdot (A\nabla F + \underline{b}F) + \bar{c}F - \lambda[F^* - F]_+^{\frac{1}{k}} = 0, \qquad (S, \delta, t) \in \Omega \times [0, T) \tag{41}$$

with the boundary and the terminal conditions Eqs (20)–(21). In what follows, we let $I_S = (0, S_{\max})$ and $I_\delta = (\delta_{\min}, \delta_{\max})$, and divide the intervals I_S and I_δ into N_S and N_δ sub-intervals, respectively:

$$I_{S_i} := (S_i, S_{i+1}), \quad I_{\delta_j} := (\delta_j, \delta_{j+1}), \qquad i = 0, 1, \cdots, N_S - 1, \quad j = 0, 1, \cdots, N_\delta - 1,$$

in which

$$0 = S_0 < S_1 < \cdots < S_{N_S} = S_{\max} \quad \text{and} \quad \delta_{\min} = \delta_0 < \delta_1 < \cdots < \delta_{N_\delta} = \delta_{\max}.$$

This defines a mesh on $I_S \times I_\delta$ with all mesh lines perpendicular to one of the axes.

Next we define another partition of $I_S \times I_\delta$ by letting

$$S_{i-\frac{1}{2}} = \frac{S_{i-1} + S_i}{2}, \; S_{i+\frac{1}{2}} = \frac{S_i + S_{i+1}}{2}, \; \delta_{j-\frac{1}{2}} = \frac{\delta_{j-1} + \delta_j}{2}, \; \delta_{j+\frac{1}{2}} = \frac{\delta_j + \delta_{j+1}}{2}$$

for $i = 1, 2, \cdots, N_S - 1$ and $j = 1, 2, \cdots, N_\delta - 1$, and $S_{-\frac{1}{2}} = 0, S_{N_S+\frac{1}{2}} = S_{\max}, \delta_{-\frac{1}{2}} = \delta_{\min}$, and $\delta_{N_\delta+\frac{1}{2}} = \delta_{\max}$. Also, we put $h_{S_i} = S_{i+\frac{1}{2}} - S_{i-\frac{1}{2}}$ and $h_{\delta_j} = \delta_{j+\frac{1}{2}} - \delta_{j-\frac{1}{2}}$ for each $i = 0, 1, \cdots, N_S$ and $j = 0, 1, \cdots, N_\delta$.

Integrate Eq (41) over $\mathcal{R}_{i,j} = [S_{i-\frac{1}{2}}, S_{i+\frac{1}{2}}] \times [\delta_{j-\frac{1}{2}}, \delta_{j+\frac{1}{2}}]$ to obtain

$$-\int_{S_{i-\frac{1}{2}}}^{S_{i+\frac{1}{2}}}\int_{\delta_{j-\frac{1}{2}}}^{\delta_{j+\frac{1}{2}}}\frac{\partial F}{\partial t}dSd\delta-\int_{S_{i-\frac{1}{2}}}^{S_{i+\frac{1}{2}}}\int_{\delta_{j-\frac{1}{2}}}^{\delta_{j+\frac{1}{2}}}\nabla\cdot(A\nabla F+\underline{b}F)dSd\delta$$

$$+\int_{S_{i-\frac{1}{2}}}^{S_{i+\frac{1}{2}}}\int_{\delta_{j-\frac{1}{2}}}^{\delta_{j+\frac{1}{2}}}\bar{c}FdSd\delta-\lambda\int_{S_{i-\frac{1}{2}}}^{S_{i+\frac{1}{2}}}\int_{\delta_{j-\frac{1}{2}}}^{\delta_{j+\frac{1}{2}}}[F^{*}-F]_{+}^{\frac{1}{k}}dSd\delta=0,$$

for $i = 1, 2, \cdots, N_S-1$, $j = 1, 2, \cdots, N_\delta-1$. Applying the mid-point quadrature rule to the above equation, we have

$$-\frac{\partial F_{i,j}}{\partial t}R_{i,j}-\int_{\mathcal{R}_{i,j}}\nabla\cdot(A\nabla F+\underline{b}F)dSd\delta+\bar{c}_{i,j}F_{i,j}R_{i,j}-\lambda[F_{i,j}^{*}-F_{i,j}]_{+}^{\frac{1}{k}}R_{i,j}=0 \tag{42}$$

for $i = 1, 2, \cdots, N_S-1$, $j = 1, 2, \cdots, N_\delta-1$, where $R_{i,j}=(S_{i+\frac{1}{2}}-S_{i-\frac{1}{2}})\times(\delta_{j+\frac{1}{2}}-\delta_{j-\frac{1}{2}})$ $\bar{c}_{i,j}=\bar{c}(S_i,\delta_j,t)$, $F_{i,j} = F(S_i, \delta_j, t)$, and $F_{i,j}^{*}=F^{*}(S_i,\delta_j,t)$.

Now we consider how to approximate the second term in Eq (42). First of all, it follows from the definition of flux A∇F+b−F and integration by parts that

$$\begin{aligned}\int_{R_{i,j}}\nabla\cdot(A\nabla F+\underline{b}F)dSd\delta &= \int_{\partial R_{i,j}}(A\nabla F+\underline{b}F)\cdot l ds\\ &= \int_{(S_{i+\frac{1}{2}},\delta_{j-\frac{1}{2}})}^{(S_{i+\frac{1}{2}},\delta_{j+\frac{1}{2}})}(a_{11}F_S+a_{12}F_\delta+b_1F)d\delta\\ &- \int_{(S_{i-\frac{1}{2}},\delta_{j-\frac{1}{2}})}^{(S_{i-\frac{1}{2}},\delta_{j+\frac{1}{2}})}(a_{11}F_S+a_{12}F_\delta+b_1F)d\delta\\ &+ \int_{(S_{i-\frac{1}{2}},\delta_{j+\frac{1}{2}})}^{(S_{i+\frac{1}{2}},\delta_{j+\frac{1}{2}})}(a_{21}F_S+a_{22}F_\delta+b_2F)dS\\ &- \int_{(S_{i-\frac{1}{2}},\delta_{j-\frac{1}{2}})}^{(S_{i+\frac{1}{2}},\delta_{j-\frac{1}{2}})}(a_{21}F_S+a_{22}F_\delta+b_2F)dS,\end{aligned} \tag{43}$$

where l denote the unit vector outward-normal to $\partial\mathcal{R}_{ij}$.

Next we deal with Eq (43) term by term. In fact, for the first term we can approximate the integral by a constant, i.e,

$$\int_{(S_{i+\frac{1}{2}},\delta_{j-\frac{1}{2}})}^{(S_{i+\frac{1}{2}},\delta_{j+\frac{1}{2}})}(a_{11}F_S+a_{12}F_\delta+b_1F)d\delta\approx(a_{11}F_S+a_{12}F_\delta+b_1F)\big|_{\left(S_{i+\frac{1}{2}},\delta_j\right)}\cdot h_{\delta_j}.$$

Clearly, we now need to derive approximations of the (A∇F+b−F)·l defined above at the mid-point, (Si+12, δ j), on the interval I_{Si} for any $i = 0, 1, \cdots, N_S-1$. Next we discuss it in detail.

Case 1: For $i \geq 1$.

According to Eq (23) we have

$$\begin{aligned} a_{11}F_S + a_{12}F_\delta + b_1 F &= \tfrac{1}{2}\sigma_s^2 S^2 F_S + \tfrac{1}{2}\rho_{sc}\sigma_s\sigma_c S F_\delta + S(k_s(n(t) - \ln S) - \delta - \lambda_s\sigma_s - \sigma_s^2)F \\ &= S(\tfrac{1}{2}\sigma_s^2 S F_S + \tfrac{1}{2}\rho_{sc}\sigma_s\sigma_c F_\delta + (k_s(n(t) - \ln S) - \delta - \lambda_s\sigma_s - \sigma_s^2)F). \end{aligned} \tag{44}$$

Following the idea in [29], we approximate the term $a_{11} F_S + b_1 F$ by solving the following two-point boundary value problem:

$$\left(\frac{1}{2}\sigma_s^2 S F_S + (k_s(n(t) - \ln S) - \delta - \lambda_s\sigma_s - \sigma_s^2)F\right)' \equiv \left(aSF_S + b_{i+\frac{1}{2},j}F\right)' = 0, \tag{45a}$$

$$F(S_i, \delta_j) = F_{i,j}, \quad F(S_{i+1}, \delta_j) = F_{i+1,j}, \tag{45b}$$

where $a = \frac{1}{2}\sigma_s^2$ and $b = k_s(n(t) - lnS) - \delta - \lambda_s\sigma_s - \sigma_s^2$, $b_{i+\frac{1}{2},j} = b(S_{i+\frac{1}{2}}, \delta_j)$, and $F_{i,j}$ and $F_{i+1,j}$ denote the values of F at (S_i, δ_j) and (S_{i+1}, δ_j), respectively. Evidently, from Eq (45a) we can obtain

$$aSF_S + b_{i+\frac{1}{2},j}F = C_1,$$

where C_1 is an arbitrary constant and can be determined by the boundary condition Eq (45b) as follows [29, 41]:

$$C_1 = b_{i+\frac{1}{2}j}\frac{S_{i+1}^{\alpha_{i,j}}F_{i+1,j} - S_i^{\alpha_{i,j}}F_{i,j}}{S_{i+1}^{\alpha_{i,j}} - S_i^{\alpha_{i,j}}},$$

where $\alpha_{i,j} = \frac{b_{i+\frac{1}{2},j}}{a}$. Then,

$$a_{11}F_S + a_{12}F_\delta + b_1 F \approx S_{i+\frac{1}{2}}\left(b_{i+\frac{1}{2}j}\frac{S_{i+1}^{\alpha_{i,j}}F_{i+1,j} - S_i^{\alpha_{i,j}}F_{i,j}}{S_{i+1}^{\alpha_{i,j}} - S_i^{\alpha_{i,j}}} + d \cdot F_\delta\right), \tag{46}$$

where $d = \frac{1}{2}\rho_{sc}\sigma_s\sigma_c$. Additionally, the derivative F_δ can be approximated by a forward difference

$$\frac{F_{i,j+1} - F_{i,j}}{h_{\delta_j}}.$$

Then, we have

$$(a_{11}F_S + a_{12}F_\delta + b_1 F)|_{(S_{i+\frac{1}{2}},\delta_j)} \cdot h_{\delta_j} \approx S_{i+\frac{1}{2}}\left(b_{i+\frac{1}{2}j}\frac{S_{i+1}^{\alpha_{i,j}}F_{i+1,j} - S_i^{\alpha_{i,j}}F_{i,j}}{S_{i+1}^{\alpha_{i,j}} - S_i^{\alpha_{i,j}}} + d_{i,j}\frac{F_{i,j+1} - F_{i,j}}{h_{\delta_j}}\right) \cdot h_{\delta_j}, \tag{47}$$

where $d_{ij} = d(S_i, \delta_j)$. In the same way, we can also approximate the second term in Eq (43) as follows:

$$(a_{11}F_S + a_{12}F_\delta + b_1F)|_{(S_{i-\frac{1}{2}},\delta_j)} \cdot h_{\delta_j} \approx S_{i-\frac{1}{2}}\left(b_{i-\frac{1}{2},j}\frac{S_i^{\alpha_{i-1,j}}F_{i,j} - S_{i-1}^{\alpha_{i-1,j}}F_{i-1,j}}{S_i^{\alpha_{i-1,j}} - S_{i-1}^{\alpha_{i-1,j}}} + d_{i,j}\frac{F_{i,j+1} - F_{i,j}}{h_{\delta_j}}\right) \cdot h_{\delta_j}. \tag{48}$$

Case 2: For $i = 0$.

As the Eq (45a) is degenerate on $(0, S_1)$, we should rethink the problem Eqs (45a)–(45b) on $(0, S_1)$ below:

$$\left(aSF_S + b_{\frac{1}{2},j}F\right)' = C_2, \tag{49a}$$

$$F(0,\delta_j) = F_{0,j}, \quad F(S_1,\delta_j) = F_{1,j}, \tag{49b}$$

where $b_{\frac{1}{2},j} = b(S_{\frac{1}{2}},\delta_j)$ and C_2 is an unknown constant to be determined next. After integrating both sides of above equation, we can obtain

$$aSF_S + b_{\frac{1}{2},j}F - C_2S + C_3.$$

Thus, we have

$$(a_{11}F_S + a_{12}F_\delta + b_1F)|_{(S_{\frac{1}{2}},\delta_j)} \cdot h_{\delta_j} \approx \left\{\frac{1}{2}\left[\left(a + b_{\frac{1}{2},j}\right)F_{1,j} - \left(a - b_{\frac{1}{2},j}\right)F_{0,j}\right] + d_{1,j}\frac{h_{\delta_j}}{F_{1,j+1} - F_{1,j}}\right\} \cdot h_{\delta_j}. \tag{50}$$

In the approximations of last two terms of Eq (43), we do not need to consider the case 2 as before, since $\delta_0 \neq 0$. We apply the similar method to the above and get the following results:

$$(a_{21}F_S + a_{22}F_\delta + b_2F)|_{(S_i,\delta_{j+\frac{1}{2}})} \cdot h_{S_i} \approx \left(\bar{b}_{i,j+\frac{1}{2}}\frac{e^{\bar{\alpha}_{i,j}\delta_{j+1}}F_{i,j+1} - e^{\bar{\alpha}_{i,j}\delta_j}F_{i,j}}{e^{\bar{\alpha}_{i,j}\delta_{j+1}} - e^{\bar{\alpha}_{i,j}\delta_j}} + \bar{d}_{i,j}\frac{F_{i+1,j} - F_{i,j}}{h_{S_i}}\right) \cdot h_{S_i}; \tag{51}$$

$$(a_{21}F_S + a_{22}F_\delta + b_2F)|_{(S_i,\delta_{j-\frac{1}{2}})} \cdot h_{S_i} \approx \left(\bar{b}_{i,j-\frac{1}{2}}\frac{e^{\bar{\alpha}_{i,j-1}\delta_j}F_{i,j} - e^{\bar{\alpha}_{i,j-1}\delta_{j-1}}F_{i,j-1}}{e^{\bar{\alpha}_{i,j-1}\delta_j} - e^{\bar{\alpha}_{i,j-1}\delta_{j-1}}} + \bar{d}_{i,j}\frac{F_{i+1,j} - F_{i,j}}{h_{S_i}}\right) \cdot h_{S_i}, \tag{52}$$

for $j = 0, 1, \cdots, N_\delta - 1$, where $\bar{\alpha}_{i,j} = \frac{\bar{b}_{i,j+\frac{1}{2}}}{\bar{a}}$, $\bar{a} = \frac{1}{2}\sigma_c^2$, $\bar{b} = k_c(\theta_c - \delta) - \lambda_c\sigma_c - \frac{1}{2}\rho_{sc}\sigma_s\sigma_c$, and $\bar{d}_{i,j} = \frac{1}{2}\rho_{sc}\sigma_s\sigma_c S_i$. Hence, we obtain the following equations by combining Eqs (43), (47), and (48) with Eqs (50)–(52):

$$-\frac{\partial F_{i,j}}{\partial t}R_{i,j} + e_{i-1,j}^{i,j}F_{i-1,j} + e_{i,j-1}^{i,j}F_{i,j-1} + e_{i,j}^{i,j}F_{i,j} + e_{i,j+1}^{i,j}F_{i,j+1}$$

$$+e_{i+1,j}^{i,j}F_{i+1,j} \quad \lambda[F_{i,j}^* \quad F_{i,j}]^{\frac{1}{k}}_{+}R_{i,j} = 0, \tag{53}$$

for $i = 1, 2, \cdots, N_S - 1$ and $j = 1, 2, \cdots, N_\delta - 1$, where

$$e_{0,j}^{1,j} = -\tfrac{1}{2} S_{\frac{1}{2}} h_{\delta_j} \left(a - b_{\frac{1}{2},j}\right), \quad e_{1,j-1}^{1,j} = -\frac{\bar{b}_{1,j-\frac{1}{2}} e^{\bar{\alpha}_{1,j-1}\delta_{j-1}} h_{S_1}}{e^{\bar{\alpha}_{1,j-1}\delta_j} - e^{\bar{\alpha}_{1,j-1}\delta_{j-1}}}, \tag{54}$$

$$\begin{aligned} e_{1,j}^{1,j} &= h_{S_1}\left(\frac{\bar{b}_{1,j+\frac{1}{2}} e^{\bar{\alpha}_{1,j}\delta_j}}{e^{\bar{\alpha}_{1,j}\delta_{j+1}} - e^{\bar{\alpha}_{1,j}\delta_j}} + \frac{\bar{b}_{1,j-\frac{1}{2}} e^{\bar{\alpha}_{1,j-1}\delta_j}}{e^{\bar{\alpha}_{1,j-1}\delta_j} - e^{\bar{\alpha}_{1,j-1}\delta_{j-1}}} + d_{1,j}\right) \\ &\quad + S_{\frac{3}{2}} h_{\delta_j} \frac{b_{\frac{3}{2},j} S_1^{\alpha_{1,j}}}{S_2^{\alpha_{1,j}} - S_1^{\alpha_{1,j}}} + \frac{1}{2} S_{\frac{1}{2}} h_{\delta_j}\left(a + b_{\frac{1}{2},j}\right) + \bar{c}_{1,j} R_{1,j}, \end{aligned} \tag{55}$$

$$e_{1,j+1}^{1,j} = -h_{S_1}\left(\frac{\bar{b}_{1,j+\frac{1}{2}} e^{\bar{\alpha}_{1,j}\delta_{j+1}}}{e^{\bar{\alpha}_{1,j}\delta_{j+1}} - e^{\bar{\alpha}_{1,j}\delta_j}} + d_{1,j}\right), \quad e_{2,j}^{1,j} = -S_{\frac{3}{2}} b_{\frac{3}{2},j} \frac{S_2^{\alpha_{1,j}} h_{\delta_j}}{S_2^{\alpha_{1,j}} - S_1^{\alpha_{1,j}}}, \tag{56}$$

for $j = 1, 2, \cdots, N_\delta - 1$, and

$$e_{i-1,j}^{i,j} = -S_{i-\frac{1}{2}} h_{\delta_j} \frac{b_{i-\frac{1}{2},j} S_{i-1}^{\alpha_{i-1,j}}}{S_i^{\alpha_{i-1,j}} - S_{i-1}^{\alpha_{i-1,j}}}, \quad e_{i,j-1}^{i,j} = -\frac{\bar{b}_{i,j-\frac{1}{2}} e^{\bar{\alpha}_{i,j-1}\delta_{j-1}} h_{S_i}}{e^{\bar{\alpha}_{i,j-1}\delta_j} - e^{\bar{\alpha}_{i,j-1}\delta_{j-1}}}, \tag{57}$$

$$\begin{aligned} e_{1,j}^{1,j} &= h_{S_i}\left(\frac{\bar{b}_{i,j+\frac{1}{2}} e^{\bar{\alpha}_{i,j}\delta_j}}{e^{\bar{\alpha}_{i,j}\delta_{j+1}} - e^{\bar{\alpha}_{i,j}\delta_j}} + \frac{\bar{b}_{i,j-\frac{1}{2}} e^{\bar{\alpha}_{i,j-1}\delta_j}}{e^{\bar{\alpha}_{i,j-1}\delta_j} - e^{\bar{\alpha}_{i,j-1}\delta_{j-1}}} + d_{i,j}\right) \\ &\quad + h_{\delta_j}\left(S_{i+\frac{1}{2}} \frac{b_{i+\frac{1}{2},j} S_i^{\alpha_{i,j}}}{S_{i+1}^{\alpha_{i,j}} - S_i^{\alpha_{i,j}}} + S_{i-\frac{1}{2}} \frac{b_{i-\frac{1}{2},j} S_i^{\alpha_{i-1,j}}}{S_i^{\alpha_{i-1,j}} - S_{i-1}^{\alpha_{i-1,j}}}\right) + \bar{c}_{i,j} R_{i,j}, \end{aligned} \tag{58}$$

$$e_{i,j+1}^{i,j} = -h_{S_i}\left(\frac{\bar{b}_{i,j+\frac{1}{2}} e^{\bar{\alpha}_{i,j}\delta_{j+1}}}{e^{\bar{\alpha}_{i,j}\delta_{j+1}} - e^{\bar{\alpha}_{i,j}\delta_j}} + d_{i,j}\right), \quad e_{i+1,j}^{i,j} = -S_{i+\frac{1}{2}} b_{i+\frac{1}{2},j} \frac{S_{i+1}^{\alpha_{i,j}} h_{\delta_j}}{S_{i+1}^{\alpha_{i,j}} - S_i^{\alpha_{i,j}}}, \tag{59}$$

for $i = 2, 3, \cdots, N_S - 1$, $j = 1, 2, \cdots, N_\delta - 1$, and $e^{i,j}_{m,n} = 0$ if $m \neq i-1, i, i+1$ and $n \neq j-1, j, j+1$. It can be easily seen that Eq (53) is an $(N_S - 1)^2 \times (N_\delta - 1)^2$ linear system of equations for

$$F = (F_{1,1}, \cdots, F_{1,N_\delta - 1}, F_{2,1}, \cdots, F_{2,N_\delta - 1}, \cdots, F_{N_S - 1,1}, F_{N_S - 1,2}, \cdots, F_{N_S - 1,N_\delta - 1})^{\top}.$$

Note that for $i = 1, 2, \cdots, N_S$ and $j = 1, 2, \cdots, N_\delta$, $F_{0,j}(t)$, $F_{i,0}(t)$, $F_{0,N\delta}(t)$ and $F_{NS},0(t)$ are equal to the given boundary conditions. Obviously, the coefficient matrix of Eq (53) is penta-diagonal.

Let

$$E_{i,j} = (0, \cdots, 0, e_{i-1,j}^{i,j}, 0, \cdots, 0, e_{i,j-1}^{i,j}, e_{i,j}^{i,j}, e_{i,j+1}^{i,j}, 0, \cdots, 0, e_{i+1,j}^{i,j}, 0, \cdots, 0)$$

for $i = 1, 2, \cdots, N_S - 1$ and $j = 1, 2, \cdots, N_\delta - 1$. Now we are in the position to discuss the time-discretization of the system Eq (53). To this purpose, we first rewrite Eq (53) as

$$-\frac{\partial F_{i,j}}{\partial t} R_{i,j} + E_{i,j}F + p(F_{i,j}) = 0, \tag{60}$$

where

$$p(F_{i,j}) = -\lambda R_{i,j}[F^*_{i,j} - F_{i,j}]_+^{\frac{1}{k}}. \tag{61}$$

Then, we select $M-1$ points numbered from t_1 to t_{M-1} between 0 and T and let $T = t_0$, $t_M = 0$ to form a partition: $T = t_0 > t_1 > \cdots > t_M = 0$. Thus, the full discrete form of Eq (60) can be obtained by applying the two-level implicit time-stepping method with a splitting parameter $\theta \in [12,1]$ to it as follows:

$$(\theta E^{m+1} + G^m)F^{m+1} + \theta D(F^{m+1}) = (G^m - (1-\theta)E^m)F^m - (1-\theta)D(F^m), \tag{62}$$

where

$$\begin{aligned} F^m &= (F^m_{1,1}, \cdots, F^m_{1,N_\delta-1}, F^m_{2,1}, \cdots, F^m_{2,N_\delta-1}, \cdots, F^m_{N_S-1,1}, \cdots, F^m_{N_S-1,N_\delta-1})^\top, \\ E^m &= (E^m_{1,1}, \cdots, E^m_{1,N_\delta-1}, E^m_{2,1}, \cdots, E^m_{2,N_\delta-1}, \cdots, E^m_{N_S-1,1}, \cdots, E^m_{N_S-1,N_\delta-1})^\top, \\ G^m &= \operatorname{diag}(-R_{1,1}/\Delta t_m, \cdots, -R_{N_S-1,N_\delta-1}/\Delta t_m)^\top, \\ D(F^m) &= (p(F^m_{1,1}), \cdots, p(F^m_{N_S-1,N_\delta-1}))^\top, \end{aligned} \tag{63}$$

for $m = 0, 1, \cdots, M-1$. Note that $\Delta t_m = t_{m+1} - t_m < 0, E^m_{i,j} = E_{i,j}(t_m)$, and F^m denotes the approximation of F at $t = t_m$.

The solution of the discrete system

The standard Newton method is employed to solve the nonlinear discrete system Eq (62). Eq (61) clearly indicates that $p'(V^m_{i,j}) \to \infty$ as $F^*_{i,j} - F_{i,j} \to 0^+$ when $k > 1$. So, we need to smooth out $p(F^m_{i,j})$ by redefining it as follows:

$$p(F^m_{i,j}) = \begin{cases} \lambda R_{i,j}[F^*_{i,j} - F_{i,j}]_+^{\frac{1}{k}}, & F^*_{i,j} - F^m_{i,j} \geq \epsilon, \\ -\lambda R_{i,j}\Big(\epsilon^{\frac{1}{k}-n+1}(n - \frac{1}{k})[F^*_{i,j} - F^m_{i,j}]_+^{n-1} & \\ +\epsilon^{\frac{1}{k}-n}(\frac{1}{k} - n + 1)[F^*_{i,j} - F^m_{i,j}]_+^n\Big), & F^*_{i,j} - F^m_{i,j} < \epsilon, \end{cases} \tag{64}$$

where $k > 0$, n is a positive integer, and $0 < \ \ll 1$ is a transition parameter. Applying Newton method to Eq (62), we can get

$$[\theta E^{m+1} + G^m + \theta J_D(\omega^{q-1})]d\omega^q = [G^m - (1-\theta)E^m]F^m - (1-\theta)D(F^m) - (\theta E^{m+1} + G^m)\omega^{q-1} - \theta D(\omega^{q-1}),$$

$$\omega^q = \omega^{q-1} + \gamma \cdot d\omega^q, \tag{65}$$

for $q = 1, 2, \cdots$, where ω^0 is a given initial guess, $J_D(\omega)$ denotes the Jacobian matrix of the column vector $D(\omega)$, and $\gamma \in (0, 1]$ is a damping parameter used to accelerate convergence. Then, we choose

$$F^{m+1} = \lim_{q\to\infty} \omega^q,$$

as the solution of Eq (65).

Remark 4. *For the European options and futures, the system* Eq (62) *degenerates to a linear system and can be solved using normal methods.*

What is more, we have the following theorem.

Theorem 3. *For* $m = 1, 2, \cdots, M-1$, *if* $|\Delta t_m|$ *is sufficiently small and* $\bar{c} \geq 0$, *, then the system matrix of* Eq (65) *is an M-matrix.*

Proof. From the definition of $D(F)$ in Eq (63), it is easy to see that its Jacobian is the following diagonal matrix:

$$J_D(\omega^l) = \text{diag}(p'(F^m_{1,1}), \cdots, p'(F^m_{N_S-1,N_\delta-1})).$$

From Eq (64) we know that $p'(F^m_{i,j}) \geq 0$ for all $i = 1, \cdots, N_S-1$ and $j = 1, \cdots, N_\delta-1$. Thus, to show that the system matrix of Eq (65) is an M-matrix, it suffices to show that $\theta E^{m+1} + G^m$ is an M-matrix.

First, we note that $e^{i,j}_{m,n} \leq 0$ for all $m \neq i$, $n \neq j$, since

$$\frac{b_{i+\frac{1}{2},j}}{S_{i+1}^{\alpha_{i,j}} - S_i^{\alpha_{i,j}}} > 0, \quad \frac{\bar{b}_{i,j+\frac{1}{2}}}{e^{\bar{\alpha}_{i,j}Y_{j+1}} - e^{\bar{\alpha}_{i,j}Y_j}} > 0 \tag{66}$$

for any i and j, and for any $\alpha = b/a$ and any $\bar{\alpha} = \bar{b}/\bar{a}$. This is because the function S^α is increasing when $b > 0$ and decreasing when $b < 0$, and the function $e^{\bar{\alpha}\delta}$ is increasing when $\bar{b} > 0$ and decreasing when $\bar{b} < 0$. Eq (66) also holds when $b_{i+\frac{1}{2},j} \to 0, \bar{b}_{i,j+\frac{1}{2}} \to 0$. Furthermore, from Eqs (54)–(56) we know that when $\bar{c}_{i,j} \geq 0$, for all $i = 1, \cdots, N_S-1$, $j = 1, \cdots, N_\delta-1$, there holds

$$\begin{aligned}(e_{i,j}^{i,j})^{m+1} &\geq |(e_{i-1,j}^{i,j})^{m+1}| + |(e_{i,j-1}^{i,j})^{m+1}| + |(e_{i,j+1}^{i,j})^{m+1}| + |(e_{i+1,j}^{i,j})^{m+1}| + \bar{c}_{i,j}^{m+1} R_{i,j} \\ &= \sum_{p=1}^{N_S-1} \sum_{q=1}^{N_\delta-1} |(e_{p,q}^{i,j})^{m+1}| + \bar{c}_{i,j}^{m+1} R_{i,j}.\end{aligned}$$

Therefore, E^{m+1} is a diagonally dominant with respect to its columns. Hence, from the above analysis, we see that for all admissible i, j, E^{m+1} is a diagonally dominant matrix with positive diagonal elements and non-positive off-diagonal elements. This implies that E^{m+1} is an M-matrix.

Second, G^m of the system matrix Eq (65) is a diagonal matrix with positive diagonal entries. In fact, when $|\Delta t_m|$ is sufficiently small, we have

$$\theta \bar{c}_{i,j} R_{i,j} + \frac{R_{i,j}}{-\Delta t_m} > 0,$$

which demonstrates that $\theta E^{m+1} | G^m$ is an M-matrix.

Here we emphasize that Theorem 3 implies that the fully discrete system Eq (65) satisfies the discrete maximum principle and the discretization is monotone, such that Eq (65) has a unique solution.

NUMERICAL RESULTS

In this section, some numerical results are presented to demonstrate the efficiency and the usefulness of the numerical method proposed in the above. Furthermore, the varieties of derivatives prices with respect to the spot price, convenience and time, are examined. These are of interest to market participants.

Test 1: The European call option with the cash-or-nothing final condition Eq (13).

Parameters: $S_{max} = 500$, $\delta_{min} = -1$, $\delta_{max} = -0.25$, $T = 1$, $r = 0.1$, $_{sc} = 0.9$, $\sigma_s = 0.3$, $\lambda_s = -0.2$, $\sigma_c = 0.3$, $\lambda_c = -0.3$, $k_s = 0.04$, $k_c = 0.02$, $\theta_c = -0.2$, $\xi = 0.03$, $C_s = 21$, $K = 200$, $B = 100$.

To solve the pricing problem numerically, we divide $(0, S_{max})$, $(\delta_{min}, \delta_{max})$, and $(0, T)$ uniformly into 50, 50, and 50 sub-intervals, respectively. The boundary conditions are given by Eqs (14), (39), and (40), and the numerical values of these boundary conditions determined by 1D initial-boundary problems are plotted in Fig 1. By means of these initial and boundary conditions, the European option values for Test 1 are computed and the intersection surfaces of different time points are plotted in Fig 2.

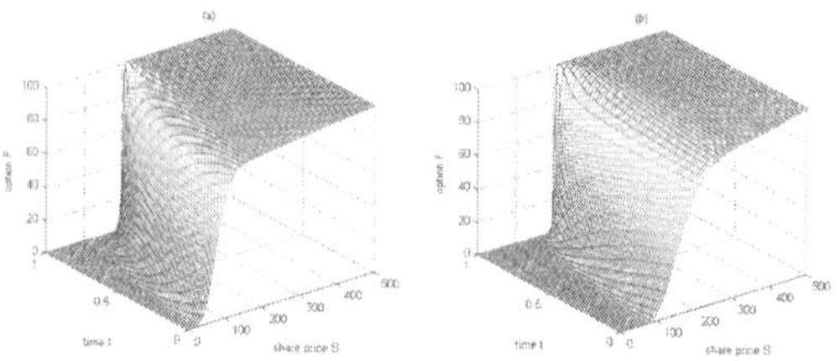

Figure 1. The boundary conditions at $\delta = \delta_{min}$ **and** $\delta = \delta_{max}$ **for Test 1.**

(a) the boundary condition $F(S, \delta_{min}, t)$; (b) the boundary condition $F(S, \delta_{max}, t)$

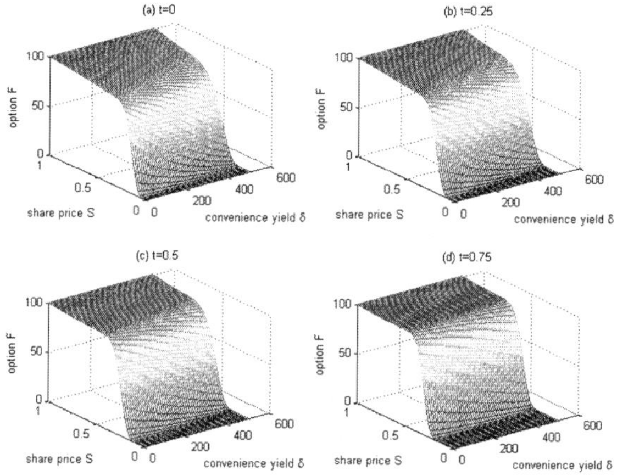

Figure 2. The European cash-or-nothing option values at the time points for Test 1.

We show some option values at some special points in Table 1.

Table 1. The European cash-or-nothing option values for some points.

t	$t = 0$	$t = 0.25$	$t = 0.5$	$t = 0.75$
(S, δ)				
(240, −1)	93.1038	93.6169	95.0070	96.9608
(240, −0.64)	90.1785	91.4336	93.3164	95.4086
(240, −0.25)	82.5970	83.9493	86.1757	89.9564

Next the convergence rates of the discretization method is gauged. To this end, we define three discrete norms $\|\cdot\|_{1,hS}$, $\|\cdot\|_{1,h\delta}$, and $\|\cdot\|_{0,h}$ as follows [42]:

$$\| v_h \|_{1,h_S}^2 = \sum_{i=1}^{N_S-1} \sum_{j=1}^{N_\delta-1} S_{i+1/2}^2 b_{i+1/2,j} h_{\delta_j} \frac{S_{i+1}^{\alpha_{i,j}} + S_i^{\alpha_{i,j}}}{S_{i+1}^{\alpha_{i,j}} - S_i^{\alpha_{i,j}}} (v_{i+1,j} - v_{i,j})^2,$$

$$\| v_h \|_{1,h_\delta}^2 = \sum_{i=1}^{N_S-1} \sum_{j=1}^{N_\delta-1} \delta_{j+1/2}^2 \bar{b}_{i+1/2,j} h_{S_i} \frac{\delta_{j+1}^{\bar{\alpha}_{i,j}} + \delta_j^{\bar{\alpha}_{i,j}}}{\delta_{j+1}^{\bar{\alpha}_{i,j}} - \delta_j^{\bar{\alpha}_{i,j}}} (v_{i,j+1} - v_{i,j})^2,$$

$$\| v_h \|_{0,h}^2 = \sum_{i=1}^{N_S-1} \sum_{j=1}^{N_\delta-1} v_{i,j}^2 R_{i,j},$$

from which we can define the following weighted discrete H^1-norm:

$$\| v_h \|_{H^1}^2 = \| v_h \|_{1,h_S}^2 + \| v_h \|_{1,h_\delta}^2 + \| v_h \|_{0,h}^2 .$$

In addition, the ratio is defined as:

$$\text{ratio} = \frac{\| F_h^{\Delta t} - F \|_\gamma}{\| F_{h/2}^{\Delta t/2} - F \|_\gamma} \quad \text{with} \quad \gamma = \infty \text{ or } H^1.$$

In Test 1, we employ the numerical solution on the mesh with $N_S = 128 = N_\delta$ and $M = 128$ as the "exact" solution F, and list the errors in the discrete L^∞-norm and the weighted discrete H^1-norm at the final time step $t = 0$ for four consecutive meshes in Table 2. Moreover, the linear regression is used to show that these data obey the basic error estimates as follows:

$$\| F - F_h \|_\infty \approx 0.3136 h^{1.0597} \quad \text{and} \quad \| F - F_h \|_{H^1} \approx 2.0922 h^{1.6493}.$$

Note that these results are reasonable because of the non-smoothness of the solution due to the pay-off function Eq (13).

Table 2. Computed errors in the L^∞-norm and the H^1-norm at t **= 0.**

mesh	L^∞-norm	ratio	H^1-norm	ratio
4×4×4	67.1832		6.2069e+003	
8×8×8	17.1719	3.9124	1.6034e+003	3.8711
16×16×16	10.8381	1.5844	712.2445	2.2512
32×32×32	8.1675	1.3270	204.2685	3.4868
64×64×64	2.4752	3.2997	57.2606	3.5673

From Figs 1, 2, and Table 1 we can conclude that

1. The European cash-or-nothing option values are higher when the option comes to maturity. This is acceptable in that the European option can be only exercised at maturity, and one cannot get more benefits when holding a European option that is not closed to maturity, which is different from the American option.
2. The greater convenience yield δ, the lower the option value. This can be explained as follows: since the convenience yield δ measures the benefits from holding the carbon emission permits rather than the derivatives contract, people will hold more carbon emissions spots when facing a higher convenience yield, which leads to the demand reduction for derivatives and the option price down.
3. As the value of carbon emission spots rises, the value of call option also rises. This is because the holder of call option will get more chances for benefits from the derivatives. This result is similar to the general call options.

Test 2: The American call option with the ramp payoff final condition Eq (12).

Parameters: $S_{max} = 500$, $\delta_{min} = -1$, $\delta_{max} = -0.25$, $T = 1$, $r = 0.1$, $_{sc} = 0.9$, $\sigma_s = 0.3$, $\lambda_s = -0.2$, $\sigma_c = 0.3$, $\lambda_c = -0.3$, $k_s = 0.04$, $k_c = 0.02$, $\theta_c = -0.2$, $\xi = 0.03$, $C_s = 21$, $K = 200$, $\lambda = 100$, $k = 16$.

To solve the American call option with the above parameters in Test 2, we divide $(0, S_{max})$, $(\delta_{min}, \delta_{max})$, and $(0, T)$ uniformly into 30, 30, and 30 sub-intervals, respectively. The final and boundary conditions are given by Eqs (20), (21), (39), and (40), and the numerical values of these boundary conditions determined by the 1D initial-value problems are plotted in Fig 3. According to these initial and boundary conditions, the American call option problem for Test 2 is solved and the intersection surfaces of different time points are depicted in Fig 4. We also show some option values at some special points in Table 3.

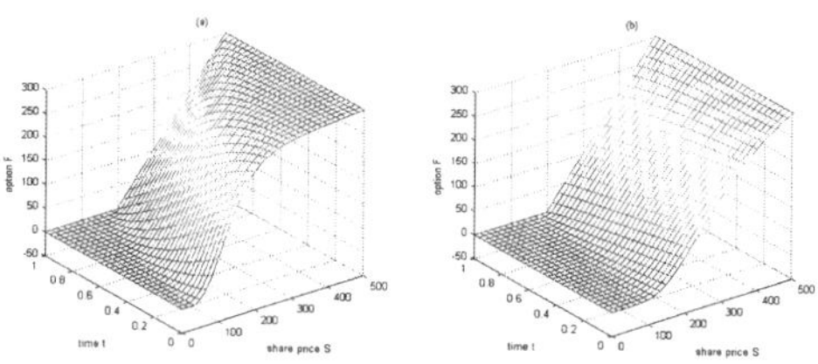

Figure 3. The boundary conditions at $\delta = \delta_{min}$ **and** $\delta = \delta_{max}$ **for Test 2.**

(a) the boundary condition $F(S, \delta_{min}, t)$; (b) the boundary condition $F(S, \delta_{max}, t)$

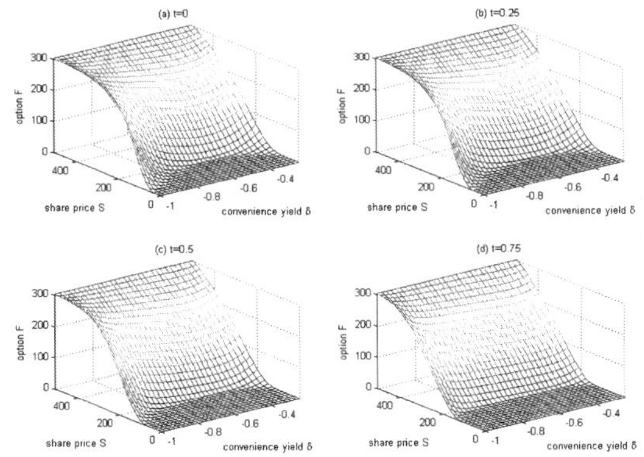

Figure 4. American call option values at time points for Test 2.

Table 3. American call option values for some points.

t / (S, δ)	$t = 0$	$t = 0.25$	$t = 0.5$	$t = 0.75$
(100, −0.6)	0	0	0	0
(200, −0.6)	48.3564	36.5279	21.7587	8.9085
(300, −0.6)	158.9589	152.3870	139.0894	115.0307
(200, −1)	240.3643	206.7208	137.2206	54.2146
(200, −0.6)	48.3564	36.5279	21.7587	10.5838
(200, −0.3)	50.8563	39.9947	26.0251	11.6181

From Figs,3, 4 and Table 3, we have the conclusions as follows:

1. The American vanilla call option values are lower when the option comes to maturity. An American option can be exercised at any time between the date of purchase and the expiration date. As the option comes to maturity, the benefits from the early exercise of option are reduced, and then the option prices decrease. This is the same as the popular financial option contracts.
2. Unlike European options, for American options the greater convenience yield δ does not mean a lower option value (see the last two lines of Table 3). As we all have known, the value of American option contains an early exercise premium. Therefore, although the greater convenience yield reduces the demand for

derivatives and decreases the option price, the early exercise premium may also increase the option price oppositely. So, there is no certain relationship between the convenience yield and the option value.

3. As the value of carbon emission spots rises, the value of the call option also goes up. This is because the holder of the call option can get more chances to benefit from it. This result is similar to general call options.

Financial engineers could pay more attention to the optimal exercise boundary compared with the option value in the case of American option. We determine the boundary $S^*(t)$ by Eq (16), and the results are plotted in Fig 5. The blue line is the optimal exercise boundary, the left of the blue line is the continuation region, and the right of the blue line is the stopping region. We can clearly see that as the convenience yield increases, the stopping region becomes smaller. Note that our pricing model is different from the classical Black-Scholes model in details, and this result should be reasonable.

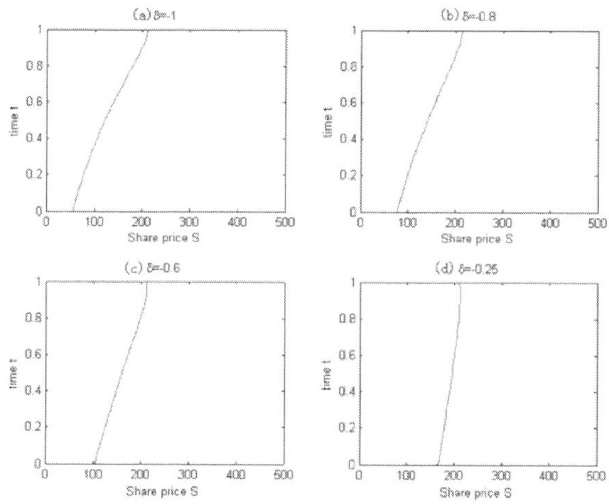

Figure 5. Optimal exercise boundaries about different δ **for Test 2.**

Test 3: The futures with the final condition Eq (17).

Parameters: $S_{max} = 500$, $\delta_{min} = -1$, $\delta_{max} = -0.25$, $T = 1$, $r = 0.1$, ${}_{sc} = 0.9$, $\sigma_s = 0.3$, $\lambda_s = -0.2$, $\sigma_c = 0.3$, $\lambda_c = -0.3$, $k_s = 0.04$, $k_c = 0.02$, $\theta_c = -0.2$, $\xi = 0.03$, $C_s = 21$.

To solve the futures with the above parameters in Test 3, we divide $(0, S_{max})$, $(\delta_{min}, \delta_{max})$, and $(0, T)$ uniformly into 50, 50, and 50 sub-intervals, respectively. The boundary conditions are given by Eqs (14), (39), and (40), and the numerical values of these boundary conditions determined by the 1D initial-value problems are plotted in Fig 6. According to these initial and boundary conditions, the futures problem for Test 3 is solved, and the intersection surfaces of different time points are depicted in Fig 7.

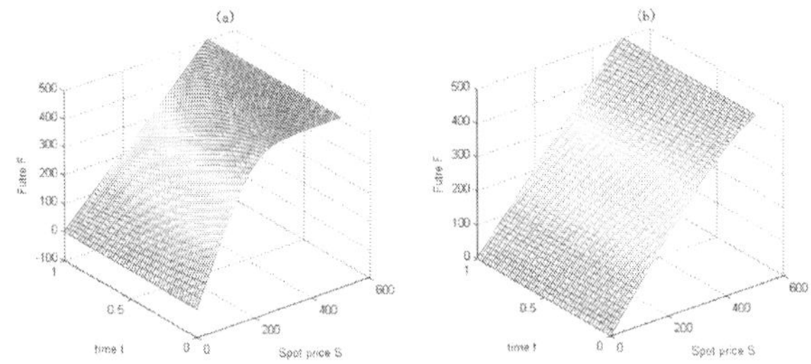

Figure 6. The boundary conditions at $\delta = \delta_{min}$ **and** $\delta = \delta_{max}$ **for Test 3.**

(a) the boundary condition $F(S, \delta_{min}, t)$; (b) the boundary condition $F(S, \delta_{max}, t)$

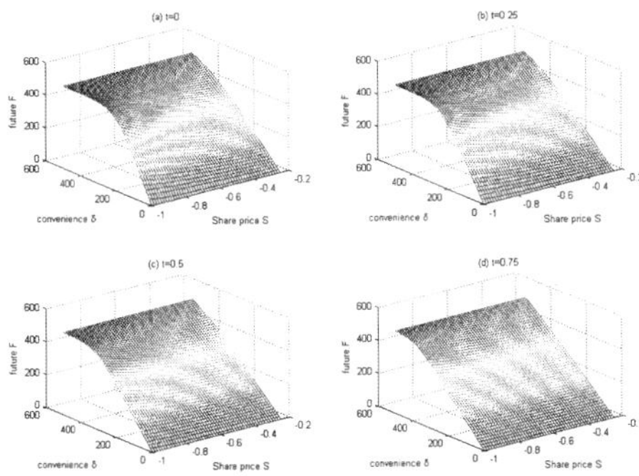

Figure 7. Futures values at time points for Test 3.

Some futures values at some special points are shown in Table 4.

Table 4. Futures values for some points.

t (S, δ)	$t = 0$	$t = 0.25$	$t = 0.5$	$t = 0.75$
(250, 1)	477.8765	451.4826	393.4370	322.6292
(250, −0.82)	410.0061	372.1902	327.5114	282.7984
(250, −0.61)	354.9645	322.1501	293.8633	267.3085

From Figs 6, 7 and Table 4, we have the conclusions as follows:

1. The futures prices are lower when the futures contract comes to the maturity. This is because the futures prices converge to the spot prices when the futures contract comes to the maturity. With time approaching to the maturity date, the uncertainty, which should increase the futures prices, is reducing.
2. The greater the convenience yield δ, the lower the futures value. This can be explained as follows: since the convenience yield δ measures the benefits from holding the carbon emission permits rather than the derivatives contract, people will hold more carbon emission spots when facing a higher convenience yield, which leads to the demand reduction for derivatives and the futures price down.
3. As the value of carbon emission spots goes up, the value of futures also rises. This is because the holder of futures will get more chances to benefit from the derivatives. This result is similar to general futures.

DISCUSSIONS

In this section, we examine the sensitivity of above results to the parameters variations. There are four parameters which are likely to have an impact on the price of emission allowance derivatives.

Discussion 1. We examine the effects of some parameters on the European cash-or-nothing option prices in Fig 8, where we fix $t = 0$ and $\delta = -1$.

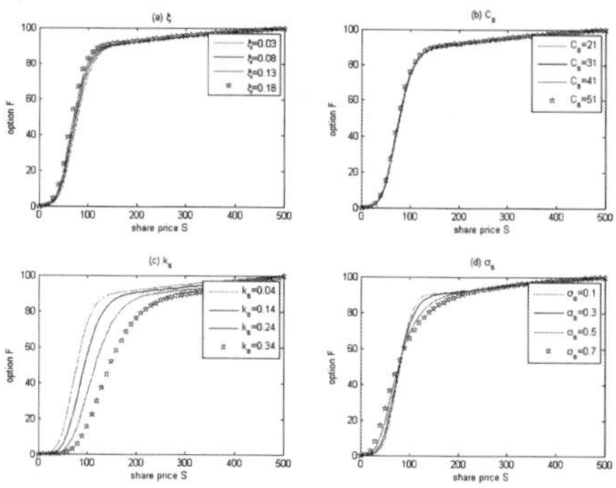

Figure 8. The effects of some parameters on the European option prices for Test 1.

From Fig 8 we can see that

a. The option price increases with the higher growing rate of marginal abatement costs. The higher growing rate of marginal abatement costs will make the abatement costs change quickly and the enterprises will live in a more uncertain world when they must choose investing on clean technology or trading the emission permits. To eliminate this uncertainty, enterprises will buy more derivatives such as options, which leads to the option value increasing.

b. The effect of the marginal abatement costs is negligible. The purpose to trade the emission permits is just to hedge the risk of marginal abatement costs, and if marginal abatement costs increase, people will hold more emission permit spots, and the spots price will increase and vice versa. So, it is natural that the option value does not change following the movement of marginal abatement costs.

c. Increasing the speed of mean-reversion k_s reduces the option prices. This is because emission permit prices have a stronger tendency towards the mean value, the price risk of emission permits decreases, and then people need not to hold many derivatives to hedge the risk. The decrease of demand will reduce the option prices.

4. The effect of the volatility σ_s is irregular compared with the classical Black-Scholes model. In the Black-Scholes world, the higher volatility of underlying assets increases the risk of investment, and the derivatives price will increase. In our model, since the stochastic convenience yield is considered, the effect of volatility on option prices may be counteracted by the stochastic convenience yield. So, there is no certain relationship between the volatility and the option price when there is a stochastic convenience yield.

Note that the above sensitivity analysis results are also satisfied by the following American option and futures, as they are all the derivatives used to hedge the price risk.

Additionally, the effects of stochasticity of convenience yield on the valuations of carbon European options are also examined in Fig 9, where $t = 0$ is fixed. Note that the solid line states the deterministic convenience yield simulation results and the dash line represents the stochastic ones.

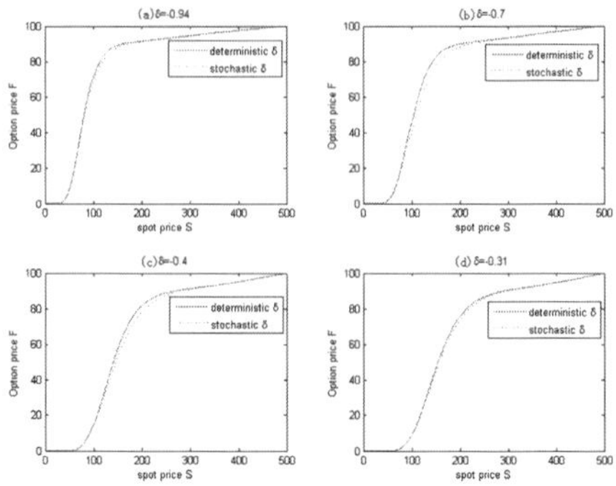

Figure 9. The effects of stochasticity of convenience yield on the European option prices for Test 1.

Discussion 2. The effects of some parameters on the American option price are examined in Fig 10, in which $t = 0$ and $\delta = -1$.

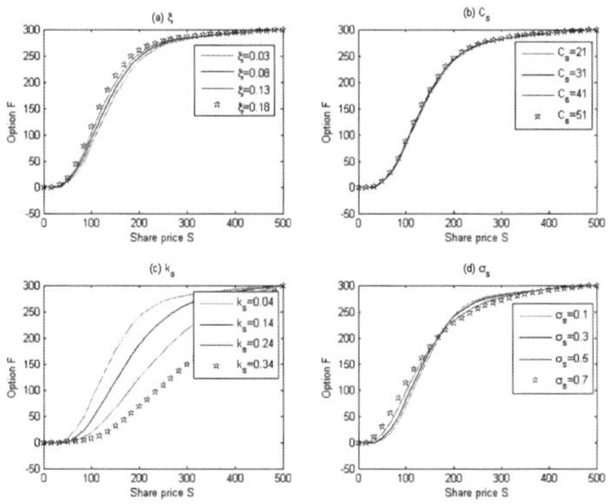

Figure 10. The effects of some parameters on the American option prices for Test 2.

From Fig 10, again we can see that

a. The option price increases with the higher growing rate of marginal abatement costs.

b. The effect of the marginal abatement costs is negligible.

c. Increasing the speed of mean-reversion k_s reduces the option prices.

d. `The effect of the volatility σ_s is irregular compared with the classical Black-Scholes model.

Similarly, the effects of stochasticity of convenience yield on the valuation of carbon American option are also examined in Fig 11, where $t =$

0 is fixed. Note that the solid line states the deterministic convenience yield simulation results and the dash line represents the stochastic ones.

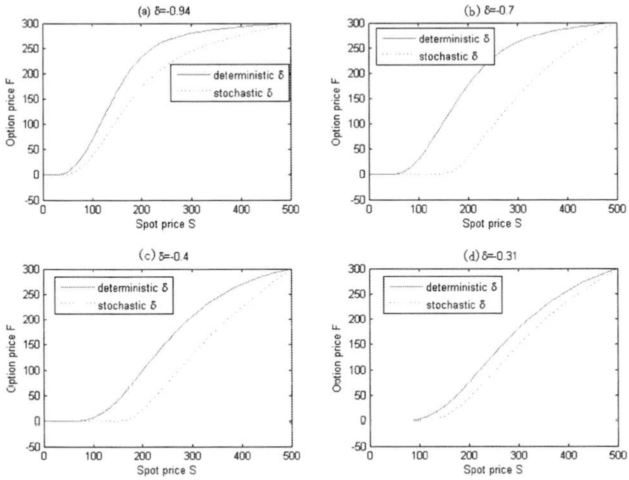

Figure 11. The effects of stochasticity of convenience yield on the American option prices for Test 2.

Discussion 3. The effects of some parameters on the futures prices are examined in Fig 12, in which $t = 0$ and $\delta = -1$.

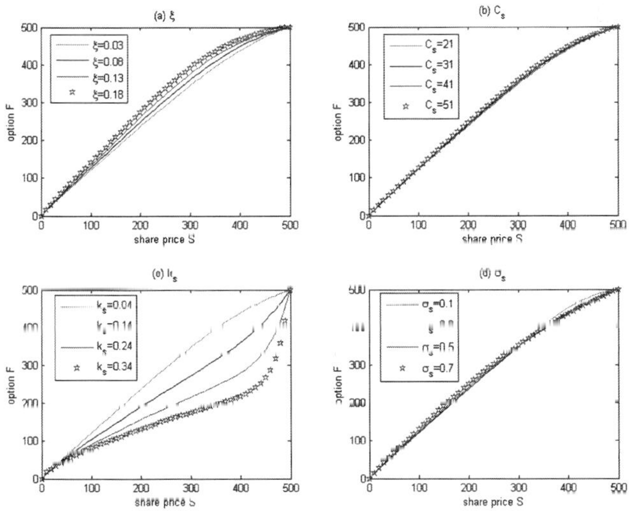

Figure 12. The effects of some parameters on the futures price for Test 3.

From Fig 12 we can see that

a. he futures price increases with the higher growing rate of marginal abatement costs, which increases the uncertainty.

b. The effect of the marginal abatement costs is negligible.

c. Increasing the speed of mean-reversion k_s reduces the futures price, because the emission permits price trends to the mean value more quickly, and the uncertainty of emission permits price would become smaller.

d. There is no certain effect of the volatility on futures prices.

Once again, the effects of stochasticity of convenience yield on the valuations of carbon futures are examined in Fig 13 with the fixed $t = 0$. Note that the solid line corresponds to the deterministic convenience yield simulation results and the dash line represents the stochastic ones.

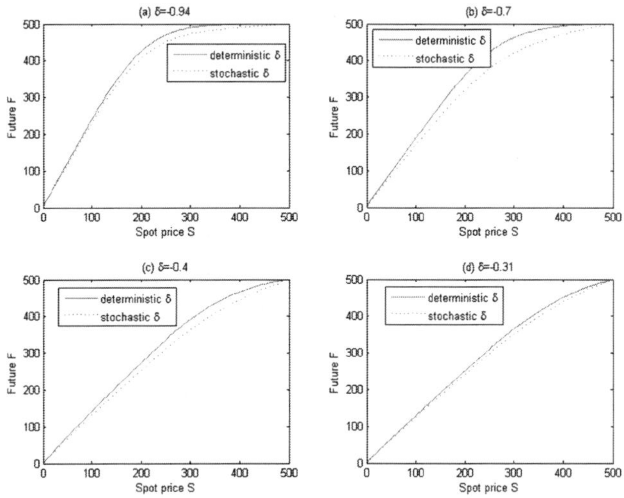

Figure 13. The effects of stochasticity of convenience yield on the futures prices for Test 3.

From the above discussions, we can summarize some experiences about risk management. Firstly, as the Kyoto Protocol came into force in 2005, several countries are expanding their investment scales on the new technologies to reduce carbon emissions, and thus the emission allowance derivatives are also becoming a risk hedging tool for market participants. Secondly, a firm may not be scarce to the emission allowance derivatives when it can adjust its marginal abatement costs well. Finally, the impact of volatility of carbon emission allowances price is difficult to measure when a stochastic convenience yield is involved.

CONCLUDING REMARKS

This paper presents a methodology for modelling and computing the valuation of carbon derivatives with stochastic convenience yields. The principle of absence of arbitrage opportunities and the stochastic calculus are used to develop the mathematical model, the partial differential equation, satisfied by carbon derivatives. For American options, we formulate the pricing problem to a linear parabolic variational inequality, and develop a power penalty method to solve it. Then, a so-called fitted finite volume method is designed to solve the nonlinear partial differential equation resulting from the power penalty method, which governs the futures, European and American option valuation.

Also, we discuss the effects of the stochastic convenience yield on the prices of carbon derivatives in details. First of all, a greater convenience yield means a lower European option price and futures price while it does not mean a lower American option price. Therefore, the property of optimal exercise boundary makes the American option quite different from the European option. Furthermore, from the sensitivity analysis we can see that (*a*) the derivatives price increases with the higher growing rate of marginal abatement costs; (*b*) the effect of the marginal abatement costs is negligible; (*c*) increasing the speed of mean-reversion k_s reduces the derivatives price; (*d*) there is no certain effect of the volatility on derivatives price compared with the classical Black-Scholes model.

The carbon market participants, such as investors, hedgers and arbitragers, may get help from our theoretical results, and they can work out better hedging strategies, adjust portfolio structures and strengthen their capabilities to manage risks. We expect that our work can emphasize the importance of the convenience yield concept for the emission allowance pricing. At the same time, we also anticipate that our derivatives pricing methodology from the perspective of partial differential equations combined with numerical methods can make a few contributions to the development of carbon market.

REFERENCES

1. Croker T, The structuring of atmospheric pollution control systems. The Economics of Air Pollution. 1966: 61–86.
2. Montgomery W, Markets in licenses and efficient pollution control programs. Journal of Economic Theory. 1972;5: 395–418. 10.1016/0022-0531(72)90049-X
3. Hahn R, Market power and transferable property rights. The Quarterly Journal of Economics. 1984;99: 753–765. 10.2307/1883124
4. Edwards T, Hutton J, Allocation of carbon permits within a country: a general equilibrium analysis of the United Kingdom. Energy Economics. 2001;23: 371–386. 10.1016/S0140-9883(00)00077-3
5. Cramton P, Kerr S, Tradeable carbon permit auctions: How and why to auction not grandfather. Energy Policy. 2002;30: 333–345. 10.1016/S0301-4215(01)00100-8
6. Klaassen G, Nentjes A, Smith M, Testing the theory of emissions trading: Experimental evidence on alternative mechanisms for global carbon trading. Ecological Economics. 2005;53: 47–58. 10.1016/j.ecolecon.2004.12.017
7. Mnif W, Davison M, Carbon emission markets. Computational Risk Management. 2011: 95–108.
8. Benz E, Truck S, Modeling the price dynamics of CO_2 emission allowances. Energy Economics. 2009;31: 4–15. 10.1016/j.eneco.2008.07.003
9. Benz E, The Price of CO_2 Allowances in the European Greenhouse Gas Emissions Trading Scheme: an Empirical, Experimental and Theoretical Study. Dissertation. 2008.
10. Daskalakis G, Psychoyios D, Markellos R, Modeling CO_2 emission allowance prices and derivatives: evidence from the European trading scheme. Journal of Banking & Finance. 2009;33: 1230–1241. 10.1016/j.jbankfin.2009.01.001
11. Chang K, Wang S, Peng K, Mean reversion of stochastic convenience yields for CO_2 emissions allowances: empirical evidence from the EU ETS. The Spanish Review of Financial Economics. 2013;11: 49–35. 10.1016/j.srfe.2013.01.001
12. Chevallier J, Modelling risk premia in CO_2 allowances spot and futures prices. Economic Modelling. 2010;27: 717–729. 10.1016/j.econmod.2010.01.012

13. Byun S, Cho H, Forecasting carbon futures volatility using GARCH models with energy volatilities. Energy Economics. 2013;40: 207–221. 10.1016/j.eneco.2013.06.017
14. Medina V, Pardo A, Pascual R, The timeline of trading frictions in the European carbon market. Energy Economics. 2014;42: 378–394. 10.1016/j.eneco.2014.01.008
15. Rudin D, A model of intertemporal emission trading, banking and borrowing. Journal of Environmental Economics and Management. 1996;31: 269–286. 10.1006/jeem.1996.0044
16. Wagner W, CO_2–Emissionszertifikate–Preismodellierung und Derivatebewertung. Universität Karlsruhe; 2007.
17. Investopedia website. Available: http://www.investopedia.com/terms/c/convenienceyield.asp.
18. Cortazar G, Schwartz E, Implementing a stochastic model for oil futures prices. Energy Economics. 2003;25: 215–238. 10.1016/S0140-9883(02)00096-8
19. Schwartzb E, The stochastic behavior of commodity prices: implications for valuation and hedging. Journal of Finance. 1997;52: 923–973. 10.1111/j.1540-6261.1997.tb02721.x
20. Hilliard J, Reis J, Valuation of commodity futures and options under stochastic convenience yields, interest rates, and jump diffusions in the spot. Journal of Financal and Quantitative Analysis. 1998;33: 61–86. 10.2307/2331378
21. Wang Z, Szolnoki A, Perc M, Optimal interdependence between networks for the evolution of cooperation. Scientific Reports. 2013;3: 1–7. 10.1038/srep02470
22. Liu C, Du W, Wang W, Particle swarm optimization with scale-free interactions. Plos One. 2014;9: 1–8.
23. Wang Z, Murks A, Du W, Rong Z, Perc M, Coveting the neighbors fitness as a means to resolve social dilemmas. Journal of Theoretical Biology. 2011;277: 19–26. 10.1016/j.jtbi.2011.02.016
24. Wang Z, Zhu Z, Arenzon J, Cooperation and age structure in spatial games. Physical Review E. 2012;85 10.1103/PhysRevE.85.011149
25. Wang Z, Xia C, Meloni S, Zhou C, Moreno Y, Impact of social punishment on cooperative behavior in complex networks. Scientific Reports. 2013;3: 1–7. 10.1155/2013/472956
26. Wang Z, Kokubo S, Tanimoto J, Fukuda E, Shigaki K, Insight into

the so-called spatial reciprocity. Physical Review E. 2013;88 10.1103/PhysRevE.88.042145

27. Jin Q, Wang Z, Wang Z, Wang Y, Strategy changing penalty promotes cooperation in spatial prisoner's dilemma game. Chaos, Solitons & Fractals. 2012;45: 359–401.
28. Szolnoki A, Wang Z, Perc M, Wisdom of groups promotes cooperation in evolutionary social dilemmas. Scientific Reports. 2012;2: 1–6. 10.1038/srep00576
29. Wang S, A novel fitted finite volume method for the Black-Scholes equation governing option pricing. IMA J. Numer. Anal. 2004;24: 699–720. 10.1093/imanum/24.4.699
30. Leveque R, Finite Volume Methods for Hyperbolic Problems. Cambridge University Press; 2004.
31. Liu J, Mu L, Ye X, An adaptive discontinuous finite volume method for elliptic problems. J. Comput. Appl. Math. 2011;235: 5422–5431. 10.1016/j.cam.2011.05.051
32. Tillmann M, Valuation and Hedging of Carbon-derivatives. Thesis, Institut für Numerische and Angewangdte Mathematik Fachbereich Mathematik, 2008.
33. Brennan M, Schwartz E, A continuous time approach to the pricing of bonds. The Journal of Banking & Finance. 1979;3: 133–155. 10.1016/0378-4266(79)90011-6
34. Black F, The pricing of commodity contracts. Journal of Financial Economics. 1976;3: 167–179. 10.1016/0304-405X(76)90024-6
35. Samuelson P, Proof that properly anticipated prices uctuate randomly. Industrial Management Review. 1965;6: 41–49.
36. Hull J, Options, Futures, and other Derivatives. Tsinghua University Press; 2011.
37. Wang S, Yang X, Teo K, Power penalty method for a linear complementarity problem arising from American option valuation. J. Optimization Theory and Applications. 2006;129: 227–254. 10.1007/s10957-006-9062-3
38. Zhang K, Wang S, Yang X, Teo K, A power penalty approach to numerical solutions of two-asset American options. Numerical Mathematics: Theory, Method and Applications. 2009;2: 202–233.
39. Wang S, Yang X, A power penalty method for linear complementarity problems. Operations Research Letters. 2008;36: 211–214. 10.1016/j.

orl.2007.06.006

40. Huang C, Wang S, A power penalty approach to a nonlinear complementarity problem. Operations Research Letters. 2010;38: 72–76. 10.1016/j.orl.2009.09.009
41. Huang C, Hung C, Wang S, A fitted volume method for the valuation of options on assets with stochastic volatilities. Computing. 2006;77: 297–320. 10.1007/s00607-006-0164-4
42. Huang C, Hung C, Wang S, On convergence of a fitted finite-volume method for the valuation of options on assets with stochastic volatilities. IMA Journal of Numerical Analysis. 2010;30: 1101–1120. 10.1093/imanum/drp016

CHAPTER

15

Variance and Dimension Reduction Monte Carlo Method for Pricing European Multi-Asset Options with Stochastic Volatilities

Yijuan Liang [1,2,] **and Xiuchuan Xu** [2]

[1]Agricultural Education and Development Research Center, Southwest University, Chongqing 400715, China

[2]School of Economics and Management, Southwest University, Chonging 400715, China

ABSTRACT

Pricing multi-asset options has always been one of the key problems in financial engineering because of their high dimensionality and the low convergence rates of pricing algorithms. This paper studies a method to accelerate Monte Carlo (MC) simulations for pricing multi-asset options

Citation: Liang, Y.; Xu, X. "Variance and Dimension Reduction Monte Carlo Method for Pricing European Multi-Asset Options with Stochastic Volatilities". Sustainability 2019, 11, 815. https://doi.org/10.3390/su11030815

with stochastic volatilities. First, a conditional Monte Carlo (CMC) pricing formula is constructed to reduce the dimension and variance of the MC simulation. Then, an efficient martingale control variate (CV), based on the martingale representation theorem, is designed by selecting volatility parameters in the approximated option price for further variance reduction. Numerical tests illustrated the sensitivity of the CMC method to correlation coefficients and the effectiveness and robustness of our martingale CV method. The idea in this paper is also applicable for the valuation of other derivatives with stochastic volatility.

Keywords:-Conditional Monte Carlo; variance reduction; multi-asset options; stochastic volatility; martingale control variate

INTRODUCTION

In the last 40 years, financial derivatives have become increasingly important in finance. They are actively traded on many exchanges throughout the world, and are entered into by financial institutions, fund managers, and corporate treasurers in the over-the-counter market. They are especially important for market anticipants because they can be used to transfer a wide range of risks in the economy from one entity to another. Efficient use of financial derivatives can certainly promote financial and social sustainability. For instance, there are many different types of energy and agricultural commodity derivatives that are designed and used to contest weather and market risks, and to protect the benefit and reduce the potential loss of anticipants. Another example is that the real options approach is very popular in valuing the real estate sustainable investment. Conversely, inappropriate use of derivatives may cause great, even global, disasters, for example, the credit crisis that started in 2007. As pointed out by Hull [1], "we have now reached the stage where those who work in finance, and many who work outside finance, need to understand how derivatives work, how they are used, and how they are priced."

The accurate and fast pricing of financial derivatives is one of the most important things in financial engineering since many of the problems in economics and finance eventually turn into the pricing of financial derivatives. For example, Kim et al. [2] decided the optimal investment timing using rainbow options valuation for economic sustainability appraisement. Yoo et al. [3] determined an optimum combination of financial models including options to achieve a sustainable profit for overseas investment projects. The pioneering work of Black and Scholes [4] and Merton [5] lay the foundations for option pricing models. It is well known that the stochastic volatility model can be used to generalize the constant volatility assumption in the Black–Scholes model to capture the character of empirical observations from financial markets, such as the observed volatility smile and the leptokurtic features of the asset return distribution [6,7]. Stochastic volatility models describe volatility behavior with another stochastic differential equation. There are many studies on stochastic volatility models, such as those of Hull and White [8], Scott [9], Stein and Stein [10], Ball and Roma [11], Heston [12], Schöbel and Zhu [13], and Hagan et al. [14]. In addition to these one-factor stochastic volatility models, Fouque et al. [15,16,17,18] proposed a multi-factor mean-reverting stochastic volatility model. A comprehensive treatment of stochastic volatility models can be found in Reference [19].

Multi-asset options refer to a wide variety of contingent claims whose payoff depends on the overall performance of more than one underlying asset. Usually, they can be grouped into three categories: rainbow options, basket options, and quanto options. The prices of rainbow options rely on price changes of underlying assets, such as exchange options, outperformance options, spread options, chooser options, max-call options, and their variations. Basket options prices are always determined by the average price of underlying assets, while the value of a quanto option depends on the performance of domestic and foreign underlying assets. Jiang [20] introduced the concepts and constructed the pricing models of multi-asset options in detail, where the volatilities were constant. The pricing problem of multi-asset options pricing is essentially equivalent to a high dimensional integral. It is challenging to compute such a high dimensional integral, especially when the number of underlying assets is large, or stochastic volatilities are considered in the model.

There are mainly three pricing methods for multi-asset options: the analytic approximation method, the fast Fourier transformation (FFT) method, and the MC simulation method. The analytic approximation method typically constructs an approximate pricing model for the original problem that results in a closed form solution, and this method is always elegantly designed to the original problem. Several studies focus on this approach, for instance, those by Turnbull and Wakeman [21], Curran [22], Milevsky and Posner [23], Ju [24], Zhou and Wang [25], Alexander and Venkatramanan [26], Datey et al. [27], Brigo et al. [28], Borovkova et al. [29], Deelstra et al. [30,31], and Li et al. [32]. The main disadvantage of the analytic approximation method is that the size of the error is unknown and there is no way to systematically reduce it. The FFT method, which was prosed by Carr and Madan [33], has successfully been used in option pricing problems with a low dimension because of its high efficiency (see Carr and Wu [34], Heston [12], Grzelak et al. [35,36,37], and He and Zhu [38]). However, the FFT method depends on the availability of a characteristic function (usually in an affine framework), which is not always promised in a general stochastic volatility model. It is also difficult to apply the FFT method to high dimensional problems due to their dimensionality. Thus, for higher dimensional options, the most practical method seems to be MC simulations. Kim et al. [2] and Yoo et al. [3] also used MC simulations to price the embedded option prices in valuation real investment projects since the high dimension of problem. MC uses the sample mean as an estimator for the expectation of a random variable. Its speed of convergence is not influenced by the dimension of the problem. In addition, it allows for a simple error bound, given by the central limit theorem.

The major drawback of an MC simulation is that its convergence rate is quite slow, that is, O(m−1/2), where *m* is the number of samples in MC simulation. As a result, often the main challenge in developing an efficient MC method is to find an effective variance reduction technique. There are a lot of studies about how to improve the efficiency of an MC simulation, and we refer the reader to Glasserman [39] and relevant references therein for a detailed discussion on various variance reduction techniques. Kemna and Vorst [40] presented one of the classical works on accelerating MC simulations. They used the geometric average option as a CV to price the arithmetic average option, which proved to be very successful. For a European multi-asset option pricing problem, Barraquand [41] proposed a "quadratic resampling" method by matching the moments of the underlying assets to reduce the variance of the MC simulation. Pellizari [42] designed a CV method called mean Monte Carlo to gain variance reduction of an MC simulation. The key of their success was that a Black–Scholes formula could be obtained when all underlying assets except for one were replaced by their mean. Borogovac and Vakili [43] proposed a "database Monte Carlo" CV method that avoids computing the expectation of CV, but the database, constructed in advance, requires huge calculations. Dingeç and Hörmann [44] exploited the property that the geometric average price was larger than the arithmetic average price to construct a CV by conditioning the payoff on the assumption that the geometric average of all prices was larger than the strike price. The expectation of their CV was computed by numerical methods, and their numerical tests for Asian options and basket options showed a great accelerating effect on the MC simulations. Liang et al. [45] designed a CV for European multi-asset options based on principal component analysis. Sun and Xu [46] used the CMC method with the importance sampling technique to accelerate MC simulations for basket options. There are some other approaches to speed up MC simulations, such as the quasi-Monte Carlo method [39,47,48,49,50,51,52], and parallelized implementations of MC simulations on CPUs/GPUs [53,54,55,56,57,58].

However, there is little research on variance reduction of MC simulations in pricing multi-asset options with stochastic volatilities. Du et al. [59] proposed a variance reduction method in multi-asset options under stochastic volatility models by matching the moments of the volatilities. Although their method shows great variance reduction of MC simulations, there are some restrictions to it. (1) All underlying assets are assumed to be driven by one stochastic volatility factor, which is not reasonable in practice. A more reasonable model is to assume that each underlying asset is driven by its own stochastic volatility factor (see Antonelli et al. [60], Shiraya and Takahashi [61], and Park et al. [62]). (2) Their moment matching technique greatly depends on the Hull–White stochastic volatility model, and is not applicable to general stochastic volatility models. (3) They only conducted numerical tests for options with two assets, which is not general enough for most multi-asset options.

In this paper, we aim to develop an efficient dimension and variance reduction method for MC simulations in pricing European multi-asset options with general stochastic volatilities. In our pricing framework, the underlying asset is assumed to be driven by its own stochastic volatility process, and full correlations between factors are allowed. The stochastic volatility model, which could be the Hull–White [8] or Heston [12] models, is quite general, such that our pricing model has a wide range of applicability. Our dimension and variance reduction method is built on the idea developed by Liang and Xu [63], who designed a CMC simulation with martingale CV to price single-asset European options with stochastic volatility. Our main contributions are: (1) A CMC pricing framework is deduced for European multi-asset options with general stochastic volatility models, which results in dimension and variance reduction. (2) A martingale CV based on a martingale representation theorem is combined with the CMC to obtain further variance reduction of the MC simulations. (3) The algorithm was tested on typical multi-asset options, such as exchange options, basket options (which can be more than two assets), and quanto options, showing the broad applicability and high efficiency of our method.

The paper is organized as follows. In Section 2, we introduce the pricing model of multi-asset options with stochastic volatilities. In Section 3, we deduce the CMC pricing framework, prove the martingale presentation theorem, and construct the martingale CV in detail. We present numerical tests and their results in Section 4, to evaluate the efficiency of our proposed method. Finally, we conclude the paper in Section 5.

PRICING MODEL

In this section, we give the pricing model of multi-asset options with stochastic volatilities. Specifically, in a risk-neutral world, let $S_i(t)$ be the price of the *i*th underlying asset $(i=1,2,\cdots,n)$ at time *t*, which we assume obeys the following stochastic differential equations:

$$\frac{dS_i(t)}{S_i(t)} = (r-\delta_i)dt + f_i(Y_i(t))dW_i(t), \tag{1}$$

$$dY_i(t) = \mu_i(t,Y_i(t))dt + g_i(t,Y_i(t))dZ_i(t), \tag{2}$$

where r is the constant risk-free interest rate and δ_i is the continuous dividend rate. Y_i is the stochastic variance, and the functions $f_i(y), \mu_i(t,y)$, and $g_i(t,y)$ determine the specific volatility model, which can be quite general. $dW_i(t)$, and $dZ_i(t)$ are standard Brownian noise terms, and the covariance between them is captured as follows:

$$\mathrm{cov}(dW_i(t), dW_j(t)) = \rho_{ij}dt, \quad i \neq j, \tag{3}$$

$$\mathrm{cov}(dW_i(t), dZ_i(t)) = \rho_i dt, \quad i = 1,2,\cdots,n, \tag{4}$$

$$\mathrm{cov}(dW_i(t), dZ_j(t)) = 0, \quad i \neq j, \tag{5}$$

$$\mathrm{cov}(dZ_i(t), dZ_j(t)) = 0, \quad i \neq j. \tag{6}$$

where the correlation coefficients ρ_{ij}, ρ_i are constant.

Equation (3) indicates that the underlying assets are correlated. Equations (4) and (5) indicate that any underlying asset is driven by only one stochastic variance factor and is not directly affected by the other stochastic variance factors. Equation (6) indicates that the random stochastic variance factors are mutually independent, but this assumption could be relaxed allowing for correlated random processes. Several popular stochastic volatility models are collected in Table 1.

Table 1. Models of stochastic volatility.

Reference	ρ_i	$f_i(y)$	$\mu_i(t,y)$	$g_i(t,y)$
Hull–White [8]	0	$\sqrt{y}$	μy	σy
Scott [9]	0	e^y	$a(\theta - y)$	σ
Stein–Stein [10]	0	$\|y\|$	$a(\theta - y)$	σ
Ball–Roma [11]	0	$\sqrt{y}$	$a(\theta - y)$	$\sigma\sqrt{y}(2a\theta > \sigma^2)$
Heston [12]	$\neq 0$	$\sqrt{y}$	$a(\theta - y)$	$\sigma\sqrt{y}(2a\theta > \sigma^2)$
Hagan et al. [14]	$\neq 0$	y	0	σy

In the following, we introduce our notations for convenience. The underlying asset vector is $S(t)=(S_1(t),\cdots,S_n(t))'$, and the stochastic variance vector is $Y(t)=(Y1(t),\cdots,Yn(t))'$, where $'$ represents the transpose of a vector or matrix. Additionally, the correlation matrix is given by $\Gamma=(\rho_{ij})$.

Suppose the payoff function of the European multi-asset option at maturity *T* is given by:

$$h(S_1(T), S_2(T), \cdots, S_n(T)) =: h(S(T)). \tag{7}$$

Denote V(t,s,y) as the value of a European multi-asset option with stochastic volatilities at time *t*; then, by the no-arbitrage pricing principle we obtain:

$$V(t,s,y) = E\left[e^{-r(T-t)}h(S(T)) \mid S(t) = s, Y(t) = y\right], \tag{8}$$

where E[·] is the expectation in a risk-neutral world. Given the initial asset price S(0) and initial variance Y(0), the European option price at the initial time is actually:

$$V(0, S(0), Y(0)) = E[e^{-rT}h(S(T))]. \tag{9}$$

MC simulation can be used to compute the option price based on Equation (9) (please see Glasserman [39]). Suppose the number of samples in MC simulation is m. Firstly, for the jth sample, we need to simulate the processes of the Brownian motions $W_i^{(j)}(t)$ and $Z_i^{(j)}(t), i = 1, 2, \cdots, n$ to get the processes of $Y^{(j)}(t)$ and $S^{(j)}(t)$ and the discounted payoff $V^{(j)} = e^{-rT}h(S^{(j)}(T)), j = 1, 2, \cdots, m$. Then, we average the samples of discounted payoff and use the sample mean $\overline{V} = \frac{1}{m}\sum_{j=1}^{m} V^{(j)}$ as an estimation of the option price. The law of large numbers guarantees the convergence of MC simulation. The central limit theorem guarantees that the standard error—the standard deviation of sample mean

$\overline{V}\cdot$—from MC simulation has a form of $Std = \frac{\text{var}(V^{(i)})}{\sqrt{m}}$. The standard error can be used to measure how far the sample mean is likely to be from the option price or to make confidence intervals of the option price, for instance, a 95% confidence interval $\overline{V}\cdot \pm 1.96$Std. It also indicates that the MC simulation has a convergence rate as O(m−1/2), which is rather low. Thus, in the next section, using a similar idea as in Liang and Xu [63], we propose our efficient CMC simulation framework with martingale CV for this option pricing problem.

If the stochastic volatility $f_i(Y_i(t))$ in Equation (**1**) is replaced by a constant volatility σ_i, we can obtain the dynamic process of an underlying asset with constant volatility as follows:

$$\frac{dS_i(t)}{S_i(t)} = (r - \delta_i)dt + \sigma_i dW_i(t), \quad i = 1, 2, \cdots, n. \tag{10}$$

The correlations between $dW_i(t)$ are defined by Equation (**3**). Jiang [**20**] carefully studied the explicit expression for a European multi-asset option value with constant volatility. Denote $V^{BS}(t,S(t);\sigma,\Gamma)$ as the corresponding price at time *t*, where the volatility vector is $\sigma=(\sigma_1,\sigma_2,\cdots,\sigma_n)'$. However, an analytic solution exists only for some specific options [**20**], such as exchange options, outer performance options, spread options, two dimension chooser options, basket options with a geometric average price, and quanto options. We give the specific expression for some of these in the numerical tests.

DIMENSION AND VARIANCE REDUCTION

In this section, we apply the acceleration methods of Liang and Xu [63] to price European multi-asset options with stochastic volatilities. The idea is that a martingale CV can be combined with the CMC method to reduce the variance of an MC simulation.

CMC Method

CMC can be used to reduce the variance of an MC simulation. Willard [64] initially put forward the CMC simulation to price options with stochastic volatilities. His method is feasible for those options that have a closed-form solution under the constant volatility model. Drimus [65] used CMC to analyze the variance products under the log-Ornstein–Uhlenbeck (log-OU) model. Boyle et al. [66] also used the CMC approach in pricing a down-and-in call option with a discretely monitored barrier. Broadie and Kaya [67] applied the CMC to accelerate exact simulations of the stochastic volatility with affine jump diffusion processes. Yang et al. [68] employed the CMC to reduce the variance of MC simulations when calculating the prices and greeks of barrier options. Dingeç and Hörmann [44] and Sun and Xu [46] combined CMC simulations and other variance reduction techniques to price basket options.

When we consider computing the expectation E[Y] of a random variable Y, the conditional expectation E[Y|X] of Y on some other variable X is also an unbiased estimator of E[Y]. This results from the double expectation formula E[Y]=E[E[Y|X]], and the variance decomposition formula [69]:

$$\mathrm{var}(Y) = \mathrm{var}\left(E[Y|X]\right) + E[\mathrm{var}(Y|X)] \geq \mathrm{var}(E[Y|X]),$$

which indicates that the variance of E[Y|X] is always smaller than that of Y. The so-called CMC method uses the conditional expectation of the random variable E[Y|X] instead of that of the original random variable Y, which can obviously reduce variance. The key is that we need to have a closed form of E[Y|X].

Now, we intend to deduce the CMC pricing formula for the pricing problem of Equation (8). The most important thing is to obtain the conditional expectation of the discounted payoff $e^{-rT}h(S(T))$ on other random variables or stochastic information. First, a Cholesky decomposition of Brownian noise dWi(t) is conducted according to Equation (4):

$$dW_i(t) = \rho_i dZ_i(t) + \sqrt{1-\rho_i^2}d\tilde{Z}_i(t), \quad (i = 1, 2, \cdots, n), \tag{11}$$

where $dZ_i(t)$ and $d\tilde{Z}_i(t)$ are independent standard Brownian noises, which means that:

$$\mathrm{cov}(dZ_i(t), d\tilde{Z}_i(t)) = 0, \quad (i = 1, 2, \cdots, n). \tag{12}$$

If we denote the vectors $d\mathbf{W}(t) = (dW_1(t), \cdots, dW_n(t))'$, $d\mathbf{Z}(t) = (dZ_1(t), \cdots, dZ_n(t))'$, and $d\tilde{\mathbf{Z}}(t) = (d\tilde{Z}_1(t), \cdots, d\tilde{Z}_n(t))'$, then Equation (**11**) can be rewritten to:

$$d\mathbf{W}(t) = diag(\rho)d\mathbf{Z}(t) + diag(q)d\tilde{\mathbf{Z}}(t). \tag{13}$$

where $\rho = (\rho_1, \cdots, \rho_n)'$, $q = \left(\sqrt{1-\rho_1^2}, \cdots, \sqrt{1-\rho_n^2}\right)'$, and diag(v) is a diagonal matrix, with diagonal values from the vector v. According to Equations (**12**) and (**13**), it is obvious that:

$$\mathrm{cov}(d\mathbf{W}(t)) = diag(\rho)^2 dt + diag(q)\mathrm{cov}(d\tilde{\mathbf{Z}}(t))diag(q). \tag{14}$$

Notice that Equation (**3**) implies:

$$\mathrm{cov}(d\mathbf{W}(t)) = \Gamma dt. \tag{15}$$

We can solve for the covariance of $d\tilde{Z}_i(t)$ by Equations (**14**) and (**15**) as:

$$\operatorname{cov}(d\tilde{\mathbf{Z}}(t)) = diag(\boldsymbol{q})^{-1}\left(\Gamma - diag(\boldsymbol{\rho})^2\right) diag(\mathbf{q})^{-1} dt =: \tilde{\Gamma} dt. \tag{16}$$

The entries of matrix $\tilde{\Gamma}$ are

$$\tilde{\Gamma}_{ii} = 1, \quad \tilde{\Gamma}_{ij} = \frac{\rho_{ij}}{\sqrt{1-\rho_i^2}\sqrt{1-\rho_j^2}} \quad (i \neq j). \tag{17}$$

To ensure the matrix $\tilde{\Gamma}$ is well-defined, the condition of correlation coefficients $| \, |\rho_i| < 1, | \, \tilde{\Gamma}_{ij}| < 1$ should be satisfied.

Then, applying the Itô formula to $\ln(S_i(t))$, with the help of Equations (1) and (11), results in:

$$d\ln(S_i(t)) = \left(r - \delta_i - \tfrac{1}{2} f_i^2(Y_i(t))\right) dt + \rho_i f_i(Y_i(t)) dZ_i(t) + \sqrt{1-\rho_i^2} f_i(Y_i(t)) d\tilde{Z}_i(t), \quad (i = 1, 2, \cdots, n). \tag{18}$$

Integrating both sides of the above equation from *t* to *T* results in:

$$S_i(T) = S_i(t)\xi_i(t,T)\exp\left((r-\delta_i)(T-t) - \tfrac{1-\rho_i^2}{2}\int_t^T f_i^2(Y_i(s))ds + \sqrt{1-\rho_i^2}\int_t^T f_i(Y_i(s))d\tilde{Z}_i(s)\right), \tag{19}$$

where

$$\xi_i(t,T) = \exp\left(-\frac{1}{2}\rho_i^2 \int_t^T f_i^2(Y_i(s))ds + \rho_i \int_t^T f_i(Y_i(s))dZ_i(s)\right). \tag{20}$$

Note that $\xi_i(t,T)$ is actually an exponential martingale with expectation $E[\xi_i(t,T)]=1$.

Let $\bar{\sigma}_i(t,T)$ denote the average volatility of underlying asset Si on the interval [t,T], which is given by:

$$\bar{\sigma}_i(t,T)^2 = \frac{1}{T-t}\int_t^T f_i^2(Y_i(s))ds. \tag{21}$$

It is observed that, given the information of stochastic processes $\{Z_1(t), Z_2(t), \cdots, Z_n(t)\}$, the quantities $\xi_i(t,T)$, and $\bar{\sigma}_i(t,T)$ $(i=1,2,\cdots,n)$ are totally determined. The pricing problem of Equation (**8**) then becomes the expectation of random variables $\{\tilde{Z}_1(t), \tilde{Z}_2(t), \cdots, \tilde{Z}_n(t)\}$. Assume there exists a Black–Scholes formula with constant volatilities. Then, calculating expectations with $\{\tilde{Z}_1(t), \tilde{Z}_2(t), \cdots, \tilde{Z}_n(t)\}$ gives us the CMC pricing formula of a European multi-asset option with stochastic volatility, as follows:

$$V(t, \boldsymbol{S}(t), \boldsymbol{Y}(t)) = E\left[V^{\mathrm{BS}}\left(t, \boldsymbol{S}(t)\cdot\boldsymbol{\xi}(t,T); \boldsymbol{q}\cdot\bar{\boldsymbol{\sigma}}(t,T), \tilde{\Gamma}\right)\right]. \tag{22}$$

where $\cdot$ is the dot product of two vectors, $\xi(t,T)=(\xi_1(t,T),\cdots,\xi_n(t,T))'$, and $\bar{\sigma}(t,T) = (\bar{\sigma}_1(t,T), \cdots, \bar{\sigma}_n(t,T))$

Compared with the MC formula in Equation (8), we now only need to simulate the n random variables {Z(t)} instead of the 2n random variables {W(t), and Z(t)}. Thus, the dimension and variance are reduced by the properties of the CMC.

Martingale Control Variate (CV)

To further reduce the simulation variance of the variable $V^{BS}(t, S(t) \cdot \xi(t,T); q \cdot \bar{\sigma}(t,T), \tilde{\Gamma})$ in Equation (22), a general martingale CV is proposed to combine with the CMC simulation. Some brief introductions about the CV method are given at first (for more details and references, please refer to Glasserman [39]).

When the CV method is used to compute the expectation E[Y], the CV estimator is:

$$Y(b) = Y - b(X - E[X]),$$

where *X* is called a CV and E[X] is the closed-form expectation of *X*. The constant *b* can be selected as b^*=cov(X,Y)/var(X) to minimize the variance of the CV estimator with an optimal variance reduction ratio R^2=var(Y)/var(Y(b^*))=1/$(1-\rho_{XY}^2)$. The efficiency of the CV method can be measured by the variance reduction ratio R2 or the standard error reduction ratio *R*. The success of the CV depends on high correlations with the naive variable *Y* and the availability of its expectation E[X]. Thus, the CV is always elegantly designed to a specific problem. In this paper, a martingale whose expectation equals to zero is used as a CV, which avoids any extra effort needed to obtain its expectation.

To construct an efficient CV in the CMC framework, a martingale representation theorem for the stochastic volatility pricing model of Equations (1) and (2) is proved in the following theorem.

Theorem 1 (Martingale Representation Theorem).

If the underlying assets satisfy Equations (1) and (2), the European multi-asset option price at the initial time V(0,S(0),Y(0)) *with payoff* h(S(T)) *can be expressed as:*

$$e^{-rT}h(S(T)) - V(0,S(0),Y(0)) = \int_0^T e^{-rt}\left(\sum_{i=1}^{n} S_i(t) f_i(Y_i(t)) \frac{\partial V}{\partial S_i} dW_i(t) + \sum_{i=1}^{n} \mu_i(t,Y_i(t)) \frac{\partial V}{\partial Y_i} dZ_i(t)\right), \quad (23)$$

which can be rewritten in vector form as:

$$e^{-rT}h(S(T)) - V(0,S(0),Y(0)) = \int_0^T e^{-rt}\left((S(t) \cdot f(Y(t)) \cdot \nabla_S V)' dW(t) + (\mu(t,Y(t)) \cdot \nabla_Y V)' dZ(t)\right), \quad (24)$$

where $f(\mathbf{Y}(t)) = (f_1(Y_1(t)), \cdots, f_n(Y_n(t)))'$, $\mu(t, \mathbf{Y}(t)) = (\mu_1(t, Y_1(t)), \cdots, \mu_n(t, Y_n(t)))$ $\nabla_S V = (\frac{\partial V}{\partial S_1}, \cdots, \frac{\partial V}{\partial S_n})'$, *and* $\nabla_Y V = (\frac{\partial V}{\partial Y_1}, \cdots, \frac{\partial V}{\partial Y_n})'$.

Proof.

Applying Itô's formula to $e^{-rt}V(t,S(t),Y(t))$ yields:

$$d\left(e^{-rt}V(t, \mathbf{S}(t), \mathbf{Y}(t))\right) = e^{-rt}LVdt + e^{-rt}\left(\sum_{i=1}^{n} S_i(t) f_i(Y_i(t)) \frac{\partial V}{\partial S_i} dW_i(t) + \sum_{i=1}^{n} \mu_i(t, Y_i(t)) \frac{\partial V}{\partial Y_i} dZ_i(t)\right),$$

where

$$\begin{aligned} LV = & \frac{\partial V}{\partial t} + \frac{1}{2}\sum_{i,j=1}^{n} \rho_{ij} S_i S_j f_i(Y_i) f_j(Y_j) \frac{\partial^2 V}{\partial S_i \partial S_j} + \sum_{i=1}^{n} \rho_i S_i f_i(Y_i) \mu_i(t, Y_i) \frac{\partial^2 V}{\partial S_i \partial Y_i} \\ & + \frac{1}{2}\sum_{i=1}^{n} g_i^2(t, Y_i) \frac{\partial^2 V}{\partial Y_i^2} + \sum_{i=1}^{n} (r - \delta_i) S_i f_i(Y_i) \frac{\partial V}{\partial S_i} + \sum_{i=1}^{n} \mu_i(t, Y_i) \frac{\partial V}{\partial Y_i} - rV. \end{aligned}$$

Furthermore, LV=0 because of the Feynman–Kac formula [69], and thus:

$$d\left(e^{-rt}V(t, \mathbf{S}(t), \mathbf{Y}(t))\right) = e^{-rt}\left(\sum_{i=1}^{n} S_i(t) f_i(Y_i(t)) \frac{\partial V}{\partial S_i} dW_i(t) + \sum_{i=1}^{n} \mu_i(t, Y_i(t)) \frac{\partial V}{\partial Y_i} dZ_i(t)\right).$$

Now, by integrating both sides of the above equation on the interval [0,T], and noticing that V(T,S(T),Y(T))=h(S(T)), we obtain the conclusion of the martingale representation theorem. □

The martingale representation theorem gives us inspiration to construct efficient CVs. For simplicity, denote:

$$X = \int_0^T e^{-rt}\left((\mathbf{S}(t) \cdot f(\mathbf{Y}(t)) \cdot \nabla_S V)' d\mathbf{W}(t) + (\mu(t, \mathbf{Y}(t)) \cdot \nabla_Y V)' d\mathbf{Z}(t)\right), \tag{25}$$

The martingale expression in Equation (23) indicates that the variance of $e^{-rT}h(S(T))$ in the MC simulation is totally determined by the martingale *X*. Thus, the martingale *X* plays the role of a perfect CV for an MC simulation. Fouque and Han [18] actually gave a similar representation in their work, and used the martingales as a CV to price single-asset options under a specific multi-factor stochastic volatility model. This can be understood in financial terminology. The martingale CV corresponds to a continuous delta hedge strategy taken by a trader who sells the option. The integrands of the martingale would, in theory, be the perfect delta hedges. Even though perfect replication by delta hedging under stochastic volatility models is impossible, the variance of replication error is directly related to the variance induced by the martingale CV method.

For the CMC pricing framework, taking the conditional expectation of both sides of Equation (**23**) based on the information {Z(t),0≤t≤T}, results in:

$$V^{BS}\left(t, S(t)\cdot\xi(t,T); q\cdot\bar{\sigma}(t,T), \tilde{\Gamma}\right) - V(0, S_0, Y_0) = \hat{X}, \tag{26}$$

where

$$\hat{X} = E\left[X|\mathbf{Z}(t), 0 \le t \le T\right]. \tag{27}$$

We can determine the expression for X^ by first substituting the Cholesky decomposition, Equation (**11**), into the expression of X in Equation (**25**). Then, we compute the expectations about $\{\tilde{Z}_1(t), \tilde{Z}_2(t), \cdots, \tilde{Z}_n(t)\}$ as:

$$\hat{X} = \int_0^T e^{-rt}(\rho\cdot F_1(t))' d\mathbf{Z}(t) + \int_0^T e^{-rt} F_2(t)' d\mathbf{Z}(t), \tag{28}$$

where

$$F_1(t) = E[\mathbf{S}(t)\cdot \boldsymbol{f}(\mathbf{Y}(t))\cdot \nabla_S V|\mathbf{Z}(s), 0 \le s \le t], \tag{29}$$

$$F_2(t) = E[\boldsymbol{\mu}(t, \mathbf{Y}(t))\cdot \nabla_Y V|\mathbf{Z}(s), 0 \le s \le t]. \tag{30}$$

Equation (**26**) shows that the variance of $V^{BS}\left(t, S(t)\cdot\xi(t,T); q\cdot\bar{\sigma}(t,T), \tilde{\Gamma}\right)$ is totally determined by the zero martingale X^, since V(0,S0,Y0) is a constant. This indicates that $\hat{X}$, theoretically, is a perfect CV for $V^{BS}\left(t, S(t)\cdot\xi(t,T); q\cdot\bar{\sigma}(t,T), \tilde{\Gamma}\right)$ in CMC simulations. It is a pity that we have no explicit expression of this perfect zero martingale, since there is no exact expression for V(t,s,y). A possible solution is that we approximate the option price V(t,s,y) with a Black–Scholes option price along with some carefully selected volatility parameters. In the following, we show our approach.

Given the information {Z(s),0≤s≤t}, the conditional expectation of S(t) can be computed by Equation (**19**) as:

$$\hat{S}(t) = E[S(t)|\mathbf{Z}(s), 0 \le s \le t] = e^{(r-\delta)t} S(0)\cdot\xi(0,t). \tag{31}$$

We intend to use V^{BS}(t,S(t);θ,Γ) as an approximation of V(t,S(t),Y(t)), where θ is a constant vector whose values should be carefully selected. The partial derivatives can thus be approximated as:

$$\nabla_S V(t, S(t), Y(t)) \approx \nabla_S V^{BS}(t, S(t); \boldsymbol{\theta}, \Gamma),$$
$$\nabla_Y V(t, S(t), Y(t)) \approx \nabla_Y V^{BS}(t, S(t); \boldsymbol{\theta}, \Gamma) = 0.$$

Now, given the information {Z(s),0≤s≤t}, by using the approximated derivatives and substituting S(t) with $\hat{S}(t)$ in Equations (**29**) and (**30**), $F_1(t)$ can be approximately expressed as:

$$\hat{F}_1(t) = \hat{S}(t) \cdot f(Y(t)) \cdot \nabla_S V^{\mathrm{BS}}(t, \hat{S}(t); \theta, \Gamma). \tag{32}$$

Notice that $\nabla_Y V^{\mathrm{BS}}(t, \hat{S}(t); \theta, \Gamma) = 0$, so $\hat{F}_2(t) \approx 0$.

In the end, we obtain our martingale CV:

$$\tilde{X} = \int_0^T e^{-rt} (\rho \cdot \hat{F}_1(t))' d\mathbf{Z}(t). \tag{33}$$

The value of the constant volatility vector θ should be determined if we want to use the martingale $\tilde{X}$ as a CV. Fouque and Han [18] illustrated a method for pricing a single-asset option with multi-factor volatility. They picked the long-term mean of the volatility as the volatility parameter in their specific multi-factor model. However, they did not offer a solution for non-mean-reverting stochastic volatility models, such as the Hull–White model.

In this paper, we set parameter θ=f(E[Y(t)]). The idea is that, on the interval [t,T], the stochastic variance Y(t) is approximated by the expectations of their initial state E[Y(t)]. This results in a corresponding approximated stochastic volatility f(E[Y(t)]). We hope that the dynamic behavior of the approximated process with such parameters is similar to the original process.

Remark 1.

It is difficult to use the CMC pricing formula of Equation (22) if the analytic solution of a European multi-asset option price under constant volatility does not exist. However, we can still construct a martingale CV for an MC simulation in those cases.

According to the martingale representation theorem (Theorem 1), the variance of $e^{-rT}h(S(T))$ *in an MC simulation is totally determined by the zero martingale X (see Equation (23)). We can select a value* Vapprox(t,S(t);θ,Γ) *with a constant volatility parameter* θ *as the approximation of option price* V(t,S(t),Y(t)) *under a stochastic volatility model. Following the idea in the CMC framework, we select* θ=f(E[Y(t)]). *Thus, the partial derivatives can be approximated as:*

$$\begin{aligned} \nabla_S V(t, S(t), Y(t)) &\approx \nabla_S V^{\text{approx}}(t, S(t); \theta, \Gamma), \\ \nabla_Y V(t, S(t), Y(t)) &\approx \nabla_Y V^{\text{approx}}(t, S(t); \theta, \Gamma) = 0. \end{aligned}$$

Furthermore, the martingale CV is:

$$\tilde{X} = \int_0^T e^{-rt} \left(\rho \cdot S(t) \cdot f(Y(t)) \cdot \nabla_S V^{\text{approx}}(t, S(t); \theta, \Gamma)\right)' dZ(t).$$

Taking the arithmetic average basket option with stochastic volatilities as an example, we can use the geometric average basket option with constant volatilities as an approximation and then construct the corresponding CV. It is expected that, for a more accurate approximated price Vapprox(t,S(t);θ,Γ), *a larger variance reduction ratio can be obtained by the corresponding martingale CV.*

NUMERICAL TESTS

In this section, we present some numerical tests designed for the typical multi-asset options— including the exchange options, basket options and quanto options—to emphasize the efficiency of our method.

Exchange Options

The exchange option, which was first studied by Margrabe [70], empowers its holder with the right to exercise it by comparing the difference between the prices or the rates of return of two underlying assets. Its payoff is:

$$h^{\text{excha}}(S_1(T), S_2(T)) = \max(S_2(T) - S_1(T), 0). \tag{34}$$

If the underlying assets evolute with constant volatilities σ_1 and σ_2, the exchange option has a pricing formula at time t, as shown by Margrabe [70] and Jiang [20] as follows:

$$V^{BS-\text{excha}}(t, S_1, S_2) = e^{-\delta_2(T-t)} S_2 N(-d_2) - e^{-\delta_1(T-t)} S_1 N(-d_1), \tag{35}$$

where

$$d_1 = \frac{\ln \frac{S_1}{S_2} + \left(\delta_2 - \delta_1 + \frac{1}{2}(\sigma_1^2 - 2\rho_{12}\sigma_1\sigma_2 + \sigma_2^2)\right)(T-t)}{\sqrt{(\sigma_1^2 - 2\rho_{12}\sigma_1\sigma_2 + \sigma_2^2)(T-t)}}, \tag{36}$$

$$d_2 = d_1 - \sqrt{(\sigma_1^2 - 2\rho_{12}\sigma_1\sigma_2 + \sigma_2^2)(T-t)}, \tag{37}$$

and $N(x) = \frac{1}{\sqrt{2\pi}}\int_{-\infty}^{x} e^{-\frac{t^2}{2}}dt$ is the cumulative distribution function of a standard normal variable. It is easy to derive the derivatives:

$$\begin{aligned} \frac{\partial}{\partial S_1} V^{\text{BS-excha}}(t, S_1, S_2) &= -e^{-\delta_1(T-t)} N(-d_1), \\ \frac{\partial}{\partial S_2} V^{\text{BS-excha}}(t, S_1, S_2) &= e^{-\delta_2(T-t)} N(-d_2). \end{aligned}$$

We assumed the stochastic volatilities obey the Heston model, for i=1,2:

$$\begin{aligned} \frac{dS_i(t)}{S_i(t)} &= (r - \delta_i)dt + \sqrt{Y_i(t)}dW_i(t), \\ dY_i(t) &= a_i(\theta_i - Y_i(t))dt + \sigma_i\sqrt{Y_i(t)}dZ_i(t). \end{aligned}$$

The parameters should satisfy the Feller condition [71] to guarantee the positiveness of variance, i.e., $2a_1\theta_1 > \sigma_1^2$ and $2a_2\theta_2 > \sigma_2^2$. We used a truncated Euler discrete scheme [39,72,73] with equal time intervals to simulate the Heston process in our tests.

At first, we wanted to examine the acceleration effect of a CMC simulation compared with a traditional MC simulation. We fixed the parameters $S_1(0)=S_2(0)=K=30$, $r=0.05$, $T=1$, $\delta_1=\delta_2=0$, $a_1=a_2$, $\sigma_1=\sigma_2=0.2$, $Y_1(0)=0.01$, $Y_2(0)=0.04$, $\theta_1=0.015$, and $\theta_2=0.05$. Additionally, we took the number of time steps N=100, and the number of simulations m= 100,000 in all numerical simulations. Note that $\left|\frac{\rho_{12}}{\sqrt{1-\rho_1^2}\sqrt{1-\rho_2^2}}\right| < 1$ should be satisfied from Equation (17). Taking $\rho_{12}=0$ for simplicity, the numerical results with different correlation coefficients are recorded in Table 2 and Table 3.

Table 2. Exchange option: Estimated prices for MC and CMC with different correlation coefficients.

$\rho_1 \mid \rho_2$	−0.75	−0.50	−0.25	0.00	0.25	0.50	0.75	
−0.75	2.8083	2.8218	2.8336	2.8436	2.8519	2.8587	2.8639	
−0.50	2.8169	2.8294	2.8402	2.8493	2.8567	2.8623	2.8663	
−0.25	2.8248	2.8362	2.8459	2.8539	2.8603	2.8649	2.8678	
0.00	2.8317	2.8420	2.8505	2.8573	2.8625	2.8661	2.8680	$\overline{V}_{MC}$
0.25	2.8372	2.8462	2.8536	2.8594	2.8635	2.8660	2.8667	
0.50	2.8413	2.8492	2.8554	2.8600	2.8630	2.8644	2.8641	
0.75	2.8440	2.8508	2.8558	2.8592	2.8612	2.8615	2.8601	
−0.75	2.8153	2.8294	2.8420	2.8528	2.8616	2.8684	2.8731	
−0.50	2.8253	2.8377	2.8486	2.8578	2.8651	2.8703	2.8735	
−0.25	2.8340	2.8446	2.8539	2.8615	2.8672	2.8709	2.8725	
0.00	2.8413	2.8503	2.8579	2.8639	2.8679	2.8700	2.8700	$\overline{V}_{CMC}$
0.25	2.8473	2.8546	2.8606	2.8649	2.8674	2.8678	2.8661	
0.50	2.8519	2.8576	2.8620	2.8647	2.8654	2.8641	2.8608	
0.75	2.8550	2.8592	2.8620	2.8630	2.8621	2.8591	2.8539	

Table 3. Exchange option: Numerical results for MC and CMC with different correlation coefficients.

$\rho_1 \| \rho_2$	−0.75	−0.50	−0.25	0.00	0.25	0.50	0.75	
−0.75	0.0132	0.0137	0.0143	0.0148	0.0153	0.0159	0.0164	
−0.50	0.0132	0.0137	0.0142	0.0148	0.0153	0.0158	0.0164	
−0.25	0.0131	0.0137	0.0142	0.0147	0.0153	0.0158	0.0164	
0.00	0.0131	0.0136	0.0141	0.0147	0.0152	0.0158	0.0163	Std_{MC}
0.25	0.0130	0.0135	0.0141	0.0146	0.0152	0.0157	0.0163	
0.50	0.0130	0.0135	0.0140	0.0146	0.0151	0.0157	0.0163	
0.75	0.0129	0.0134	0.0140	0.0145	0.0151	0.0156	0.0162	
−0.75	0.0088	0.0065	0.0049	0.0045	0.0058	0.0084	0.0119	
−0.50	0.0081	0.0055	0.0036	0.0032	0.0049	0.0077	0.0115	
−0.25	0.0076	0.0048	0.0025	0.0020	0.0042	0.0073	0.0112	
0.00	0.0074	0.0045	0.0019	0.0011	0.0039	0.0071	0.0110	Std_{CMC}
0.25	0.0075	0.0046	0.0020	0.0012	0.0039	0.0072	0.0111	
0.50	0.0078	0.0050	0.0027	0.0021	0.0043	0.0074	0.0113	
0.75	0.0084	0.0057	0.0036	0.0031	0.0048	0.0078	0.0116	
−0.75	1.5014	2.1268	2.9394	3.2711	2.6283	1.8898	1.3762	
−0.50	1.6377	2.4911	3.9677	1.6222	3.1339	2.0450	1.4323	
−0.25	1.7336	2.8222	5.5991	7.3682	3.6300	2.1571	1.4677	
0.00	1.7692	2.9943	7.3944	13.0340	3.9300	2.2070	1.4805	$R = \frac{Std_{MC}}{Std_{CMC}}$
0.25	1.7385	2.9276	7.0000	11.8935	3.8805	2.1900	1.4714	
0.50	1.6551	2.6807	5.2352	7.0184	3.5503	2.1170	1.4433	
0.75	1.5402	2.3701	3.9081	4.7380	3.1223	2.0066	1.4001	

Table 2 records the estimated option values calculated by the MC and CMC simulations, which are denoted as $\bar{V}_{MC}$ and $\overline{V}_{CMC}$, respectively. The upper part of Table 3 records the standard errors of an MC simulation, denoted as StdMC. The standard errors are almost the same for various correlation coefficients, and increase slightly with correlation ρ_2 while decreasing with ρ_1. The exchange option can be seen as a call option on asset S_2 for fixed S_1; a higher correlation ρ_2 implies a larger variation in the price of asset S_2, thus resulting in a larger value of the option price and a larger simulation variance. Similar analysis can be conducted with respect to correlation ρ_1 by regrading the exchange option as a put option on asset S_1 for fixed S_2.

The middle part of Table 3 records the standard errors of a CMC simulation, denoted as Std_{CMC}. Obviously, the standard CMC errors are always smaller than MC. It is interesting that a standard CMC error rapidly declines as correlation coefficient ρ_1 or ρ_2 tends to zero. Thus, the ratio of the standard errors of a CMC simulation to an MC simulation reduces. We denote this ratio as $R=Std_{MC}/Std_{CMC}$ and present its values at the bottom of Table 3. *R* becomes larger when the correlation coefficient is getting closer to the original point, and decays rapidly in the opposite direction. For example, for $\rho_1=\rho_2=0,0.25,0.5$, and 0.75, the reduction ratios of the standard error are 13.0340, 3.8805, 2.1170, and 1.4001, respectively. This can be explained by Equation (11); the CMC simulation removes the randomness that is independent from the stochastic variances Y_1,Y_2, and its quantity is proportional to $\sqrt{1-\rho_1^2}$, or $\sqrt{1-\rho_2^2}$. In other words, a larger variance reduction ratio is promised when the absolute value of ρ_1 or ρ_2 is smaller. This property indicates that a CMC simulation is more competitive when the correlation between the underlying asset and stochastic volatility is weak.

We also investigated the computational costs of the MC and CMC methods. The computational platform for this paper was an Intel i5-6200U CPU, 2.30 GHz, 8 GB memory, and the software environment was Matlab R2018a for Windows 10. It took 50.88 s to calculate all of the data in the upper part of Table 3 and 25.85 s for the middle part, which means that the time cost of a CMC simulation is almost half that of an MC simulation. This is because the MC method needs to simulate four random variables, {W_1(t),W_2(t),Z_1(t), and Z_2(t)}, while the CMC method only needs to simulate two random variables, {Z_1(t), and Z_2(t)}.

Taking the variance reduction ratio into consideration, the speed up ratio of a CMC simulation to an MC simulation is defined as $\frac{Std_{MC}^2}{Std_{CMC}^2}\cdot\frac{t_{MC}}{t_{CMC}}$. Thus, when correlation ρ_1=ρ_2=0, the speed up ratio of the CMC is 13.03402·50.8825.85=334.38. Even for the case of a larger correlation ρ_1=ρ_2=0.75, the speed up ratio of the CMC is $1.4001^2\cdot\frac{50.88}{25.85}=3.86$, which improves the efficiency of the MC simulation by roughly 75%. In summary, the CMC simulation enjoys the advantages of saving time and having a great variance reduction ratio, especially when the correlation coefficients are small.

We next tested the efficiency of our martingale CV method. As a contrast, we constructed another CV for the stochastic model, as suggested by Ma and Xu [74]. Consider dummy assets whose prices $\tilde{S}_i(t), (i=1,2,\cdots,n)$ satisfy the following stochastic differential equations:

$$\frac{d\tilde{S}_i(t)}{\tilde{S}_i(t)}=(r-\delta_i)dt+\tilde{\sigma}_i(t)dW_i(t),$$

where $\tilde{\sigma}_i(t)$ is a determined function. The covariance of dW_i(t) is given by Equation (**3**). It can be computed by matching the first two moments of the underlying asset prices as $\tilde{\sigma}_i^2(t)=E[f_i^2(Y_i(t))]$. In the case of a Heston stochastic volatility model:

$$\tilde{\sigma}_i^2(t)=E[Y_i(t)]=\theta_i+(Y_i(0)-\theta_i)e^{-a_it},\quad i=1,2.$$

We used the payoff hexcha $(\tilde{S}_1(T), \tilde{S}_2(T))$ as a CV to the MC method, and we called this CV method a function CV method. The corresponding exchange option price can be computed using Equation (35) by replacing $\sigma_1^2 + \sigma_2^2 - 2\rho_{12}\sigma_1\sigma_2$ in Equations (36) and (37) with the average volatility on the interval [0,T] given by:

$$\frac{1}{T}\int_0^T E[Y_1(t)] + E[Y_2(t)] - 2\rho_{12}\sqrt{E[Y_1(t)]E[Y_2(t)]}dt.$$

We changed the values of the correlation coefficients and kept the other parameters fixed as before. Remember that $\left|\frac{\rho_{12}}{\sqrt{1-\rho_1^2}\sqrt{1-\rho_2^2}}\right| < 1$. The detailed results are shown in Table 4.

Table 4. Exchange option: Numerical results for CVs with different correlation coefficients.

ρ_{12}	$\rho_1 = \rho_2$	Std_{MC}	Std_{Mar}	Std_{Fun}	R_1	R_2
−0.5	−0.6	0.0160	0.0015	0.0039	10.3761	4.1166
	−0.4	0.0163	0.0013	0.0040	12.2233	4.0626
	−0.2	0.0166	0.0012	0.0041	13.9129	4.0540
	0.0	0.0170	0.0011	0.0042	15.0719	4.0821
	0.2	0.0173	0.0013	0.0042	13.2504	4.1410
	0.4	0.0176	0.0016	0.0042	11.1433	4.2271
	0.6	0.0180	0.0019	0.0041	9.5880	4.3348
0.0	−0.75	0.0132	0.0014	0.0039	9.6978	3.4035
	−0.50	0.0137	0.0012	0.0040	11.2209	3.4318
	−0.25	0.0142	0.0010	0.0041	14.6458	3.4829
	0.00	0.0147	0.0009	0.0041	16.2942	3.5500
	0.25	0.0152	0.0011	0.0042	13.3596	3.6291
	0.50	0.0157	0.0015	0.0042	10.2764	3.7178
	0.75	0.0162	0.0017	0.0043	9.6550	3.8097
0.5	−0.6	0.0106	0.0010	0.0038	10.9374	2.8023
	0.4	0.0110	0.0009	0.0040	12.9115	2.7755
	−0.2	0.0115	0.0006	0.0041	17.7294	2.7910
	0.0	0.0119	0.0006	0.0042	19.8106	2.8424
	0.2	0.0123	0.0008	0.0042	14.7523	2.9203
	0.4	0.0127	0.0012	0.0042	10.7868	3.0531
	0.6	0.0131	0.0012	0.0041	10.6183	3.2264

In Table 4, Std_{MC}, Std_{Mar}, and Std_{Fun} are the standard errors from the MC simulation, the martingale CV method, and the function CV method, respectively. $R_1=Std_{MC}/Std_{Mar}$ is the standard error reduction ratio of the martingale CV method compared to the MC simulation, and $R_2=Std_{MC}/Std_{Fun}$ is the standard error reduction ratio of the function CV method compared to the MC simulation.

It is obvious that the standard error reduction ratio of the CMC is much larger than that of the function CV method, the former falling in 9–20 while the latter being about 3 or 4. Table 4 also shows that, for a fixed $\rho_1=\rho_2$, the standard errors of the MC simulation, martingale CV method, and function CV method decrease with the correlation value of ρ_{12}. For a fixed ρ_{12}, the standard errors of the MC simulation and function CV method increase with the value of $\rho_1=\rho_2$ while the martingale CV method decreases with the absolute value of $\rho_1=\rho_2$, which is mainly caused by the properties of the CMC. Thus, the standard error reduction ratio of the martingale CV method also decreases with the absolute value of $\rho_1=\rho_2$.

The computing times for all values of the MC, the martingale CV, and the function CV methods are 22.33, 22.26, and 25.50 s, respectively. The time costs of the MC method and the martingale CV method are almost the same, while the function CV method is slightly slower. Thus, the martingale CV method proposed in our paper is superior to the function CV method, when considering the variance reduction ratio and the time cost.

Fixing the parameters ρ12=0, and ρ1=ρ2=0.5, we next examined the effects of the volatility parameters for the stochastic volatility. In the Heston stochastic volatility model, the Feller condition should be satisfied [**71**]; thus, $\sigma_1 < \sqrt{2a_1\theta_1} = \sqrt{2 \cdot 2 \cdot 0.015}$ =0.2449, and $\sigma_2 < \sqrt{2a_2\theta_2} = \sqrt{2 \cdot 2 \cdot 0.05} = 0.4472$. Numerical results of these tests are shown in Table 5.

Table 5. Exchange option: Numerical results for CVs with different volatilities of the stochastic volatilities.

σ_1	σ_2	Std_{MC}	Std_{Mar}	Std_{Fun}	R_1	R_2
0.2	0.1	0.0151	0.0013	0.0030	11.7123	4.9767
	0.2	0.0157	0.0015	0.0042	10.2764	3.7178
	0.3	0.0163	0.0018	0.0057	9.0825	2.8678
	0.4	0.0169	0.0021	0.0073	8.1624	2.3279
0.05	0.2	0.0157	0.0014	0.0034	11.2239	4.5840
0.10		0.0157	0.0014	0.0036	10.9406	4.3545
0.15		0.0157	0.0015	0.0039	10.6170	4.0447
0.20		0.0157	0.0015	0.0042	10.2764	3.7178

As shown in Table 5, the standard errors of the three simulation methods all increase with increasing volatilities of stochastic volatilities. Standard error reduction ratios also decline with the volatility of the stochastic volatilities. However, our martingale CV method is much more efficient than the function CV method, especially in the case of large volatility.

Basket Options

The payoff of the basket option at maturity depends on the average price of the underlying assets. Since the basket option with arithmetic average price does not have a closed-form price, even with constant volatility, we considered the geometric average basket option whose payoff at time T is:

$$h^{\text{GeomBasket}}(\mathbf{S}(T)) = \max\left(\prod_{i=1}^{n} S_i^{\alpha_i}(T) - K, 0\right), \tag{38}$$

where n is the number of underlying assets, $\alpha_i \geq 0$ are the weights of each underlying asset with $\sum_{i=1}^{n} \alpha_i = 1$, and K is the strike price.

The geometric average basket option has a closed-form solution if the underlying assets have constant volatilities as $\sigma_1, \cdots, \sigma_n$. Denote:

$$\hat{\sigma}^2 = \sum_{i,j=1}^{n} \alpha_i \alpha_j \sigma_i \sigma_j \rho_{ij},$$
$$\hat{\delta} = \sum_{i=1}^{n} \alpha_i \left(\delta_i + \frac{\sigma_i^2}{2}\right) - \frac{\hat{\sigma}^2}{2}.$$

The geometric average basket option price at time t is given by Jiang [20]:

$$V^{\text{BS-GeomBasket}}(t, \mathbf{S}) = S_1^{\alpha_1} \cdots S_n^{\alpha_n} e^{-\hat{\delta}(T-t)} N(d_1) - Ke^{-r(T-t)} N(d_2), \tag{39}$$

where

$$d_1 = \frac{\ln \frac{S_1^{\alpha_1} \cdots S_n^{\alpha_n}}{K} + \left(r - \hat{\delta} + \frac{1}{2}\hat{\sigma}^2\right)(T-t)}{\hat{\sigma}\sqrt{T-t}},$$
$$d_2 = d_1 - \hat{\sigma}\sqrt{T-t}$$

Thus, the derivatives are:

$$\frac{\partial}{\partial S_j} V^{\text{BS-GeomBasket}}(t, \mathbf{S}) = e^{-\hat{\delta}(T-t)} N(d_1) \alpha_j S_j^{-1} \prod_{i=1}^{n} S_i^{\alpha_i}, \quad j = 1, 2, \quad , n.$$

For a basket option with n underlying assets, we still used the Heston stochastic volatility model and function CV method as a comparison. The expectation of the corresponding CV can be calculated using Equation (**39**) by substituting $\sigma_i \sigma_j$ with $\frac{1}{T}\int_0^T \sqrt{E[Y_i(t)]E[Y_j(t)]}dt$. We fixed the parameters r=0.05,T=1,K=30, Si(0)=30,δ_i=0,a_i=2, and σ_i=0.2(i=1,2,···,n). We allocated equal weights for the underlying assets, which means that αi=1/n. For the initial value of the stochastic volatility, we took a linear interpolation between 0.12 and 0.32 for the *n* assets. In other words, the initial variance vector was Y(0)=(0.1²,0.3²)′ for n=2 and Y(0)=(0.1²,0.15²,0.2²,0.25²,0.3²)′ for n=5. We took the long-term mean of stochastic variance as $\theta_i = (\sqrt{Y_i(0)} + 0.05)^2$, which was θ=(0.15²,0.35²)′ for n=2, for example. For the correlations between Brownian noises, we took $\rho_{ij}=\rho_0(i \neq j)$ for simplicity. To guarantee the positive definiteness of the matrix $\Gamma=(\rho_{ij})$, the parameter ρ0 should satisfy −1/(n−1)<ρ0<1. In addition, $\frac{\rho_{ij}}{\sqrt{1-\rho_i^2}\sqrt{1-\rho_j^2}}$ <1 is needed for the proper definition of $\tilde{\Gamma}$. Thus, we set $\rho_{ij}=\rho_0=0(i \neq j)$ at first. We fixed the number of time steps to N=100 and the number of simulations to m= 100,000. We tested the acceleration effects of the CVs for different correlation coefficients ρ_i. Numerical results are shown in Table 6.

Table 6. Geometric average basket option: Numerical results for CVs with different correlation coefficients.

n	ρ_i	Std_{MC}	Std_{Mar}	Std_{Fun}	R_1	R_2
2	−0.75	0.0110	0.0012	0.0021	9.2740	5.2542
	−0.50	0.0112	0.0009	0.0021	12.8705	5.2650
	−0.25	0.0115	0.0004	0.0022	27.4754	5.2935
	0.00	0.0117	0.0004	0.0022	30.5615	5.3375
	0.25	0.0120	0.0005	0.0022	23.5176	5.3968
	0.50	0.0122	0.0010	0.0022	11.9347	5.4719
	0.75	0.0124	0.0013	0.0022	9.2236	5.5629
5	−0.75	0.0068	0.0009	0.0013	7.6660	5.1178
	−0.50	0.0069	0.0006	0.0013	11.7653	5.1320
	−0.25	0.0069	0.0002	0.0013	27.9824	5.1434
	0.00	0.0070	0.0001	0.0014	110.3248	5.1544
	0.25	0.0071	0.0003	0.0014	25.1755	5.1663
	0.50	0.0071	0.0006	0.0014	11.1250	5.1785
	0.75	0.0072	0.0010	0.0014	7.4879	5.1952
10	−0.75	0.0049	0.0008	0.0010	6.3077	5.0690
	−0.50	0.0049	0.0005	0.0010	10.4984	5.0820
	−0.25	0.0049	0.0002	0.0010	26.5073	5.0928
	0.00	0.0050	1.7×10^{-5}	0.0010	291.4214	5.1023
	0.25	0.0050	0.0002	0.0010	24.2257	5.1104
	0.50	0.0050	0.0005	0.0010	10.0253	5.1192
	0.75	0.0050	0.0008	0.0010	6.1413	5.1310

Table 6 again shows that Std_{Mar}, the standard error of the martingale CV method, decreases as the correlation coefficients ρ_i tends to zero, resulting in a greater standard error reduction ratio R1 in those cases. For example, R_1 goes from 30 to 9 when $|\rho_i|$ goes from 0 to 0.75. On the other hand, the simulation error Std_{Fun} and, thus, the corresponding reduction ratio R2 of the function CV method are not sensitive to the correlation coefficient. The reduction ratio is around 5 in all cases. Considering the number of underlying assets, the reduction ratio of the martingale CV slightly decreases as the number of assets n becomes larger, except for the ρ_i=0 cases. For example, the reduction ratios of the martingale CV method are 9.2740, 7.6660, and 6.3077 for n=2,5, and 10, respectively, and for ρ_i=−0.75. On the other hand, the ratios are 30.5615, 110.3248, and 291.4214 for the ρ_i=0 case. As a contrast, the performance of the function CV method is more stable with different n. It is obvious that our martingale CV is much more efficient than the function CV method.

Next, we fixed ρ_i=0.5(i=1,2,⋯,n) and changed the value of σ_i, the volatility of the stochastic volatility. For convenience, we took an equal σ_i for every underlying asset. The results are recorded in Table 7.

Table 7. Geometric basket option: Numerical results for CVs with different volatilities of the stochastic volatility.

n	σ_i	Std_{MC}	Std_{Mar}	Std_{Fun}	R_1	R_2
2	0.1	0.0177	0.0013	0.0011	13.7479	15.6830
	0.2	0.0179	0.0013	0.0023	13.3158	7.9626
	0.3	0.0182	0.0014	0.0034	12.8095	5.4211
	0.4	0.0184	0.0015	0.0044	12.2580	4.1894
5	0.1	0.0093	0.0007	0.0006	12.7129	14.3708
	0.2	0.0093	0.0007	0.0013	12.4417	7.2688
	0.3	0.0094	0.0008	0.0019	12.1388	4.9147
	0.4	0.0094	0.0008	0.0025	11.8127	3.7571
10	0.1	0.0060	0.0005	0.0004	11.8384	13.8295
	0.2	0.0060	0.0005	0.0009	11.6365	6.9848
	0.3	0.0060	0.0005	0.0013	11.4192	4.7179
	0.4	0.0060	0.0005	0.0017	11.1910	3.5979

As shown in Table 7, the standard errors of the three simulation methods increase with the volatilities of the stochastic volatility at fixed n, and decrease with the number of underlying assets for a fixed σ_i. The standard error reduction ratios of the two CV methods decrease with increasing volatility of the stochastic volatility and increasing number of assets. However, the martingale CV method is more robust for different volatilities compared to the function CV method. For example, for the case of n=2, the standard error reduction ratio of the martingale CV method decreases from 13.7479 to 12.2580 when σ_i increases from 0.1 to 0.4, while that of the function CV method sharply decreases from 15.6830 to 4.1894. The results suggest that our martingale CV method is especially efficient in high volatility cases, while the function CV method has some advantages in a low volatility environment.

Quanto Options with Real Data

The quanto option is a contract written when someone invests money in foreign securities. Usually, its risk depends on the volatility of the securities' prices and the change of the foreign currency rate. Its main purpose is to provide exposure to a foreign asset without taking the corresponding exchange rate risk. We applied our method to price a quanto option. Park et al. [62] used a power series expansion method to obtain an analytic approximation value for the quanto option price under the Hull–White stochastic volatility model.

First, we give the quanto option pricing model with Hull–White stochastic volatility, as shown in Park et al. [62]. Let S(t) be a stock price in foreign currency, and F(t) be a foreign exchange (FX) rate, that is the amount of domestic currency value per one foreign currency value. In a risk-neutral world, they are assumed to obey the following stochastic differential equations:

$$\begin{aligned} dS(t)/S(t) &= (r_f - \rho_{12}\sigma_1(t)\sigma_2(t))dt + \sigma_1(t)dW_1(t), \\ dF(t)/F(t) &= (r_d - r_f)dt + \sigma_2(t)dW_2(t), \\ d\sigma_1(t)/\sigma_1(t) &= \mu_1 dt + \xi_1 dZ_1(t), \\ d\sigma_2(t)/\sigma_2(t) &= \mu_2 dt + \xi_2 dZ_2(t), \end{aligned}$$

where rd is a risk-free domestic interest rate and r_f is a risk-free foreign interest rate.

The correlations among the Brownian noises are given by $\text{cov}(dW_1(t),dW_2(t))=\rho_{12}dt$, $\text{cov}(dW_1(t),dZ_1(t))=\rho_1 dt$, and $\text{cov}(dW_2(t),dZ_2(t))=\rho2dt$. Additionally, $\sigma_1(t)$ and $\sigma_2(t)$ are the stochastic volatilities of the stock price and the FX rate, respectively. This form of the Hull–White stochastic volatility is a little different from that in Table 1 (for more details, please see Park et al. [62]). The parameters μ_1,μ_2,ξ_1, and ξ_2 are constants.

Park et al. [62] considered a specific quanto option with payoff:

$$h^{\text{Quanto}}(\mathbf{S}(T)) = F_0 \max(S(T) - K, 0), \quad (40)$$

where F_0 is a predetermined FX rate, and *K* is the strike price. A more general quanto option payoff would be $\max(F_0,F(T))\max(S(T)-K,0)$ (see Jiang et al. [20]). When volatilities $\sigma_1(t)$ and $\sigma_2(t)$ take constant values σ_1 and σ_2, respectively, the authors gave the Black–Scholes quanto option price as:

$$V^{\mathrm{BS-Quanto}}(t,\mathbf{S}) = F_0 e^{-r_d(T-t)} \left(S e^{(r_f-\rho_{12}\sigma_1\sigma_2)(T-t)} N(d_1) - K N(d_2) \right), \tag{41}$$

where

$$\begin{aligned} d_1 &= \frac{\ln\frac{S}{K} + \left(r_f - \rho_{12}\sigma_1\sigma_2 + \frac{1}{2}\sigma_1^2\right)(T-t)}{\sigma_1\sqrt{T-t}}, \\ d_2 &= d_1 - \sigma_1\sqrt{T-t}. \end{aligned}$$

It is easy to obtain the derivative

$$\frac{\partial}{\partial S} V^{\mathrm{BS-Quanto}}(t,\mathbf{S}) = F_0 e^{(r_f-r_d-\rho_{12}\sigma_1\sigma_2)(T-t)} N(d_1).$$

The authors [62] supposed a quanto European call option of the S&P500 index with 1200 strike and a predetermined FX rate of 1100 (KRW/USD). The model parameters shown in Table 8 were observed on 13 October 2010. Furthermore, we assume that the contract multiplier of the S&P500 option is 100 and the maturity is 13 October 2011. Without loss of generality, we set the unobserved values as zeros.

Table 8. Market dataset parameters.

View Date	13 October 2010
S&P500	1169.77
FX Rate (KRW/USD)	1127
Volatility of S&P500	18.58%
Volatility of FX Rate	11.83%
Correlation between S&P500 and FX Rate	−0.2297
Correlation between S&P500 and its volatility	−0.55
Volatility of volatility of S&P500	11.72%
Volatility of volatility of FX Rate	16.8%
USD LIBOR(1Y)	0.77%
KRW Treasury Rate(1Y)	2.91%

We changed the values of the correlation between the S&P500 and the FX rate and fixed all other parameters. The number of time steps was set to N=100 and the number of simulations was set to m= 100,000. The numerical results for these models are recorded in Table 9.

Table 9. Quanto option: Numerical results with different correlation coefficients.

ρ_{12}	V_{Appro} (10^6)	V_{MC} (10^6)	Std_{MC} (10^6)	V_{Mar} (10^6)	Std_{Mar} (10^6)	V_{Fun} (10^6)	Std_{Fun} (10^6)	R_1	R_2
−0.6	9.1393	9.1082	0.0462	9.1223	0.0038	9.1162	0.0047	12.1373	9.7825
−0.4	8.8325	8.8187	0.0454	8.8316	0.0037	8.8258	0.0046	12.3813	9.8215
−0.2	8.5311	8.5357	0.0447	8.5473	0.0035	8.5416	0.0045	12.6122	9.8533
0.0	8.2352	8.2592	0.0439	8.2696	0.0034	8.2638	0.0044	12.8264	9.8768
0.2	7.9448	7.9890	0.0432	7.9984	0.0033	7.9924	0.0044	13.0198	9.8906
0.4	7.6599	7.7253	0.0424	7.7335	0.0032	7.7276	0.0043	13.1884	9.8948
0.6	7.3807	7.4679	0.0417	7.4750	0.0031	7.4691	0.0042	13.3285	9.8900

In **Table 9**, ρ_{12} stands for the correlation between the S&P500 and FX rate. VAppro is the approximated value obtained by the series expansion method in Park et al. [62]. V_{MC}, V_{Mar}, and V_{Fun} are the estimated values of the MC simulation, the martingale CV method, and the function CV method, respectively. Std_{MC}, Std_{Mar}, and Std_{Fun} are the standard errors of the MC simulation, the martingale CV method, and the function CV method, respectively. $R1=Std_{MC}/Std_{Mar}$ is the standard error reduction ratio of the martingale CV method compared to the MC simulation, and $R_2=Std_{MC}/Std_{Fun}$ is the standard error reduction ratio of the function CV method compared to the MC simulation. For the function CV method, the expectation of the corresponding CV can be calculated by using Equation (41) and substituting $\sigma_i\sigma_j$ with $\frac{1}{T}\int_0^T E[\sigma_i(t)]E[\sigma_j(t)]dt$, where $E[Y_i(t)] = Y_i(0)e^{\mu_i t}, i = 1,2$. It is obvious that our martingale CV method has a larger standard reduction ratio than the function CV method. This, again, shows the efficiency and robustness of our method.

CONCLUSIONS

In the context of European multi-asset options with stochastic volatilities, we propose a dimension and variance reduction Monte Carlo method. A conditional Monte Carlo pricing formula is deduced, and then the martingale representation theorem is proved. A martingale control variate is combined with the conditional Monte Carlo simulation.

Numerical tests on typical multi-asset options—including exchange options, basket options, and currency options—showed that this method yields considerable variance reduction, not only when compared to a traditional Monte Carlo simulation, but also with respect to the function control variate in Ma and Xu [74].

For future research, it would be interesting and challenging to extend the framework in this paper to price more options with stochastic volatilities, not only European options but also exotic options such as American options or barrier options. Furthermore, it would be interesting to study jump diffusion models with stochastic volatilities. Another important approach is to use this framework in empirical financial studies and risk management. After model parameters are calibrated with real market data, our method can be used to accurately and quickly value option prices which can be widely used in areas of economics and finance. We would also like to extend this method to other areas like risk management and civil engineering.

ABBREVIATIONS

The following abbreviations are used in this manuscript:

MC	Monte Carlo
CV	control variate
CMC	conditional Monte Carlo
FFT	fast Fourier transformation
FX	foreign exchange

REFERENCES

1. Hull, J.C. *Options, Futures and Other Derivatives*, 10th ed.; Pearson Education: New York, NY, USA, 2018.
2. Kim, Y.; Shin, K.; Ahn, J.; Lee, E.-B. Probabilistic Cash Flow-Based Optimal Investment Timing Using Two-Color Rainbow Options Valuation for Economic Sustainability Appraisement. *Sustainability* 2017, *9*, 1781.
3. Yoo, J.-I.; Lee, E.-B.; Choi, J.-W. Balancing Project Financing and Mezzanine Project Financing with Option Value to Mitigate Sponsor's Risks for Overseas Investment Projects. *Sustainability* 2018, *10*, 1498.
4. Black, F.; Scholes, M. The Pricing of Options and Corporate Liabilities. *J. Political Econ.* 1973, *81*, 637–654.
5. Merton, R.C. Theory of Rational Option Pricing. *Bell J. Econ. Manag. Sci.* 1973, *4*, 141–183.
6. Rubinstein, M. Nonparametric tests of alternative option pricing models using all reported trades and quotes on the 30 most active CBOE option classes from August 23, 1976 through August 31, 1978. *J. Financ.* 1985, *40*, 455–480.
7. Andersen, L.; Andreasen, J. Jump-diffusion processes: Volatility smile fitting and numerical methods for option pricing. *Rev. Deriv. Res.* 2000, *4*, 231–262.
8. Hull, J.; White, A. The pricing of options on assets with stochastic volatilities. *J. Financ.* 1987, *42*, 281–300.
9. Scott, L.O. Option pricing when the variance changes randomly: Theory, estimation, and an application. *J. Financ. Quant. Anal.* 1987, *22*, 419–438.
10. Stein, E.M.; Stein, J.C. Stock price distributions with stochastic volatility: An analytic approach. *Rev. Financ. Stud.* 1991, *4*, 727–752.
11. Ball, C.A.; Roma, A. Stochastic volatility option pricing. *J. Financ. Quant. Anal.* 1994, *29*, 589–607.
12. Heston, S.L. A closed solution for options with stochastic volatility with applications to bond and currency options. *Rev. Financ. Stud.* 1993, *6*, 327–343.
13. Schöbel, R.; Zhu, J. Stochastic volatility with an Ornstein-Uhlenbeck process: An extension. *Eur. Financ. Rev.* 1999, *3*, 23–46.
14. Hagan, P.S.; Kumar, D.; Lesniewski, A.S.; Woodward, D.E. Managing

smile risk. *Wilmott Mag.* 2002, *1*, 249–296.

15. Fouque, J.P.; Papanicolaou, G.; Sircar, K.R. Mean-reverting stochastic volatility. *Int. J. Theor. Appl. Financ.* 2000, *3*, 101–142.
16. Fouque, J.P.; Tullie, T.A. Variance reduction for Monte Carlo simulation in a stochastic volatility environment. *Quant. Financ.* 2002, *2*, 24–30.
17. Fouque, J.P.; Han, C.H. Variance reduction for Monte Carlo methods to evaluate option prices under multi-factor stochastic volatility models. *Quant. Financ.* 2004, *4*, 597–606.
18. Fouque, J.P.; Han, C.H. A martingale control variate method for option pricing with stochastic volatility. *ESAIM Probab. Stat.* 2007, *11*, 40–54.
19. Fouque, J.P.; Papanicolaou, G.; Sircar, R. *Derivatives in Financial Markets with Stochastic Volatility*; Cambridge University Press: Cambridge, UK, 2000
20. Jiang, L.S. *Mathematical Models and Methdods for Options Prcing*, 2nd ed.; China Higher Education Press: Beijing, China, 2014.
21. Turnbull, S.; Wakeman, L. A quick algorithm for pricing European average options. *J. Financ. Quant. Anal.* 1991, *26*, 377–389.
22. Curran, M. Valuing Asian and portfolio options by conditioning on the geometirc mean price. *Manag. Sci.* 1994, *40*, 1705–1711.
23. Milevsky, M.A.; Posner, S.E. A closed-form approximation for valuing basket options. *J. Deriv.* 1998, *5*, 54–61.
24. Ju, N. Pricing Asian and basket options via Taylor expansion. *J. Comput. Financ.* 2002, *5*, 79–103.
25. Zhou, J.; Wang, X. Accurate closed-form approximation for pricing Asian and basket options. *Appl. Stoch. Model. Bus. Ind.* 2008, *24*, 343–358.
26. Alexander, C.; Venkatramanan, A. Analytic approximations for multi-asset option pricing. *Math. Financ.* 2012, *22*, 667–689.
27. Datey, J.Y.; Gauthier, G.; Simonato, J.G. The performance of analytical approximations for the computation of Asian quanto-basket option prices. *Multinat. Financ.* 2003, *7*, 55–82.
28. Brigo, D.; Mercurio, F.; Rapisarda, F.; Scotti, R. Approximated momentmatching dynamics for basket-options pricing. *Quant. Financ.* 2004, *4*, 1–16.
29. Borovkova, S.; Permana, F.J.; Weide, H.V.D. A closed form approach to the valuation and hedging of basket and spread option. *J.*

Deriv. 2007, *14*, 8–24.

30. Deelstra, G.; Liinev, J.; Vanmaele, M. Pricing of arithmetic basket options by conditioning. *Insur. Math. Econ.* 2004, *34*, 55–77.
31. Deelstra, G.; Diallo, I.; Vanmaele, M. Moment matching approximation of Asian basket option prices. *J. Comput. Appl. Math.* 2010, *234*, 1006–1016.
32. Li, M.; Deng, S.; Zhou, J. Closed-Form Approximations for Spread Option Prices and Greeks. 2010. Available online: https://ssrn.com/abstract=952747 (accessed on 30 December 2018).
33. Carr, P.; Madan D, B. Option valuation using the fast Fourier transform. *J. Comput. Financ.* 1999, *2*, 61–73.
34. Carr, P.; Wu, L. Time-changed Lévy processes and option pricing. *J. Financ. Econ.* 2004, *71*, 113–141.
35. Grzelak, L.A.; Oosterlee, C.W.; Weeren, S.V. The affine Heston model with correlated gaussian interest rates for pricing hybrid derivatives. *Quant. Financ.* 2011, *11*, 1647–1663.
36. Grzelak, L.A.; Oosterlee, C.W. On cross-currency models with stochastic volatility and correlated interest rates. *Appl. Math. Financ.* 2012, *19*, 1–35.
37. Grzelak, L.A.; Oosterlee, C.W. On the Heston model with stochastic interest rates. *SIAM J. Financ. Math.* 2012, *2*, 255–286.
38. He, X.J.; Zhu, S.P. A closed-form pricing formula for European options under the Heston model with stochastic interest rate. *J. Comput. Appl. Math.* 2018, *335*, 323–333.
39. Glasserman, P. *Monte Carlo Methods in Financial Engineering*; Springer: New York, NY, USA, 2004.
40. Kemna, A.G.Z.; Vorst, A.C.F. A pricing method for options based on average asset values. *J. Bank. Financ.* 1990, *14*, 113–129.
41. Barraquand, J. Numerical valuation of high dimensional multivariate European securities. *Manag. Sci.* 1995, *41*, 1882–1891.
42. Pellizari, P. Efficient Monte Carlo pricing of European options using mean value control variates. *Decis. Econ. Financ.* 2001, *24*, 107–126.
43. Borogovac, T.; Vakili, P. Control variate technique: A constructive approach. In Proceedings of the 2008 Winter Simulation Conference, Miami, FL, USA, 7–10 December 2008; IEEE: Piscataway, NJ, USA, 2008; pp. 320–327.

44. Dingeç, K.D.; Hörmann, W. Control variates and conditional Monte Carlo for basket and Asian options. *Insur. Math. Econ.* 2013, *52*, 421–434.

45. Liang, Y.; Xu, C.; Ma, J. An efficient pricipal component Monte Carlo method for pricing European options with multi-factors. *J. Southwest Univ.* 2018, *40*, 88–97.

46. Sun, Y.; Xu, C. A hybrid Monte Carlo acceleration method of pricing basket options based on splitting. *J. Comput. Appl. Math.* 2018, *342*, 292–304.

47. Fang, K.; Hickernell, F.J.; Niederreiter, H. *Monte Carlo and Quasi-Monte Carlo Methods*; Springer: Berlin, Germany, 2000.

48. Dahl, L.O.; Benth, F.E. Valuation of Asian basket options with Quasi-Monte Carlo techniques and singular value decomposition. *Pure Math.* 2001, *5*, 1–21.

49. Wang, X. Variance reduction techniques and quasi-Monte Carlo methods. *J. Comput. Appl. Math.* 2001, *132*, 309–318.

50. Wang, X.; Fang, K.T. The effective dimension and quasi-Monte Carlo integration. *J. Complex.* 2003, *19*, 101–124.

51. Wang, X. On the effects of dimension reduction techniques on some high-dimensional problems in finance. *Oper. Res.* 2006, *54*, 1063–1078.

52. Wang, X. Dimension reduction techniques in quasi-Monte Carlo methods for option pricing. *INFORMS J. Comput.* 2009, *21*, 488–504.

53. Abbas-Turki, L.A.; Lapeyre, B. American options pricing on multi-core graphic cards. In Proceedings of the Second International Conference on Business Intelligence and Financial Engineering (BIFE2009), Beijing, China, 24–26 July 2009; IEEE: Piscataway, NJ, USA, 2009; pp. 307–311.

54. Abbas-Turki, L.A.; Vialle, S.; Lapeyre, B.; Mercier, P. High dimensional pricing of exotic European contracts on a GPU cluster, and comparison to a CPU cluster. In Proceedings of the IEEE International Symposium on Parallel and Distributed Processing (IPDPS 2009), Rome, Italy, 23–29 May 2009; IEEE: Piscataway, NJ, USA; 2009; pp. 1–8.

55. Lee, M.; Jeon, J.; Bae, J.; Jang, H.S. Parallel implementation of a financial application on a GPU. In Proceedings of the 2nd International Conference on Interaction Sciences: Information Technology, Culture and Human, Seoul, Korea, 24–26 November 2009; ACM: New York,

NY, USA, 2009; pp. 1136–1141.

56. Surkov, V. Parallel option pricing with Fourier space time-stepping method on graphics processing units. *Parallel Comput.* 2010, *36*, 372–380.
57. Dang, D.M.; Christara, C.C.; Jackson, K.R. An efficient graphics processing unit-based parallel algorithm for pricing multi-asset American options. *Concurr. Comp. Pract. E* 2012, *24*, 849–866.
58. Hu, Y.; Li, Q.; Cao, Z.; Wang, J. Parallel simulation of high-dimensional American option pricing based on CPU versus MIC. *Concurr. Comp. Pract. E* 2015, *27*, 1110–1121.
59. Du, K.; Liu, G.; Gu, G. Accelerating Monte Carlo method for pricing multi-asset options under stochastic volatility models. *IAENG Int. J. Appl. Math* 2014, *44*, 62–70.
60. Antonelli, F.; Ramponi, A.; Scarlatti, S. Exchange option pricing under stochastic volatility: A correlation expansion. *Rev. Deriv. Res.* 2010, *13*, 45–73.
61. Shiraya, K.; Takahashi, A. Pricing multiasset cross-currency options. *J. Futures Mark.* 2014, *34*, 1–19.
62. Park, J.; Lee, Y.; Lee, J. Pricing of quanto option under the Hull and White stochastic volatility model. *Commun. Korean Math. Soc.* 2013, *28*, 615–633.
63. Liang, Y.; Xu, C. Efficient accelerating method of conditional Monte-Carlo simulation for two-factor option pricing model. *J. Tongji Univ.* 2014, *42*, 645–650.
64. Willard, G.A. Calculating prices and sensitivities for path-independent derivative securities in multifactor models. *J. Deriv.* 1997, *5*, 45–61.
65. Drimus, G.G. Options on realized variance in Log-OU models. *Appl. Math. Financ.* 2012, *19*, 477–494.
66. Boyle, P.; Broadie, M.; Glasserman, P. Monte Carlo methods for security pricing. *J. Econ. Dyn. Control* 1997, *21*, 1267–1321.
67. Broadie, M.; Kaya, Ö. Exact simulation of stochastic volatility and other affine jump diffusion processes. *Oper. Res.* 2006, *54*, 217–231.
68. Yang, Y.; Ma, J.; Liang, Y. The Research on the Calculation of Barrier Options under Stochastic Volatility Models Based on the Exact Simulation. *IAENG Int. J. Appl. Math.* 2018, *48*, 112–124.
69. Ross, S. *A First Course in Probability*, 9th ed.; Pearson: Boston, MA,

USA, 2014.

70. Margrabe, W. The value of an option to exchange one asset for another. *J. Financ.* 1978, *33*, 177–186.
71. Feller, W. Two singular diffusion problems. *Ann. Math.* 1951, *54*, 173–182.
72. Andersen, L.B.G. Efficient simulation of the Heston stochastic volatility model. *J. Comput. Financ.* 2008, *11*, 1–42.
73. Lord, R.; Koekkoek, R.; Dijk, D.V. A comparison of biased simulation schemes for stochastic volatility models. *Quant. Financ.* 2010, *10*, 177–194.
74. Ma, J.; Xu, C. An efficient control variate method for pricing variance derivatives. *J. Comput. Appl. Math.* 2010, *235*, 108–119.

CHAPTER

16

A Comparative Study on Barrier Option Pricing using Antithetic and Quasi Monte-Carlo Simulations

Nneka Umeorah and Phillip Mashele

Unit for Business Mathematics and Informatics, North-West University, Potchefstroom, South Africa

ABSTRACT

Monte-Carlo simulations have been utilized greatly in the pricing of derivative securities. Over the years, several variance reduction techniques have been developed to curb the instability, as well as, increase the simulation efficiencies of the Monte-Carlo methods. Our approach in this research work will consider the use of antithetic variate techniques to estimate the fair prices of barrier options. Next, we use the quasi-Monte Carlo method,

Citation: Nneka Umeorah and Phillip Mashele "A Comparative Study on Barrier Option Pricing using Antithetic and Quasi Monte-Carlo Simulations" Journal of Mathematics and Statistics 2018, Volume 14: 94.106. http://dx.doi.org/10.3844/jmssp.2018.94.106

together with Sobol sequence to estimate the values of the same option. An extended version of the Black-Scholes model will serve as basis for the exact prices of these exotic options. The resulting simulated prices will be compared to the exact prices. The research concludes by showing some results which proves that when random numbers are generated via low discrepancy sequences in contrast to the normal pseudo-random numbers, a more efficient simulation method is ensued. This is further applicable in pricing complex derivatives without closed formsolutions.

Keywords: Black-Scholes Model, Barrier Options, Monte-Carlo, Antithetic Monte-Carlo, Quasi Monte-Carlo, Sobol Sequence, 2010 Mathematics Subject Classification: 91G60, 91G20, 91G30

INTRODUCTION

The path-dependent nature of barrier options classified them as exotic options and this becomes difficult during delta-hedging. However, notwithstanding their complicated nature of pricing, investors and corporate firms had used barrier options in portfolio insurance framework and bond trading, thereby improving efficiency and eliminating possible transaction costs (Rich, 1994, p. 295). With regards to bond trading, Cox and Rubinstein (1985, p. 408-411) explained that bonds which are fixed with short Downand-Out (DO) barrier call options can be used to minimize the costs related to writing and imposing expensive bond agreements. Barrier options, as exotic options are traded both on the standardized exchange and in the over-the-counter market. As important financial instruments, Luenberger and Luenberger (1999) explained that barrier options trading account for “50% of the volume of all exotic options and 10% volume of all traded securities”. Barrier options are options whose payoffs depend on the barrier levels that the underlying prices attain to during the life of the contract. These options can be classified into up(down) options if the barrier level is positioned above(below) the underlying price respectively. Research works have been done in the class of barrier options, but relatively no such work has been carried out on the comparative study focusing on the use of antithetic and quasi Monte-Carlo simulation method to price zero-rebate barrier call option with European knock-out features. Hence, this research seeks to address the gap.

One of the earliest research on barrier option pricing can be found in Snyder (1969). He described the DO options as limited risk special options, in which the holder agrees (with a favorable price in return) to limit the risk experienced by the writer by making the option void supposing “the price of the stock declines during the life of the contract to a specific point below the striking price called the expiration price". Furthermore, Merton (1973) priced the DO barrier options using the Partial Differential Equations (PDE) approach. Pricing discrete barrier options are also found in works of literature (Kou, 2003; Broadie et al., 1997). Zvan et al. (2000) presented an implicit method for solving PDE models in relation to barrier options. An explicit formula for obtaining the knock-out discount for barrier options can be obtained in Musiela and Rutkowski (2006). On the conditional Monte-Carlo simulations, Brandimarte (2013) analyzed the problem and applied it to DO put options. Different variance reduction techniques applied in the Monte-Carlo methods on option pricing were reviewed by Bolia and Juneja (2005). They employed different recent Monte-Carlo techniques like the regression-based, random tree and the stochastic mesh methods specifically on Bermudan and American options. Rich (1994) presented a mathematical foundation, mostly the key density function approach which is in connection to the solution of the FokkerPlank PDE, to value the barrier options. He derived closed form solutions of the European barrier options possessing a rebate and a fixed barrier level. A rebate is positive discount paid to the option holder in case of an a knock-out or knock-in and its presence increases the value of the barrier option, even though it has no effect on its payoff. Furthermore, Alzubaidi (2016) employed the concept of antithetic variate, together with the Brownian bridge to improve the efficiency of the MonteCarlo methods in the valuation of rebate barrier options.

Practical Applicability of the Monte-Carlo Methods

The MCS methods are techniques for obtaining numerical solutions to certain problems using some computational algorithm that focuses on series of iterative random sampling. Kroese et al. (2014) explained that Monte Carlo methods are applied mainly in three welldefined problem classes, which are sampling, estimation and optimization. Consider the evolution of stock prices in the financial market, for example, they follow a randomized system. Thus, the Monte-Carlo methods are introduced artificially to solve deterministic functions by which corresponding samples are drawn from a given probability distribution. Depending on the complexity of the dynamical system models, large number of samples are required in the simulation using the Monte-Carlo methods. These samples duly depend on large computation time. The concept of variance reduction method results in greater efficiency, with respect to the convergence process, as well as to an acceleration of the simulation techniques (Roboam, 2012). Variance reduction techniques include but not limited to antithetic variates, correlated sampling, importance sampling, control variates and stratified sampling.

Application of Antithetic Monte-Carlo Simulation (AMCS) methods can be seen in the fields of science, finance and engineering. Carmona et al. (2012) explained that Monte-Carlo methods are applicable to credit risk analysis, dynamic parameter derivatives and option pricing in the field of finance. It is also used in the estimation of fixed income instruments and interest rate derivatives which are defined in diverse term structure models. These are mostly used in situations where the closed form solutions of these securities are non-existent. Risk and uncertainties of certain projects can be managed by the aid of the Monte-Carlo methods and this had greatly improved its use in risk management. Glasserman (2004) asserted that the AMCS had played a significant role in the measurement of risks that are associated with a given portfolio of assets, as well as in the measurement of profit and loss distribution of a portfolio.

Certain complex problems in finance, statistics, technology, economics and probability theorems which cannot be resolved by the normal mathematical methods can be resolved using the quasi Monte Carlo simulation methods (QMCS). The QMCS combines methodologies and results from algebra, combinatorics and number theory to quantitatively generate number points which can be used for simulation purposes. Furthermore, the multidimensional numerical integration which are evident in financial problems are handled properly when the QMCS method is used. Fang (2002) discussed some of the applications of the QMCS methods to statistics and they were linked to the resolution of problems in statistical inference, geometric probability, experimental designs, as well as, Bayelsian statistics. Caisch (1998) analyzed the applications of QMCS to the field of scientific computing and it was observed that the rate of convergence for the QMCS is asymptotically faster than the normal MCS. Au Yeung (2010) focused on Asian option pricing using the quasi Monte-Carlo methods, in connection with MATLAB.

The significance of this research is to conduct a comparative study on zero rebate knock-out barrier option pricing using the MCS, AMCS and QMCS methods. We further seek generalize which Monte-Carlo methods show higher efficiency in valuing the pathdependent barrier options so as to avoid underpricing. This research, however, will focus on how financial risk can be hedged using exotic option pricing and these prices will be obtained using the techniques of the MCS, AMCS and QMCS methods. It will further explain how the method can be incorporated in evaluating financial derivatives, like the path-dependent barrier options. The organization of this research is as follows: Section 1 introduced the study, with brief literature studies on barrier option valuation using numerical approximations. It further discussed the practical applicability of the Monte-Carlo methods. Section 2 considers the theoretical valuation for both the DO and the up-and-out (UO) non-dividend barrier call options.

The closed form solutions of these existing formulas are properly referenced. Section 3 discusses the numerical approaches to the knock-out barrier option valuation, which in this context refers to the ordinary MCS, AMCS and QMCS methods. Section 4 provides the numerical results and the corresponding comparative statics, whereas Section 5 concludes the study.

VALUATION OF ZERO-REBATE KNOCK-OUT BARRIER OPTIONS

Generally, a barrier option with a knock-out feature refers to an option which ceases to exist once the barrier level is being hit by the price of an underlying asset. Thus, the option holder receives a rebate (positive discount) immediately or at maturity if the barrier is triggered within the contract's life, but receives the value of European option if the underlying never touches the barrier.

Definition 2.1. (Zero-Rebate Knock-Out Barrier Options) (Zhang, 1998)

A zero-rebate knock-out barrier option pays nothing if the barrier is reached but pays the value of the European option if otherwise. It is also referred to as an option whose payoff at time T is defined by.

$$f(S(T))(1-\mathbb{I}_{\tau<T}).$$

Here, τ which is a function of the underlying S and the barrier level B, denotes the stopping time for the barrier; f is the payoff function, depending on the type of option and finally, λ I is an indicator function which pays 1 when the event λ (breaching of the barrier) occurs or 0 otherwise.

Terminal and Boundary Conditions for ZeroRebate Knock-Out Barrier Options

Let f(T, S) be the payoff function of the DO barrier call option and V (t, S) be the value of the non-dividend DO barrier call option, then the option value V (t, S) satisfies the Black-Scholes PDE given below:

$$\frac{\partial V(t,S)}{\partial t}+rS\frac{\partial V(t,S)}{\partial S}$$
$$+\frac{S^2\sigma^2}{2}\frac{\partial^2 V(t,S)}{\partial S^2}-rV(t,S)=0, \qquad (21.)$$

with the terminal and boundary conditions given as:

$$V(T,S)=f(T,S)=\max\{S(T)-K,0\} \qquad (2.2)$$

$$V(t,B)=0 \qquad (2.3)$$

$$V(t,\infty)\sim S-Ke^{-r(T-t)}. \qquad (2.4)$$

Here T refers to the maturity time; t, the current time; r, the risk-free interest rate; σ, the volatility rate; S, the current price of the underlying asset; S(T), the price of the underlying asset at expiry; K, the strike price and B, the barrier level. Equation 2.2 gives the payoff or the terminal value of the option, the condition in Equation 2.3 occurs when the minimum value of the underlying asset price coincides with the barrier level and Equation 2.4 is when the underlying price attains its maximum.

The UO call option knocks-out once the maximum value of the underlying asset price coincides with the barrier level within the contract's lifespan. Let V * (t, S) denote its value at time t. Then, the option can be priced using the same PDE above but the terminal and boundary conditions become:

$$V^*(T,S)=\max\{S(T)-K,0\} \qquad (2.5)$$

$$V^*(t,B)=0 \qquad (2.6)$$

$$V^*(t,0)\sim 0. \qquad (2.7)$$

The PDE in Equation 2.1 can be solved using the hedging analysis of Black and Scholes. The equation is converted to heat equation, whereby the valuations of vanilla options can be related to the flow of heat in an infinite bar and this is solved by the method of images with respect to the standard European options (Buchen, 2012; Wilmott et al., 1995).

Valuation of Zero-Rebate Down-and-Out Call Option

The domain considered when valuating the DO call options is given by:

$$D = \{(t, S) : B < S < \infty, t \in [0, T]\}.$$

Corollary 2.2. (Buchen, 2012)

Let v(t, S) be the value of a European call option, then the image of the function v(t, S) with respect to assumed barrier level S = B and the Black-Scholes differential operator is given by:

$$v^*(t, S) = \left(\frac{B}{S}\right)^{\alpha} v\left(t, \frac{B^2}{S}\right)$$

where, $\alpha = \frac{2r}{\sigma^2} - 1$

Theorem 2.3. (Buchen, 2012)

Let v(t, S) be the solution of the terminal value problem de-scribed as:

$$\mathcal{L}v(t, S) = 0$$
$$v(T, S) = f(T, S)\mathbb{I}_{\{S>B\}},$$

in the domain {S > 0, t < T}, where Lv t S (,) is the BlackScholes differential operator found in Equation 2.1. f(T, S) is the payoff function at Equation 2.2 and $\mathbb{I}_{\{S>B\}}$ is an indicator function which outputs 1 when S>B and 0, otherwise. Then, the DO call value:

$$V(t, S) = v(t, S) - v^*(t, S)$$

solves the terminal boundary value problem for the DO barrier option in the domain {S > B, t < T}.

According to Hull (2006), the extended BlackScholes equation which gives the exact or the theoretical value for DO at time t = 0 when B>K is given by (For the put option values, see Hull (2006)):

$$V(t, S) = S(0)N(x_1) - Ke^{-rt}N(x_1 - \sigma\sqrt{T})$$
$$-S(0)\left(\frac{B}{S(0)}\right)^{2\lambda} N(y_1)$$
$$+Ke^{-rT}\left(\frac{B}{S(0)}\right)^{2\lambda-2} N(y_1 - \sigma\sqrt{T}).$$

where:

$$x_1 = \frac{\ln\left(\frac{S(0)}{B}\right)}{\sigma\sqrt{T}} + \lambda\sigma\sqrt{T},$$

$$y_1 = \frac{\ln\left(\frac{B}{S(0)}\right)}{\sigma\sqrt{T}} + \lambda\sigma\sqrt{T},\ \lambda = \frac{r + \frac{\sigma^2}{2}}{\sigma^2}$$

For B ≤ K, the price becomes:

$$\begin{aligned} V(t,S) &= S(0)N(d_1) - Ke^{-rT}N(d_2) \\ &- S(0)\left(\frac{B}{S(0)}\right)^{2\lambda} N(y) \\ &+ Ke^{-rT}\left(\frac{B}{S(0)}\right)^{2\lambda-2} N\left(y - \sigma\sqrt{T}\right). \end{aligned} \tag{2.9}$$

Where:

$$d_1 = \frac{\ln\left(\frac{S(0)}{K}\right)}{\sigma\sqrt{T}} + \lambda\sigma\sqrt{T},$$

$$d_2 = d_1 - \sigma\sqrt{T},$$

$$y = \frac{\ln\left(\frac{B^2}{S(0)K}\right)}{\sigma\sqrt{T}} + \lambda\sigma\sqrt{T}.$$

Valuation of Zero-Rebate UO Call Option

The domain under consideration for this class of knock-out options is:

$$D = \{(t,S) : 0 \leq S < B, t \in [0,T]\}.$$

According to Hull (2006), the extended BlackScholes equation which gives the exact value for U_O at time t = 0 when B>K is given by (For the put option values, see Hull (2006)):

$$\begin{aligned} V^*(t,S) &= S(0)N(d_1) - Ke^{-rT}N(d_2) \\ &- \left[\begin{array}{l} S(0)N(x_1) - Ke^{-rT}N\left(x_1 - \sigma\sqrt{T}\right) - S(0)\left(\frac{B}{S(0)}\right)^{2\lambda} \\ \left[N(-y) - N(-y_1)\right] + Ke^{-rT}\left(\frac{B}{S(0)}\right)^{2\lambda-2} \\ \left[N\left(-y + \sigma\sqrt{T}\right) - N\left(-y_1 + \sigma\sqrt{T}\right)\right] \end{array}\right] \end{aligned} \tag{2.10}$$

Lastly, when B≤K, we have that:

$$V^*(t,S) = 0. \tag{2.11}$$

Equation 2.11 follows from the fact that the strike price is normally positioned below the barrier. But supposing that it is fixed above the barrier level, the option becomes worthless by the time it hits the barrier. Thus, the chances of the option expiring in the money become negligible.

We note that in Equations 2.9 and 2.10, the first two terms denote the Black-Scholes formula for the European call options. Likewise the first two terms in Equation 2.8, but with B = K. Furthermore, the remaining terms denote that extra condition with regards to the barrier feature, that is, the risk of the option being knocked-out with the sudden loss of the premium.

NUMERICAL APPROXIMATIONS

This section gives an overview of the simulation techniques employed by the standard MCS, AMCS and the QMCS methods to barrier option pricing.

Antithetic Monte-Carlo Simulation

Generally, the MCS is a statistical estimation method which is based on the generation of random numbers. Boyle et al. (1997) first used the MCS to price the European options under the Black-Scholes assumptions (Black and Scholes, 1973). The MCS of option pricing under the Black-Scholes framework involves the generation of N sample independent asset price movements and then the estimation of their payoffs. These payoffs are averaged and then discounted at a riskfree interest rate. One of the major setbacks of using the MCS is seen in its slow rate of convergence at which the estimated values tend to the true solution. This can be explained by the large variances obtained after the simulation. It is equally observed that the confidence interval obtained using the MCS is greatly influenced by the ratio of the standard deviation to the square root of the number of simulations. To further reduce the width of the interval, the deviation or the variance has to be reduced which in turn gives a better estimate.

The AMCS was introduced in this study to improve the flaws of the MCS. According to Glasserman (2004), the AMCS focuses on the symmetric properties of normal distribution to reduce the variance of the results been simulated and this is achieved by introducing negatively correlated random variables. Using a set of random numbers which consists of a positive and a negative number, say x_1 and x_2 respectively, the option value can be estimated. Let X be a random variable whose estimate is unknown. Also, let x_1 and x_2 be two negatively correlated variables with the same mean μ and variance σ^2. Define:

$$X = \frac{1}{2}\left(x_1 + x_2\right),$$

then the mean and the variance are given respectively as:

$$\mathbb{E}[X] = \mathbb{E}\left[\frac{1}{2}\left(x_1 + x_2\right)\right] = \frac{1}{2}\left(\mathbb{E}\left[x_1\right] + \mathbb{E}\left[x_2\right]\right) = \mu.$$

$$\begin{aligned} \operatorname{var}[X] &= \operatorname{var}\left[\frac{1}{2}\left(x_1 + x_2\right)\right] \\ &= \frac{1}{4}\left(\operatorname{var}\left[x_1\right] + \operatorname{var}\left[x_2\right] + 2\operatorname{cov}\left[x_1, x_2\right]\right) \\ &= \frac{1}{2}\left(\operatorname{var}\left[x_1\right] + \operatorname{cov}\left[x_1, x_2\right]\right). \end{aligned}$$

Thus, we have that:

$$\operatorname{var}[X]\begin{cases} = \dfrac{\sigma^2}{2} \textit{ if } x_1 \textit{ and } x_2 \textit{ are identically independent } (\operatorname{cov} = 0) \\ < \dfrac{\sigma^2}{2} \textit{ if } \operatorname{cov}\left[x_1, x_2 < 0\right] \end{cases}$$

It is essential that if the cov[x_1, x_2] < 0 or cov[x_1, x_2] = 0, then the variance is reduced. For the outputs of the antithetic variates to be negatively correlated, it is essential that the inputs are negatively correlated. The mapping between them should be monotone. Non-monotonic functions result in a non-negative correlation and this could increase the variance, instead of reducing as expected.

Definition 3.1. (Monotone Functions)

Let A be a subset of $\mathbb{R}$ and define a function g: A → $\mathbb{R}$ and x, y ∈ A. Then g is monotonic if x < y implies g(x) ≤ g(y) or if x < y implies g(x) ≥ g(y). The following corollaries help in the generation of negatively correlated random variables which are uniformly and normally distributed.

Corollary 3.2. (Chan and Wong, 2015)

If g(X_1,X_2, · · ·, X_n) is a monotone function of each of its argument. Then the following holds for a set η_1,· · ·, η_m of independent and identically distributed uniformly random numbers on (0, 1):

$$\operatorname{cov}\left[g\left(\eta_1, \cdots, \eta_m\right), g\left(1-\eta_1, 1-\eta_2, \cdots, 1-\eta_m\right)\right] \leq 0.$$

Let η_1,· · ·, η_m and β_1 = 1-η_1,· · ·,β_m = 1-ηm be uniformly random numbers with the properties that the pair (η_1, β_1) are negatively correlated. Define a function g to be monotone. If X_1 = g(η_1,· · ·,η_m), then X_2 = g(β_1,· · ·,

β_m) must be defined in the same distribution as X_1. Thus, from Corollary 3.2, we have that $cov(X_1, X_2) \leq 0$. Hence, X_1 and X_2 are negatively correlated.

Corollary 3.3. (Chan and Wong, 2015)

If $g(X_1, X_2, \cdots, X_n)$ is a monotone function of each of its argument. Then the following holds for a set $\eta_1, \cdots, \eta_m$ of independent and identically distributed normal random numbers on (0, 1):

$$\text{cov}\left[g(\eta_1, \cdots, \eta_m), g(-\eta_1, -\eta_2, \cdots, -\eta_m)\right] \leq 0.$$

Let $X_i \sim N(\mu, \sigma^2)$ and $Yi = 2\mu - Xi$. Then from the linearity condition of the normal random variable, we see that $Y_i \sim N(\mu, \sigma^2)$. Thus, X_i and Y_i are negatively correlated. This follows from the definition of covariance:

$$\begin{aligned}
&\text{cov}\left[X_i(2\mu - X_i)\right] = \mathbb{E}\left[X_i(2\mu - X_i)\right] \\
&-\mathbb{E}\left[X_i\right]\mathbb{E}\left[2\mu - X_i\right] \\
&= \mathbb{E}\left[X_i\right]\left(\mathbb{E}\left[2\mu - X_i\right] - \mathbb{E}\left[2\mu - X_i\right]\right) \\
&= 0.
\end{aligned}$$

Knock-Out Barrier Option Pricing using the MCS Methods

Consider the asset price dynamics described by the stochastic differential equation below:

$$dS(t) = S(t)\left(rdt + \sigma dB(t)\right), \tag{3.1}$$

where, B(t) is the Brownian motion defined in the riskneutral probability measure and r the risk-free interest rate. Using Ito's lemma to solve the above 3.1, we have:

$$S(t) = S(0)\exp\left(\left(r - \frac{\sigma^2}{2}\right)t + \sigma B(t)\right). \tag{3.2}$$

From Equation 3.2, we have that $\sqrt{t\epsilon} \Rightarrow B(\Delta t) \sim \sqrt{\Delta t}\epsilon$ and follows a standard normal distribution with the parameters ~ N(0, 1). Next, we consider the discredited underlying asset price process defined by:

$$S^*(t + \Delta t) = S(t)\exp\left(\left(r - \frac{\sigma^2}{2}\right)\Delta t + \sigma\left(\sqrt{\Delta t}\right)\epsilon\right). \tag{3.3}$$

The payoffs are introduced next which would be discounted at a risk-free interest rate and this depends on the type of knock-out call options being considered. For the zero rebate DO barrier call options, the discounted payoff is:

$$v^{+}(t+\Delta t)=e^{-r\Delta t}\max\left\{S^{+}(t+\Delta t)-K,0\right\}. \tag{3.4}$$

Equation 3.4 above occurs when the barrier is not triggered (that is, when max{S^{+} (t +Δt)}>B) but pays zero when it knocks out (that is, when max{S^{+} (t +Δt)} ≤ B). Finally, the option value is constructed by dividing the sum of the discounted payoff by the number of simulations. Thus, the value of the Monte-Carlo simulated DO call option is given by:

$$V_{M}=\frac{1}{M}\sum_{i=1}^{M}v^{+}(t+\Delta t).$$

NB: Denote the Monte-Carlo simulated UO call value to be V^{*}_{M}, which follows the same process as discussed above. The only difference is that equation (3.4) occurs when max{S^{+} (t +Δt)}< B and zero when max{S^{+} (t +Δt)}≥ B.

Knock-Out Barrier Option Pricing using the AMCS Methods

For the AMCS concept of option pricing, another set of asset price process is equally simulated so that the two processes would be negatively correlated. An average would be taken and this serves as the final simulated asset price. A more accurate result is obtained from the use of these antithetic pairs since the true option value is bounded by the positive and the negative pairs. Also, the generation of the pairs is computationally cheaper. Thus, consider two discretized underlying asset processes defined by:

$$S^{+}(t+\Delta t)=S(t)\exp\left(\left(r-\frac{\sigma^{2}}{2}\right)\Delta t+\sigma\left(\sqrt{\Delta t}\right)\epsilon\right) \tag{3.5}$$

And:

$$S^{-}(t+\Delta t)-S(t)\exp\left(\left(r-\frac{\sigma^{2}}{2}\right)\Delta t-\sigma\left(\sqrt{\Delta t}\right)\epsilon\right). \tag{3.6}$$

Next, using the risk-neutrality condition of asset pricing, these payoffs are discounted and this depends on the type of options being considered. For the zero rebate DO barrier call options, the discounted payoffs are:

$$v^{+}(t+\Delta t)=e^{-r\Delta t}\max\left\{S^{+}(t+\Delta t)-K,0\right\} \tag{3.7}$$

And

$$v^{-}(t+\Delta t)=e^{-r\Delta t}\max\left\{S^{-}(t+\Delta t)-K,0\right\}. \tag{3.8}$$

Equations 3.7 and 3.8 above occur when the barrier is not triggered (that is, when $\max\{S^+(t+\Delta t)\} > B$) but the option pays zero when it knocks out (that is, when $\max\{S^+(t+\Delta t)\} \leq B$). The UO call option on the other hand pays the above when $\max\{S^+(t+\Delta t)\} < B$ and zero when $\max\{S^+(t+\Delta t)\} \geq B$. Denote its value by V^*_A. Finally, the mean estimator which is the required option value is constructed by taking the average of the discounted payoffs. Thus, the value of the AMCS DO call option is given by:

$$V_A = \frac{1}{M}\sum_{i=1}^{M}\frac{1}{2}\left(v^+(t+\Delta t)+v^-(t+\Delta t)\right).$$

Quasi Monte-Carlo Simulation

Suppose we aim at calculating:

$$\mathbb{E}\left[f(u_1,\cdots,u_d)=\int_{[0,1]^d} f(m)\,dm,\right]$$

where, u_i^s are sequences of independent uniformly distributed random variables. The QMCS entails using the approximation:

$$\int_{[0,1]^d} f(m)\,dm \approx \frac{1}{n}\sum_{i=1}^{n} f(x_i),$$

with the points $x_i'^s$ obtained deterministically in the dimensional closed unit hypercube $[0, 1]^d$. The QMCS are based on the generation of low-discrepancy sequences such as Halton sequence, Faure sequence, Sobol sequence and this is in contrast to the pseudorandom numbers generated by the MCS. The error bound for QMCS is of the form $O(N^{-1}(\log N)^{d-1})$ and it is in contrast to the standard MCS which is of the form $O\left(N^{-\frac{1}{2}}\right)$. The nodes N are chosen to ensure that there exists low degree of regularity in the evaluation of the integral and the nodes are obtained from either of the sequences listed above (McLeish, 2000). Thus according to Glasserman (2004), the error bound satisfies the Koksma-Hlawka inequality which is written as:

$$\varepsilon = \left|\int_{[0,1]^d} f(m)\,dm - \frac{1}{n}\sum_{i=1}^{n} f(x_i)\right| < v(f)\,D_n(A,p),$$

where, $v(f)$ is the Hardy-Krause variation of function of f and $D_n(A, p)$ is defined below.

Definition 3.4

Let A be non-empty Lebesgue measurable subsets of $[0, 1]^d$. The discrepancy of the point sets $p = \{x_1, \cdots, x_n\}$ relative to A is defined as:

$$D_n(A, p) = \sup\left\{\frac{\#\{x_i \in A\}}{n} - \lambda_d(A)\right\}.$$

where, $\#\{x_i \in A\}$ is the number of points in A and $\lambda_d(A)$ is the Lebesgue measure of A. Extreme discrepancy and star discrepancy are obtained if A takes the form:

$$\prod_{i=1}^{d}[a_i, b_i) \text{ and } \prod_{i=1}^{d}[0, a_i) \text{ respectively, for } 0 \le a_i < b_i \le 1.$$

Discrepancy measures the uniformity deviation of specific sequence of points in $[0, 1]^d$ and deterministic sequences tend to fill up the given space uniformly. The choice of low discrepancy sequences helps in speeding up the rate of convergence from $O(1\sqrt{n})$ to $O(1/n)$ and thus, the analysis of the discrepancy can help in measuring the error of the approximation method (See Niederreiter (1992) for further detailed description on analysis of low discrepancy sequences.). Further analysis of the error bound shows that small errors arise if $x_i'^s$ with small star or extreme discrepancies are used and this gives the low discrepancy sequences. For the purpose of this research, we consider the Sobol sequence.

The Sobol sequence forms discrete uniform patterns in a specific hypercube using base two form. The construction of Sobol sequence can be found in (Jäckel, 2002) and in the generation of our results, we used the computer program (Ace option pricier (Beta) (http://www.acenumerics.com/optionpricer.html)). For the set of initial direction, this program used the methodology found in (Joe and Kuo, 2008) (Also see http://web.maths.unsw.edu.au/~fkuo/sobol/).

The knock-out barrier option pricing using the QMCS method follows the same procedure as the standard MCS. The only difference being that the QMCS uses deterministic sequences in the interval [0,1] instead of the pseudo-random numbers resulting to the presence confidence intervals. Furthermore, either the Box-Muller method or the inverse transform method can be used to transform the low discrepancy sequences to normally distributed sequences. Thus, these sequences replace the normal pseudo-random numbers as found in Equation 3.3 so as to ensure faster rate of convergence using error bounds which are already known.

COMPUTATIONAL RESULTS AND ANALYSIS

This section considers some of the findings obtained during the implementation of the above methodologies. For the knock-out barrier options, we consider the nondividend DO and UO call options with a zero-rebate when the option is knocked out. First, we review the results presented in Wang and Wang (2011) where they considered different Monte-Carlo methods in the valuation of barrier options. The parameters K = 50, r = 0.1, $\sigma = 0.2$, T = 1 were used to value the DO put options. We compare their results on AMCS and the normal MCS using M = 10000, with our QMCS value using $2^{13} = 8192$ paths. The following table is obtained.

Table 1 compares the MCS, AMCS and the QMCS in the valuation of the DO put option with zero dividend. We chose the paths for the QMCS which are relatively lower than the number of simulations of other Monte-Carlo methods and yet, we observe that the QMCS performed outstandingly better in comparison with others, followed by the AMCS. The efficiency of the AMCS can be seen in the reduction of the variances of the MCS and this actually improved the results of the simulation. The standard errors are depicted in the brackets and we observed that the standard error for the MCS are much higher when compared to the AMCS and the QMCS. Infact, we observed that the QMCS need quite a few number of steps (as small as $2^{10} = 1024$ steps) for convergence to ensue. For the computation time, the MCS took an average of 0.016 sec for a simulation number of 10000, to output its results. The AMCS took an average of 0.031 sec, which is about twice that of the MCS. This was because the AMCS had to reduce the variances first before the actual result was obtained. Finally, the QMCS used an average of 0.015 sec in conjunction with $2^{10} = 1024$ steps, to output its results. Thus, the QMCS most a times offer higher accuracy with lesser computation time and this outperforms the results of the MCS.

Results on Zero-Rebate DO Barrier call Options with at-the-Money (ATM) Features

We consider the values obtained using MCS, AMCS and QMCS for the above-named option which pays no dividend. Here, we consider the parameters: S = K = 150, B = 125, r = 0.05, $\sigma = 0.25$, T = 0.5. The extended

Black-Scholes formula in Equation 2.9 is used to obtain the exact price and the value is 12.1861. Figure 1 and 2 display the results obtained.

Table 1: Simulated values for DO barrier put option

S	B	Accurate	MCS	AMCS	QMCS
50	40	0.6264	0.7101 (0.0171)	0.6953 (0.0119)	0.6264 (0.0000)
55	40	0.4192	0.4607 (0.0140)	0.4227 (0.0099)	0.4192 (0.0000)
45	40	0.6024	0.7306 (0.0174)	0.7266 (0.0123)	0.6024 (0.0000)
50	45	0.0629	0.0899 (0.0046)	0.0828 (0.0032)	0.0629 (0.0000)
50	35	1.4404	1.5063 (0.0288)	1.4966 (0.0203)	1.4404 (0.0000)
50	30	1.8136	1.8689 (0.0343)	1.7969 (0.0237)	1.8136 (0.0000)

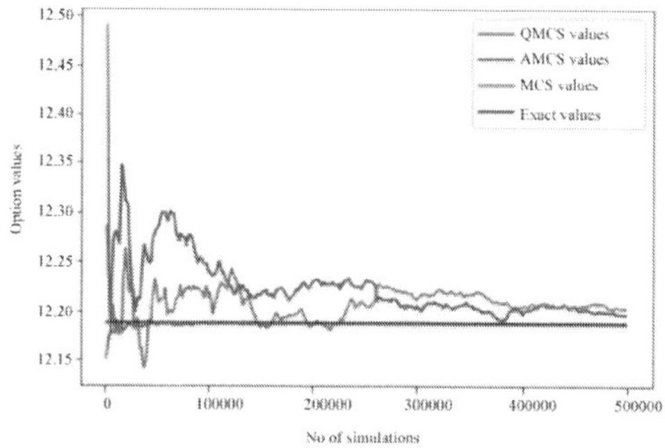

Figure. 1: Option values

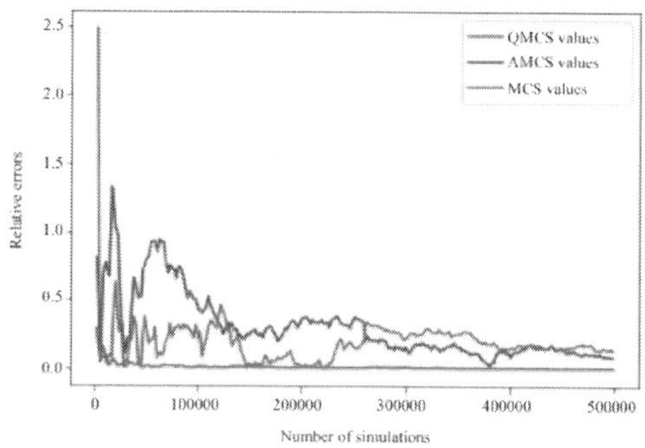

Figure. 2: Relative errors

Figure 1 shows convergence of the values of barrier option prices when the asset price process is being simulated using various Monte-Carlo forms. The exact value remains unchanged, as it is unaffected by the number of simulations. The QMCS performed extremely well, as convergence is achieved faster and this was in connection to the use of Sobol sequences. This is followed by the performance of the AMCS, which reduced the variance of the ordinary MCS to achieve its accuracy. Figure 2 represent the plot of the relative error in percentage between the simulated values and the exact Black-Scholes price. It is calculated by:

$$Rerror = 100\% \frac{|SIV - EBS|}{EBS}, \tag{4.1}$$

where, EBS denotes the value of the exact Black-Scholes prices values and SIV is the Monte-Carlo simulated values (which could be the MCS, AMCS, or the QMCS values). We observed that the QMCS is more efficient since more consistency is achieved in the deviation of the simulated prices from the exact price.

Results on the UO Call Options using MCS and AMCS Methods

Consider the non-dividend UO call option with parameters: S = 50, K = 60, B = 80, r = 0.05, σ = 0.45, T = 0.5. The exact value of the option using the extended Black-Scholes pricing formula is 0.8657. We used different time points N = 100, N = 1000 and N = 10000. The number of simulations was also considered when $M = 10^1, 10^2, \cdots, 10^5$ for each of the time points. The outputs generated are depicted graphically in the Fig. 3 and 4.

Table 2: Monte-Carlo simulated values for the knock-out call options

	Option values							
	DO values				UO values			
Asset (S)	Exact	MCS	AMCS	QMCS	Exact	MCS	AMCS	QMCS
10	0.0082	0.0080 (2.439)	0.0081 (1.220)	0.0082 (0.000)	0.0082	0.0080 (2.439)	0.0081 (1.220)	0.0082 (0.000)
13	0.3064	0.3070 (0.196)	0.3065 (0.033)	0.3064 (0.000)	0.3064	0.3070 (0.196)	0.3065 (0.033)	0.3064 (0.000)
16	1.7403	1.7419 (0.092)	1.7408 (0.029)	1.7403 (0.000)	1.7398	1.7413 (0.086)	1.7403 (0.029)	1.7398 (0.000)
19	4.2567	4.2582 (0.035)	4.2569 (0.005)	4.2567 (0.000)	4.2325	4.2349 (0.057)	4.2331 (0.014)	4.2525 (0.000)
21	6.1818	6.1837 0.031)	6.1819 (0.002)	6.1818 (0.000)	6.0303	6.0337 (0.056)	6.0310 (0.012)	6.0303 (0.000)
24	9.1588	9.1611 (0.023)	9.1589 (0.001)	9.1588 (0.000)	8.0730	8.0739 (0.011)	8.0723 (0.009)	8.0730 (0.000)
27	12.1564	12.1591 (0.022)	12.1565 (0.001)	12.1563 (0.001)	8.1537	8.1478 (0.071)	8.1524 (0.016)	8.1537 (0.000)
30	15.1561	15.1591 (0.020)	15.1563 (0.001)	15.1561 (0.000)	5.8175	5.8146 (0.050)	5.8148 (0.046)	5.8175 (0.000)
33	18.1561	18.1594 (0.018)	18.1563 (0.001)	18.1561 (0.000)	2.2253	2.2263 (0.045)	2.2266 (0.058)	2.2253 (0.000)

Figure 3 and 4 denote the MCS values and the AMCS values respectively, for different discretisation time points N. The points in the graph are the centre of our discussion, as they denote the option values at each of the specified simulation number. The points are interpolated using spline curve fittings. The straight line which is unaffected by the simulation numbers denotes the option value (0.8657). In each of the figures above, we compare the estimated values obtained when the time points are N = 100, 1000 and

10000. We observed that increase in the number of simulations and increase in the time points would result to a high rate of convergence of the estimated values to the exact value, for both the MCS and the AMCS values. But the rate of convergence of the AMCS is faster when compared to the MCS since at M = 10^5 and N = 10000, the AMCS converged to the true solution.

Figures 5 and 6 denote respectively the plot of the relative errors with respect to an incremental change in the number of simulations, for both the MCS and the AMCS values. The relative errors from both simulations are reducing with an increase in the number of simulations and time points. The relative errors obtained using the AMCS are generally lesser when compared to that of MCS. The exact Black-Scholes price for the zerorebate barrier options is obtained based on continuous monitoring time. Our approach entails solving the pricing formula using the discrete time framework. This price can be achieved when the time steps (ΔT) are reduced drastically and this follows from the increment in the number of the discretisation time points N. Hence, the discreet points $\max_{0 \leq j \leq N} S(j) < B$. tends towards the continuous points $\max_{0 \leq t \leq T} S(t) < B$.

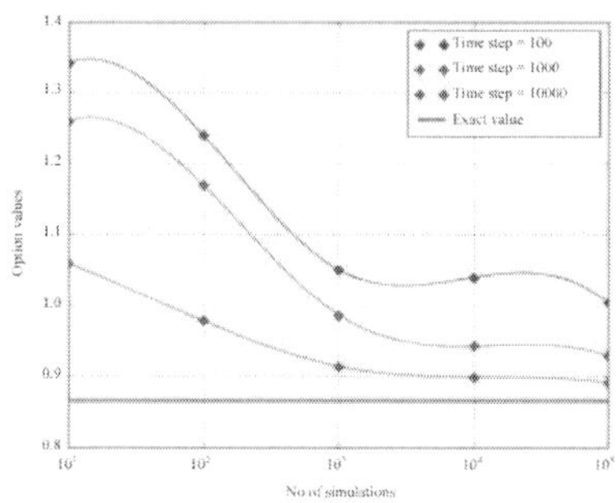

Figure. 3: MCS values with different N

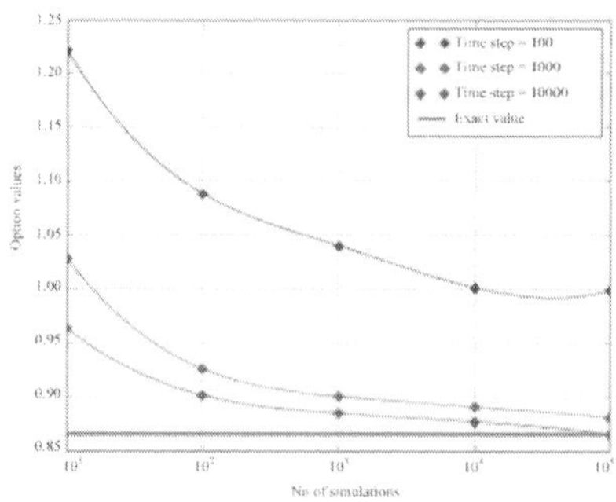

Figure. 4: AMCS values with different N

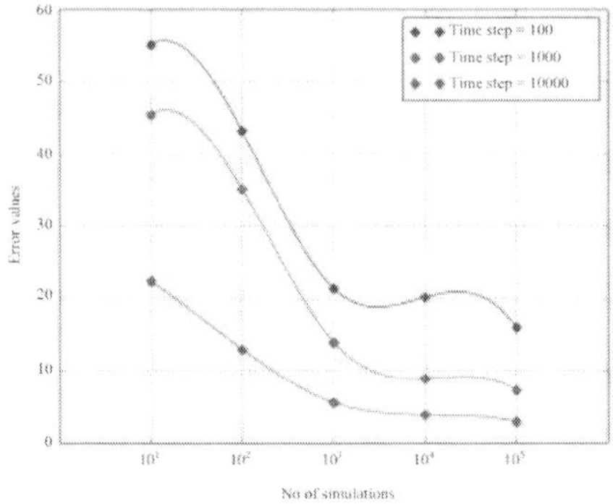

Figure. 5: Error values for MCS

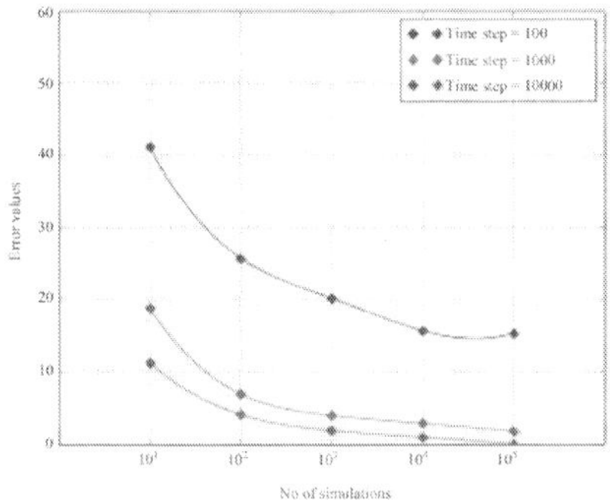

Figure. 6: Error values for AMCS

Monte-Carlo Simulation Values on Knock-Out Barrier Call Options

Table 2 shows the comparison of the AMCS, MCS and the QMCS on the DO barrier options and the UO barrier call options. The parameters used include: $K = 16$, $r = 0.1$, $T = 0.75$, $\sigma = 0.2$, upper barrier for UO = 35 and lower barrier for DO = 5. The following results were obtained:

From Table 2 above, we observed that the exact values for the DO options increased without bound as the underlying asset price moved away from the barrier level. Thus, the probability of the option being knocked-out was slim. Whereas, the UO call resulted in a non-linear function, as the option values increased and declined at some points. The reduction in the option values is often due to the fact that if the volatility is sufficiently large, the chances of the option being knocked out is increased. Hence, this increased probability results in the decline of the option value. The value for the UO barrier call option is always less than the difference between the barrier level and the strike price and hence it has limited upside profit potential. The values obtained using the MCS, AMCS and QMCS as depicted in the same table above show similar characteristics. The table also shows the deviation of the simulated values from the exact values. For the MCS, AMCS for both UO and DO, we consider the number of simulations to be M = 600000 and 2^{19} = 524288 paths. The result shows that the AMCS is really an improved version of the MCS, with regards to their deviation from the exact value. However, when the low discrepancy method like the QMCS with sobol sequences is used to output the same option value, a highly improved version and more effective result is obtained. The relative errors are further showed in the brackets and they are expressed in percentage. It follows that the errors for the AMCS is quite smaller than that of the MCS and high rate of convergence is seen in the results obtained by the QMCS. Also, the average computation time for the QMCS to simulate 2^{19} = 524288 paths is 0.187 secs. The MCS takes an average of 0.172 secs to simulate M = 600000 paths and AMCS takes about 0.203 secs (Recall, we used the computer program (Ace option pricier (Beta)). Hence, the QMCS outperforms the normal MCS and the AMCS.

CONCLUSION

A comparative study had been conducted on the valuation of non-dividend knock-out barrier options with respect to the Black-Scholes theoretical values. This research work had compared the prices of the exact Black-Scholes prices of the zero-rebate knock-out barrier options to Monte-Carlo simulated values. We have discussed results from the MCS, AMCS and the QMCS method and we observed that the ordinary MCS underpriced the option, as can be seen in the high discrepancies of the simulated values in comparison to the exact prices. It was also observed that using the AMCS resulted in the reduction of variance estimate when compared to the independent simulations obtained from the standard MCS method. This was made possible since the antithetic variables used are negatively correlated and thus, increasing one variable led to a simultaneous decrease in the other variable. Hence, the method gave a better and fairly accurate result when compared to the MCS. Furthermore, the rate of convergence is accelerated when the simulation nodes are chosen deterministically as observed by the use of low discrepancy sequences. Thus, the use of QMCS together with the

Sobol sequences resulted to a more efficient method for valuing complex derivative, like the barrier options with continuous monitoring period.

Future work will extend this result to barrier options with moving barrier, as well barrier options with double barrier.

ACKNOWLEDGEMENT

The authors wish to acknowledge the financial support from the Deutscher Akademischer Austauschdienst (DAAD) in collaboration with the African Institute for Mathematical Sciences (AIMS) and the Centre for Business Mathematics and Informatics, North-West University, Potchefstroom. We also thank the anonymous referees for their helpful comments.

REFERENCES

1. Alzubaidi, H., 2016. Efficient Monte Carlo algorithm using antithetic variate and Brownian bridge techniques for pricing the barrier options with rebate payments. J. Math. Stat., 12: 1-11. DOI: 10.3844/jmssp.2016.1.11
2. Au Yeung, J., 2010. Path-dependent options pricing: A quasi Monte Carlo simulation approach with MATLAB.
3. Black, F. and M. Scholes, 1973. The pricing of options and corporate liabilities. J. Political Economy, 81: 637-654. DOI: 10.1086/260062
4. Bolia, N. and S. Juneja, 2005. Monte Carlo methods for pricing financial options. Sadhana, 20: 347-385. DOI: 10.1007/BF02706251
5. Boyle, P., M. Broadie and P. Glasserman, 1997. Monte Carlo methods for security pricing. J. Economic Dynam. Control, 21: 1267-1321. DOI: 10.1016/S0165-1889(97)00028-6
6. Brandimarte, P., 2013. Numerical Methods in Finance and Economics: A MATLAB-Based Introduction. 2nd Edn., John Wiley and Sons, New York, ISBN-10: 1118625579, pp: 696.
7. Broadie, M., P. Glasserman and S. Kou, 1997. A continuity correction for discrete barrier options. Math. Finance, 7: 325-349. DOI: 10.1111/1467-9965.00035
8. Buchen, P., 2012. An Introduction to Exotic Option Pricing. 1st Edn., CRC Press, ISBN-10: 142009100X, pp: 296.
9. Carmona, R., P. Del Moral, P. Hu and N. Oudjane, 2012. An Introduction to Particle Methods with Financial Applications. In: Numerical Methods in Finance, Carmona, R., P.D. Moral, P. Hu and N. Oudjane (Ed.), Springer Berlin Heidelberg, ISBN-10: 3642257461, pp: 3-49.
10. Caisch, R.E., 1998. Monte Carlo and quasi-Monte Carlo methods. Acta Numerica, 7: 1-49. DOI: 10.1017/S0962492900002804
11. Chan, N. and H. Wong, 2015. Simulation Techniques in Financial Risk Management. 2nd Edn., John Wiley and Sons, Hoboken, ISBN-10: 1118735994, pp: 232.
12. Cox, J.C. and M. Rubinstein, 1985. Options Markets. 1st Edn., Prentice Hall, Englewood Cliffs, ISBN-10: 0136382053, pp: 498.
13. Fang, K.T., 2002. Some Applications of Quasi-Monte Carlo Methods in Statistics. In: Monte Carlo and QuasiMonte Carlo Methods 2000, Fang, K.T., H. Niederreiter and F.J. Hickernell (Eds.), Springer, Berlin,

Heidelberg, ISBN-10: 3642560466, pp: 10-26.

14. Glasserman, P., 2004. Monte Carlo Methods in Financial Engineering. 1st Edn., Springer Science and Business Media, New York, ISBN-10: 0387004513, pp: 596.
15. Hull, J., 2006. Options, Futures and other Derivatives. 1st Edn., Pearson Education, India, Nueva York, ISBN-10: 8131722996, pp: 822.
16. Jäckel, P., 2002. Monte Carlo Methods in Finance. 1st Edn., J. Wiley, Chichester, ISBN-10: 047149741X, pp: 238.
17. Joe, S. and F.Y. Kuo, 2008. Constructing Sobol sequences with better two-dimensional projections. SIAM J. Scientific Comput., 30: 2635-2654. DOI: 10.1137/070709359
18. Kroese, D.P., T. Brereton, T. Taimre and Z.I. Botev, 2014. Why the Monte Carlo method is so important today. Wiley Interdisciplinary Rev.: Comput. Stat., 6: 386-392. DOI: 10.1002/wics.1314
19. Kou, S., 2003. On pricing of discrete barrier options. Statistica Sinica, 13: 955-964.
20. Luenberger, D. and R. Luenberger, 1999. Pricing and hedging barrier options. Investment practice, Stanford University, EES-OR
21. Merton, R., 1973. Theory of rational option pricing. Bell J. Econom. Manage. Sci. 4: 141-183. DOI: 10.2307/3003143
22. Musiela, M. and M. Rutkowski, 2006. Martingale Methods in Financial Modelling. 2nd Edn., Springer Science and Business Media, New York, ISBN-10: 3540266534, pp: 638.
23. McLeish, D., 2000. Monte Carlo Simulation in Finance. 1st Edn., John Wiley and Sons.
24. Niederreiter, H., 1992. Random number generation and quasi-Monte Carlo methods. Society for Industrial and Applied Mathematics.
25. Rich, D.R., 1994. The mathematical foundations of barrier option-pricing theory. Adv. Futures Opt. Res., 7: 267-311.
26. Roboam, X., 2012. Systemic Design Methodologies for Electrical Energy Systems: Analysis, Synthesis and Management. 1st Edn., John Wiley and Sons, London, ISBN-10: 1118569679, pp: 352.
27. Snyder, G., 1969. Alternative forms of options. Financial Anal. J., 25: 93-99. DOI: 10.2469/faj.v25.n5.93
28. Wang, B. and L. Wang, 2011. Pricing barrier options using Monte Carlo methods.

29. Wilmott, P., S. Howison and J. Dewynne, 1995. The Mathematics of Financial Derivatives: A Student Introduction. 1st Edn., Cambridge University Press, New York, ISBN-10: 0521497892, pp: 317
30. Zhang, P., 1998. Exotic Options: A Guide to Second Generation Options. 2nd Edn., World Scientific, Singapore, ISBN-10: 9810235216, pp: 692.
31. Zvan, R., K. Vetzal and P. Forsyth, 2000. PDE methods for pricing barrier options. J. Economic Dynam. Control, 24: 1563-1590. DOI: 10.1016/S0165-1889(00)00002-6

INDEX

C

D

N

O

P

Q

R

S

T

U

V

Z